Leukocyte Typing II

VOLUME 3
Human Myeloid and Hematopoietic Cells

Leukocyte Typing II

Leukocyte Typing II

VOLUME 3
Human Myeloid and Hematopoietic Cells

Edited by
Ellis L. Reinherz   Barton F. Haynes
Lee M. Nadler   Irwin D. Bernstein

With 86 Illustrations

Springer-Verlag
New York Berlin Heidelberg Tokyo

Ellis L. Reinherz, M.D., Division of Tumor Immunology, Dana-Farber Cancer Institute, Boston, MA 02115 U.S.A.

Barton F. Haynes, M.D., Department of Medicine, Duke University School of Medicine, Durham, NC 27710 U.S.A.

Lee M. Nadler, M.D., Division of Tumor Immunology, Dana-Farber Cancer Institute, Boston, MA 02115 U.S.A.

Irwin D. Bernstein, M.D., Program in Pediatric Oncology, Fred Hutchinson Cancer Research Center, Seattle, WA 98104 U.S.A.

Library of Congress Cataloging in Publication Data
Main entry under title:
Leukocyte typing II.
  Papers presented at the Second International
Workshop on Human Leukocyte Differentiation Antigens,
held in Boston, Sept. 17–20, 1984.
  Includes bibliographies and indexes.
  Contents: v. 1. Human T lymphocytes—v. 2. Human
B lymphocytes—v. 3. Human myeloid and hematopoietic
cells.
  1. Leucocytes—Classification—Congresses.
2. Histocompatibility testing—Congresses. 3. Tissue
specific antigens—Analysis—Congresses. I. Reinherz,
Ellis L. II. International Workshop on Human
Leukocyte Differentiation Antigens (2nd : 1984 : Boston,
Mass.) III. Title: Leukocyte typing 2. IV. Title:
Leukocyte typing two.
QR185.8.L48L48 1985   616.07′9   85-22229

Typeset by Bi-Comp Inc., York, Pennsylvania.
Printed and bound by Halliday Lithograph, West Hanover, Massachusetts.
Printed in the United States of America.

9 8 7 6 5 4 3 2 1

ISBN 0-387-96177-1 Springer-Verlag New York Berlin Heidelberg Tokyo
ISBN 3-540-96177-1 Springer-Verlag New York Berlin Heidelberg Tokyo

Preface

The Second International Workshop on Human Leukocyte Differentiation Antigens was held in Boston, September 17–20, 1984. More than 350 people interested in leukocyte differentiation agreed to exchange reagents and participate in this joint venture. All in all, in excess of 400 antibodies directed against surface structures on T lymphocytes, B lymphocytes, and myeloid-hematopoietic stem cells were characterized. Because of the enormous quantity of serologic, biochemical, and functional data, *Leukocyte Typing II* has been divided into three volumes.

These books represent the written results of workshop participants. They should be helpful to both researchers and clinicians involved in scientific endeavors dealing with these broad fields of immunobiology. To those who delve into the various sections of the volumes, it will become evident that the work speaks for itself.

I am deeply indebted to the section editors, Barton F. Haynes, *Volume 1, Human T Lymphocytes,* Lee M. Nadler, *Volume 2, Human B Lymphocytes,* and Irwin D. Bernstein, *Volume 3, Human Myeloid and Hematopoietic Cells* for their major contributions in planning, executing, and summarizing the workshop, as well as council members John Hansen, Alain Bernard, Laurence Boumsell, Walter Knapp, Andrew McMichael, Cesar Milstein, and Stuart F. Schlossman. I would also like to thank the National Institutes of Health, World Health Organization, and International Union of Immunological Societies for making this meeting possible. Needless to say, I am most grateful to all of my colleagues who contributed to this effort and helped to accelerate the characterization of human immunobiology through their endeavors.

Ellis L. Reinherz, M.D.

Contents

Contributors

M.C. Alonso Department of Biochemistry, Faculty of Medicine, University of Cordoba, Cordoba, Spain

Donald C. Anderson Departments of Pediatrics, Microbiology and Immunology, Baylor College of Medicine, Houston, Texas 77030, U.S.A.

M. Amin Arnaout Divisions of Nephrology and Cell Biology, Children's Hospital Medical Center, Harvard Medical School, Boston, Massachusetts 02115, U.S.A.

David Askew Terry Fox Laboratory, B.C. Cancer Research Centre, Vancouver, British Columbia, Canada

Robert C. Atkins Department of Nephrology, Prince Henry's Hospital, Melbourne, Australia

Gail Bartley Division of Tumor Immunology, Dana-Farber Cancer Institute, Boston, Massachusetts 02115, U.S.A.

J.G.J. Bauman Radiobiological Institute TNO, Rijswijk, The Netherlands

Irwin D. Bernstein Program in Pediatric Oncology, Fred Hutchinson Cancer Research Center, Seattle, Washington 98104, U.S.A.

E. Berti First Department of Dermatology, University of Milan, Milan, Italy

P. Bettelheim First Medical Department, University of Vienna, Lazarettgasse 14, A-1090 Vienna, Austria

M.J.E. Bos Department of Immunohaematology, Central Laboratory of the Netherlands Red Cross Blood Transfusion Service, 1006 AK Amsterdam, The Netherlands

Herbert Braunsteiner Department of Internal Medicine, University of Innsbruck, Innsbruck, Austria

Angelo Cantù-Rajnoldi Laboratorio di Ematologia e Sezione di Citogenetica, Istituti Clinici di Perfezionamento, 20122 Milano, Italy

Giorgio Cattoretti   National Cancer Institute of Milan, Milan, Italy

O.R. Colamonici   Laboratory of Pathology, NCI, NIADDKD, National Institutes of Health, Bethesda, Maryland 20205, U.S.A.

F. de Braud   National Cancer Institute of Milan, Milan, Italy

Domenico Delia   National Cancer Institute of Milan, Milan, Italy

Giorgio Lambertenghi DeLiliers   Istituto di Clinica Medica I, 20122 Milano, Italy

P. Dubreuil   Centre d'Immunologie Luminy, Marseille, France

Bo Dupont   Human Immunogenetics Laboratory, Memorial Sloan-Kettering Cancer Center, New York, New York 10021, U.S.A.

Allen C. Eaves   Terry Fox Laboratory, B.C. Cancer Research Centre, Vancouver, British Columbia, Canada

C.P. Engelfriet   Department of Immunohaematology, Central Laboratory of the Netherlands Red Cross Blood Transfusion Service, Amsterdam, The Netherlands

Elisabeth Faustmann   Institute of General and Experimental Pathology, University of Vienna, Vienna, Austria

Alois Fellinger   Institute of General and Experimental Pathology, University of Vienna, Vienna, Austria

Maurizio Ferrari   Laboratorio di Ematologia e Sezione di Citogenetica, Istituti Clinici di Perfezionamento, 20122 Milano, Italy

H. Festenstein   Department of Immunology, London Hospital Medical Center, London E1 2AD, U.K.

Neal Flomenberg   Human Immunogenetics Laboratory, Memorial Sloan-Kettering Cancer Center, New York, New York 10021, U.S.A.

Shu Man Fu   Cancer Research Program, Oklahoma Medical Research Foundation, Oklahoma City, Oklahoma 73104, U.S.A.

Dietmar Geissler   Department of Internal Medicine, University of Innsbruck, Innsbruck, Austria

A.H.M. Geurts van Kessel   Department of Immunohaematology, Central Laboratory of the Netherlands Red Cross Blood Transfusion Service, 1006 AK Amsterdam, The Netherlands

Melvyn F. Greaves   Imperial Cancer Research Fund, London WC2A 3PX, U.K.

James D. Griffin   Division of Tumor Immunology, Dana-Farber Cancer Institute, Boston, Massachusetts 02115, U.S.A.

Kurt Grünewald   Department of Internal Medicine, University of Innsbruck, Innsbruck, Austria

O. Haas   St. Anna Childrens Hospital, Vienna, Austria

Sen-itiroh Hakomori   Division of Biochemical Oncology, Fred Hutchinson Cancer Research Center, Seattle, Washington 98104, U.S.A.

Wayne W. Hancock   Department of Nephrology, Prince Henry's Hospital, Melbourne, Australia

Toshiro Hara   Cancer Research Program, Oklahoma Medical Research Foundation, Oklahoma City, Oklahoma 73104, U.S.A.

Thierry Hercend   Division of Tumor Immunology, Dana-Farber Cancer Institute, Boston, Massachusetts 02115, U.S.A.

Michael A. Horton   Haemopoiesis Research Group, Department of Haematology, St. Bartholomew's Hospital, London EC1A 7BE, U.K.

Anna Janowska-Wieczorek   Department of Medicine, Cross Cancer Institute, Edmonton, Alberta T6G 1Z2, Canada

N. Joustra-Maas   Department of Immunohaematology, Central Laboratory of the Netherlands Red Cross Blood Transfusion Service, Amsterdam, The Netherlands

Fay E. Katz   Imperial Cancer Research Fund, London WC2A 3PX, U.K.

Nancy A. Kernan   Human Immunogenetics Laboratory, Memorial Sloan-Kettering Cancer Center, New York, New York 10021, U.S.A.

R.D. Klausner   Laboratory of Biochemistry and Metabolism, NIADDKD, National Institutes of Health, Bethesda, Maryland 20205, U.S.A.

W. Knapp   Institute of Immunology, Vienna, Austria

Robert W. Knowles   Human Immunogenetics Laboratory, Memorial Sloan-Kettering Cancer Center, New York, New York 10021, U.S.A.

Günther Konwalinka   Department of Internal Medicine, University of Innsbruck, Innsbruck, Austria

Dietrich Kraft   Institute of General and Experimental Pathology, University of Vienna, Vienna, Austria

Norbert Kraft   Department of Nephrology, Prince Henry's Hospital, Melbourne, Australia

M.J. Krantz   Department of Medicine, Cross Cancer Institute, Edmonton, Alberta T6G 1Z2, Canada

Lewis L. Lanier   Becton Dickinson Monoclonal Center, Inc., Mountain View, California 94043, U.S.A.

P.M. Lansdorp   Central Laboratory of the Netherlands Red Cross Blood Transfusion Service, 1006 AK Amsterdam, The Netherlands

J.G. Levy   Department of Microbiology, University of British Columbia, British Columbia, Canada

Denise Lewis   Haemopoiesis Research Group, Department of Haematology, St. Bartholomew's Hospital, London EC1A 7BE, U.K.

L. Linklater   Department of Pathology, University of Alberta, Edmonton, Alberta, Canada

Kristof Liszka   Institute of Immunology, University of Vienna, Vienna, Austria

P. M. Logan   Department of Microbiology, University of British Columbia, British Columbia, Canada

V. Lum   Department of Microbiology, University of British Columbia, British Columbia, Canada

D. Lutz   Hanusch Hospital, Vienna, Austria

O. Majdic   Institute of Immunology, Vienna, Austria

P. Mannoni   Department of Pathology, University of Alberta, Edmonton, Alberta, Canada

C. Mawas   Centre d'Immunologie Luminy, Marseille, France

Brian E. McMaster   Department of Pediatric Oncology, Fred Hutchinson Cancer Research Center, Seattle, Washington 98104, U.S.A.

Katrina McNulty   Haemopoiesis Research Group, Department of Haematology, St. Bartholomew's Hospital, London EC1A 7BE, U.K.

Frank Miedema   Central Laboratory of the Netherlands Red Cross Blood Transfusion Service, Amsterdam, The Netherlands

Cornelis J.M. Melief   Central Laboratory of the Netherlands Red Cross Blood Transfusion Service, 1006 AK Amsterdam, The Netherlands

Éva Monostori   Institute of Genetics, Biological Research Center, Hungarian Academy of Sciences, Szeged, Hungary

S. Naiman   Department of Pathology, Division of Hematopathology, Vancouver General Hospital, Vancouver, British Columbia, Canada

C. Navarrete   Department of Immunology, London Hospital Medical Center, London E1 2AD, U.K.

L.M. Neckers   Laboratory of Pathology, NCI, NIADDKD, National Institutes of Health, Bethesda, Maryland 20205, U.S.A.

D. Olive   Centre d'Immunologie Luminy, Marseille, France

M.G. Paindelli   First Department of Dermatology, University of Milan, Milan, Italy

Luciana Parola   Laboratorio di Ematologia e Sezione di Citogenetica, Istituti Clinici di Perfezionamento, 20122 Milano, Italy

C. Parravicini   Fifth Department of Pathology, University of Milan, Milan, Italy

D. Pearson   Department of Microbiology, University of British Columbia, British Columbia, Canada

J. Pena   Department of Biochemistry, Faculty of Medicine, University of Cordoba, Cordoba, Spain

Christian Peschel   Department of Internal Medicine, University of Innsbruck, Innsbruck, Austria

Joseph H. Phillips   Becton Dickinson Monoclonal Center, Inc., Mountain View, California 94043, U.S.A.

S. Pittaluga   Laboratory of Pathology, NCI, NIADDKD, National Institutes of Health, Bethesda, Maryland 20205, U.S.A.

Nicoletta Polli   Istituto di Clinica Medica I, 20122 Milano, Italy

R. Ramirez   Department of Biochemistry, Faculty of Medicine, University of Cordoba, Cordoba, Spain

Jerome Ritz   Division of Tumor Immunology, Dana-Farber Cancer Institute, Boston, Massachusetts 02115, U.S.A.

Lorenza Romitti   Laboratorio di Ematologia e Sezione di Citogenetica, Istituti Clinici di Perfezionamento, 20122 Milano, Italy

Helmut Rumpold   Institute of General and Experimental Pathology, University of Vienna, Vienna, Austria

Kert D. Sabbath   Division of Tumor Immunology, Dana-Farber Cancer Institute, Boston, Massachusetts 02115, U.S.A.

Raffaella Schirò   Laboratorio di Ematologia e Sezione di Citogenetica, Istituti Clinici di Perfezionamento, 20122 Milano, Italy

Stuart F. Schlossman   Division of Tumor Immunology, Dana-Farber Cancer Institute, Boston, Massachusetts 02115, U.S.A.

Reinhold E. Schmidt   Division of Tumor Immunology, Dana-Farber Cancer Institute, Boston, Massachusetts 02115, U.S.A.

Steve Self   Department of Pediatric Oncology, Fred Hutchinson Cancer Research Center, Seattle, Washington 98104, U.S.A.

Giuseppe Simoni   Laboratorio di Ematologia e Sezione di Citogenetica, Istituti Clinici di Perfezionamento, 20122 Milano, Italy

R. Solana   Department of Biochemistry, Faculty of Medicine, University of Cordoba, Cordoba, Spain

Timothy A. Springer   Laboratory of Membrane Immunochemistry, Dana-Farber Cancer Institute, Harvard Medical School, Boston, Massachusetts 02115, U.S.A.

Renate Steiner   Institute of General and Experimental Pathology, University of Vienna, Vienna, Austria

Hannes Stockinger   Institute of Immunology, University of Vienna, Vienna, Austria

Gabriele Stückler   Institute of General and Experimental Pathology, University of Vienna, Vienna, Austria

Robert Sutherland   Imperial Cancer Research Fund, London WC2A 3PX, U.K.

Frank W. Symington   Division of Biochemical Oncology, Fred Hutchinson Cancer Research Center, Seattle, Washington 98104, U.S.A.

László Takács   2nd Department of Anatomy, Semmelweis University of Medicine, Budapest H-1095 Tuzolto u. 58, Hungary

Masatoshi Takaishi   Cancer Research Program, Oklahoma Medical Research Foundation, Oklahoma City, Oklahoma 73104, U.S.A.

Fumio Takei   Terry Fox Laboratory, B.C. Cancer Research Centre, Vancouver, British Columbia V5Z 1L3, Canada

Fokke G. Terpstra   Central Laboratory of the Netherlands Red Cross Blood Transfusion Service, Amsterdam, The Netherlands

P.A.T. Tetteroo   Central Laboratory of the Netherlands Red Cross Blood Transfusion Service, 1006 AK Amsterdam, The Netherlands

Robert W. Tindle Beatson Institute for Cancer Research, Bearsden, Glasgow, Scotland, U.K.

Robert F. Todd III Division of Hematology and Oncology, Simpson Memorial Research Institute, Department of Internal Medicine, University of Michigan Medical School, Ann Arbor, Michigan 48109-0010, U.S.A.

A. Torres Department of Haematology, "Reina Sofia" Hospital, Cordoba, Spain

J.B. Trepel NCI, Navy Medical Oncology Branch, Naval Hospital, Bethesda, Maryland 20205, U.S.A.

J.F. Tromp Department of Immunohaematology, Central Laboratory of the Netherlands Red Cross Blood Transfusion Service, 1006 AK Amsterdam, The Netherlands

J.M. Turc CRC Blood Transfusion Service, Edmonton, Alberta, Canada

A.R. Turner Department of Medicine, Cross Cancer Institute, Edmonton, Alberta T6G 1Z2, Canada

Costante Valeggio Laboratorio di Ematologia e Sezione di Citogenetica, Istituti Clinici di Perfezionamento, 20122 Milano, Italy

C.M. van der Plas-van Dalen Department of Immunohaematology, Central Laboratory of the Netherlands Red Cross Blood Transfusion Service, Amsterdam, The Netherlands

M.B. van 't Veer Department of Immunohaematology, Central Laboratory of the Netherlands Red Cross Blood Transfusion Service, Amsterdam, The Netherlands

F.J. Visser Department of Immunohaematology, Central Laboratory of the Netherlands Red Cross Blood Transfusion Service, 1006 AK Amsterdam, The Netherlands

A.E.G. Kr. von dem Borne Department of Immunohaematology, Central Laboratory of the Netherlands Red Cross Blood Transfusion Service, 1006 AK Amsterdam, The Netherlands

Guy Werner Department of Immunohaematology, Central Laboratory of the Netherlands Red Cross Blood Transfusion Service, Amsterdam, The Netherlands

S. Whitney Department of Microbiology, University of British Columbia, British Columbia, Canada

B. Winkler-Lowen Department of Pathology, University of Alberta, Edmonton, Alberta, Canada

W.P. Zeijlemaker Laboratory for Experimental and Clinical Immunology, University of Amsterdam, Amsterdam, The Netherlands

CHAPTER 1

Joint Report of the Myeloid Section of the Second International Workshop on Human Leukocyte Differentiation Antigens

Irwin D. Bernstein and Steve Self

The purpose of the myeloid workshop was to continue to define clusters of antibodies identifying myeloid-associated antigens. This workshop also examined antibodies thought to react with natural killer (NK) cells, and with a group of heterodimeric molecules associated with specific leukocyte functions and referred to as the LFA family. The classification of myeloid antigens has advanced at a slower pace than the identification of T or B cell-associated antigens. This is because of the multiple cell populations, including granulocytes, monocytes, platelets, and red cells, and because of the complex biochemical analyses required to define the target antigens as many of the antigens appears to be carbohydrate determinants. Nonetheless, in the present workshop it was possible to define a number of new clusters in which antigens were grouped based on their distribution on hematopoietic cells and their biochemical nature.

One hundred and twenty antibodies were submitted by 41 laboratories in the form of ascites fluid in most cases. Of these, four were of insufficient quantity to be analyzed and 116 antibodies were analyzed in detail, two of which were withdrawn. Each participating laboratory* evaluated the panel of antibodies using one or more of the six myeloid protocols which are listed in Table 1.1. Participants were given information on antibody isotype, complement and protein A binding properties, and biochemical specificity provided by the contributors. In addition, antibodies thought to have anti-NK and LFA activity were designated. The submitted antibodies, their code numbers, and the submitting laboratories are in Table 1.2, and the characteristics of each antibody as designated by the contributing laboratory are in Table 1.3.

---

* The names of investigators participating in the myeloid workshop are in Table 1.2 and those contributing data reported in this chapter are in Table 1.4.

Table 1.1. Protocols at the second myeloid antigen workshop.

I. Serological studies of normal hematopoietic cells[a]

    A. Peripheral blood cells. 3 specimens of each of the following:

        1. Purified E-rosette-positive lymphocytes ("T cells").

        2. Purified B cells (E-rosette-negative, adherent cell-depleted lymphocytes).

        3. Purified adherent cells (monocytes).

        4. Purified granulocytes.

        5. Red blood cells.

        6. Platelets.

    B. Whole bone marrow (where possible)

II. Malignant cells[a]

    A. Malignant cell lines: HL-60, K562, KG1, and U 937 cells were evaluated. In addition, at least one malignant cell line of T cell origin and one cell line of B cell or pre–B cell origin were also evaluated (total of 6 lines).

    B. Fresh malignant hematopoietic cells: 4–5 acute myeloid leukemia cell specimens (FAB M1, M2, M3, M4, M5, or M6), one T-ALL, and one non-T ALL sample (6–7 specimens) are evaluated. Specimens were fresh or fresh frozen with DMSO in liquid nitrogen.

III. Biochemistry

    Studies of the glycolipid and/or glycoprotein nature of each of the antibodies were performed using techniques available in each laboratory. Information on the glycolipid or glycoprotein nature of the target antigen provided by contributors was distributed to the participants.

IV. Functional studies

    For this purpose the influence of antibodies on committed (CFU-GM, CFU-E, BFU-E) and multipotent (CFU-GEMM) hematopoietic progenitor cells was evaluated. In addition, studies on monocyte and/or granulocyte function were optional.

V. Antigen expression in tissue section

    As per description of participants.

VI. Other studies

    A. Anti–NK cell antibodies: A few of the antibodies were designated to have anti-NK activity by the contributor and the NK cell reactivity of each of these antibodies was evaluated. Evaluation of the anti-NK activity of the remaining antibodies in the panel was optional.

    B. Anti-LFA antibodies: A few investigators submitted antibodies against the LFA family of molecules and wished to focus on the biochemistry of these antigens. Antibodies said to have this activity were designated for those investigators.

---

[a] The serological studies in protocols I and II were performed using an indirect immunofluorescent test (FACS or microscope). According to the protocol, a developing reagent that recognizes *all* mouse immunoglobulin classes was used, and ascites fluids were used in a final dilution of 1 : 400 to stain $\leq 10^6$ cells in a volume ≥ 0.2 ml. Results were reported as percent of cells showing fluorescence greater than background. Percent purity of cells under study of $\geq 95\%$ (e.g., % E-rosette positive cells in T cell preparations) was required. Since the stability of many antibodies was not known, there was a recommendation to dilute the antibodies and to maintain them in frozen form in suitably sized aliquots and not subject them to repeated freezing and thawing. The antibodies provided were not sterile and some contained $\leq 0.1\%$ azide in the undiluted specimen.

Table 1.2. Antibody panel of the second myeloid antigen workshop.

Code no.	Antibody[a]	First author	Reference
WM-2 1	NKH2	Hercend, T.	*J. Clin. Invest.* (submitted)
WM-2 2	NKH1A	Hercend, T.	*J. Clin. Invest.* (submitted)
WM-2 3	82H3	Mannoni, P.	
WM-2 4	81H5	Mannoni, P.	
WM-2 5	82H5	Mannoni, P.	
WM-2 6	86H1	Mannoni, P.	
WM-2 7	3BB10	Clark, E.	
WM-2 8	CRIS-6	Vilella, R.	
WM-2 9	94-3D1	Vilella, R.	
WM-2 10	EDU3	Vilella, R.	(1984) *Thromb. Haemostas.* **51**:93
WM-2 11	JOAN-1	Vilella, R.	
WM-2 12	Mo5	Todd, R.	
WM-2 13	1G10	Bernstein, I.D.	(1982) *J. Immunol.* **128**:876
		Urdal, D.L.	(1983) *Blood* **62**:1022
		Andrews, R.G.	(1984) In: *Leucocyte typing*, A. Bernard *et al.*, eds. Springer-Verlag, Berlin, Heidelberg, pp. 396–404
WM-2 14	5F1	Bernstein, I.D.	(1982) *J. Immunol.* **128**:876
		Andrews, R.G.	(1984) In: *Leucocyte typing*, A. Bernard *et al.*, eds. Springer-Verlag, Berlin, Heidelberg, pp. 396–404
WM-2 15	T5A7	Andrews, R.G.	(1983) *Blood* **62**:124
		Symington, F.	(1984) *J. Biol. Chem.* **259**:6008
WM-2 16	L1B2	Andrews, R.G.	(1983) *Blood* **62**:124
WM-2 17	L4F3	Andrews, R.G.	(1983) *Blood* **62**:124
WM-2 18	TM.2.26	Maeda, H.	This report
WM-2 19	HLC5	Girardet, C.	
WM-2 20	BW242/408	Kurrle, R.	
WM-2 21	BW252/104	Kurrle, R.	(1984) *Behring Inst. Mitt.* **74**:49
WM-2 22	BW243/41	Kurrle, R.	
WM-2 23	CLBery2	von dem Borne, A.E.G. Kr.	

Table 1.2. *Continued*

Code no.	Antibody[a]	First author	Reference
WM-2 24	MHM 31	McMichael, A.	(1983) *Ann. N.Y. Acad. Sci.* **420**:251
WM-2 25	VIMC6	Stockinger, H.	(1984) *J. Natl. Cancer Inst.* **73**:7
		Knapp, W.	(In press) *Med. Oncol. Tumor Pharmacother.*
		Knapp, W.	(1984) In: *Research monographs in immunology*, Vol 3, C.P. Engelfriet *et al.*, eds. Elsevier Biomed Press, Amsterdam, p. 322
WM-2 26	VIM12	Knapp, W.	(In press) *Med. Oncol. Tumor Pharmacother.*
		Knapp, W.	(In press) In: *Handbook of monoclonal antibodies*, M.P. Dierich and S. Ferrone, eds. Noyes Publications, Park Ridge, N.J.
WM-2 27	VIM8	Knapp, W.	(In press) *Med. Oncol. Tumor Pharmacother.*
		Knapp, W.	(In press) In: *Handbook of monoclonal antibodies*, M.P. Dierich and S. Ferrone, eds. Noyes Publications, Park Ridge, N.J.
		Bettelheim, P.	(In press) *Brit. J. Haematol.*
		Majdic, O.	(1984) *Int. J. Cancer* **33**:617
WM-2 28	VIM2	Knapp, W.	(1983) *Ann. N.Y. Acad. Sci.* **420**:251
		Knapp, W.	(1984) In: *Leucocyte typing*, A. Bernard *et al.*, eds. Springer-Verlag, Berlin, Heidelberg, pp. 564–573
		Knapp, W.	(1983) *Berh. Dtsch. Ges. Path.* **67**:54
		Knapp, W.	(In press) In: *Histopathology of the bone marrow*, K. Lennert, ed. Fisher-Verlag
		Knapp, W.	(1984) In: *Research monographs in immunology*, Vol 3, C.P. Englefreit *et al.*, eds. Elsevier Biomed Press, Amsterdam, p. 322
		Knapp, W.	(In press) In: *Flow cytometry and monoclonal antibodies for therapy monitoring: Quo vadis?* P. Gros *et al.*, eds. Sanofi Group, Montpellier
		Knapp, W.	(In press) *Med. Oncol. Tumor Pharmacother.*
WM-2 29	VIM3	Knapp, W.	(In press) *Med. Oncol. Tumor Pharmacother.*
WM-2 30	VIM10	Knapp, W.	(In press) *Med. Oncol. Tumor Pharmacother.*
WM-2 31	VIM13	Knapp, W.	(In press) In: *Handbook of monoclonal antibodies*, M.P. Dierich and S. Ferrone, eds. Noyes Publications, Park Ridge, N.J.

WM-2 32	E2A7	Thompson, J.	(1984) In: *Advances in immunobiology: Blood cell antigens and bone marrow transplantation*, J.S. Thompson *et al.*, eds. Alan Liss, Inc., New York, pp. 169–187.
WM-2 33	G7C5	Thompson, J.	(1984) In: *Advances in immunobiology: Blood cell antigens and bone marrow transplantation*, J.S. Thompson *et al.*, eds. Alan Liss, Inc., New York, pp. 169–187.
WM-2 34	M5E2	Thompson, J.	(1984) In: *Advances in immunobiology: Blood cell antigens and bone marrow transplantation*, J.S. Thompson *et al.*, eds. Alan Liss, Inc., New York, pp. 169–187.
WM-2 35	AN51	McMichael, A.	(1984) *Brit. J. Haemat.* **49**:501
WM-2 36	M101	Knowles, R.W.	Unpublished results
WM-2 37	MA5	Ando, I.	(1983) *Lancet* **24**:1500
		Ando, I.	(1984) In: *Tissue culture and RES*, P. Rohlich and E. Bacsy, eds. Akademiai Kiado, Budapest, pp. 241–246
WM-2 38	JD2	Ando, I.	(1983) *Lancet* **24**:1500
		Ando, I.	(1984) In: *Tissue culture and RES*, P. Rohlich and E. Bacsy, eds. Akademiai Kiado, Budapest, pp. 241–246
WM-2 39	GB3	Ando, I.	(1983) *Lancet* **24**:1500
		Ando, I.	(1984) In: *Tissue culture and RES*, P. Rohlich and E. Bacsy, eds. Akademiai Kiado, Budapest, pp. 241–246
WM-2 40	MB1	Ando, I.	(1983) *Lancet* **24**:1500
		Ando, I.	(1984) In: *Tissue culture and RES*, P. Rohlich and E. Bacsy, eds. Akademiai Kiado, Budapest, pp. 241–246
WM-2 41	CC1.7	Ando, I.	(1983) *Lancet* **24**:1500
		Ando, I.	(1984) In: *Tissue culture and RES*, P. Rohlich and E. Bacsy, eds. Akademiai Kiado, Budapest, pp. 241–246
WM-2 42	D12	Clement, L.	(In press) *Leuk. Res.*
WM-2 43	NHL30.5	Askew, D.	
WM-2 44	J.15	McMichael, A.	
WM-2 45	TA-60b	Ueda, R.	(In press) *Cancer Res.*
WM-2 46	SHCL3	Schwarting, R.	
WM-2 47	SJ-1D1	Mirro, J.	

Table 1.2. *Continued*

Code no.	Antibody[a]	First author	Reference
WM-2 48	MY-9	Griffin, J.	(1984) *Leuk. Res.* **8**:521
WM-2 49	G7C10	Thompson, J.	(1984) *J. Am. Ger. Soc.* **32**:274
WM-2 50	KD3	Ando, I.	(1983) *Lancet* **24**:1500
		Ando, I.	(1984) In: *Tissue culture and RES*, P. Rohlich and E. Bacsy, eds. Akademiai Kiado, Budapest, pp. 241–246
WM-2 51	BL-5	Lebacq-Verheyden, A.-M.	
WM-2 52	HG-1	Tai, P.C.	(Submitted) *Clin. Exp. Immunol.*
WM-2 53	UCHL1	Beverley, P.	
WM-2 54	G022b	Thompson, J.	(1984) *J. Am. Ger. Soc.* **32**:274
WM-2 55	MHM23	Hildreth, J.E.K.	(1983) *Eur. J. Immunol.* **13**:202
WM-2 56	MHM24	Hildreth, J.E.K.	(1983) *Eur. J. Immunol.* **13**:202
WM-2 57	CLBFcRgran1	Miedema, F.	(1984) *Eur. J. Immunol.* **14**:518
WM-2 58	VEP13	Rumpold, H.	(1982) *J. Immunol.* **129**:1458
		Rumpold, H.	(1983) *Immunobiol.* **164**:51
		Schuller-Petrovic, S.	(1983) *Nature* **306**:179
		Rumpold, H.	(1984) In: *Protides of the biological fluids*, Vol 31, H. Peeters, ed. pp. 859–862
		Rumpold, H.	(In press) *Clin. Exp. Immunol.*
		Perussia, B.	(1984) *J. Immunol.* **133**:180
		Ziegler-Heitbrock, H.W.L.	(In press) *Clin. Exp. Immunol.*
		Ziegler-Heitbrock, H.W.L.	(In press) *Blood*
WM-2 59	BW209/2	Kurrle, R.	
WM-2 60	BW227/19	Kurrle, R.	(1984) *Proc. Am. Pancr. Assoc. Cancer Project*
WM-2 61	29	Hogg, N.	
WM-2 62	MoU26	Winchester, R.	(1983) *J. Immunol.* **130**:145
WM-2 63	MoS1	Winchester, R.	(1983) *J. Clin. Invest.* **71**:1633
WM-2 64	R17	Winchester, R.	(1984) *Blood* **64**:237
WM-2 65	MoU48	Winchester, R.	(1984) *Med. Oncol. Tumor Pharmacother.* **1**:263
WM-2 66	MoU28	Winchester, R.	

WM-2 No.	Antibody	Author	Reference
WM-2 67	MoP9	Winchester, R.	
WM-2 68	MoP15	Winchester, R.	
WM-2 69	MoS39	Winchester, R.	
WM-2 70	GA-1	Hiraiwa, A.	
WM-2 71	Ki-M4b	Radzun, H.J.	(1983) *Cell. Immunol.* **82:**174
		Parwaresch, M.R.	(1983) *Blood* **62:**585
		Parwaresch, M.R.	(1983) *J. Immunol.* **141:**2719
		Radzun, H.J.	(In press) *Am. J. Pathol.*
WM-2 73	CLB-LFA1/1	Miedma, F.	(1984) *Eur. J. Immunol.* **14:**518
WM-2 74	CIPAN	de Kretser, T.A.	(Submitted) *Hybridoma*
WM-2 75	60.3	Beatty, P.G.	(1983) *J. Immunol.* **131:**2913
WM-2 76	60.1	Beatty, P.G.	(1983) *J. Immunol.* **131:**2913
WM-2 77	G9F9	Thompson, J.	(1984) In: *Advances in immunobiology: Blood cell antigens and bone marrow transplantation,* J.S. Thompson et al., eds. Alan Liss, Inc., New York, pp. 169–187.
WM-2 78	M3D11	Thompson, J.	(1984) In: *Advances in immunobiology: Blood cell antigens and bone marrow transplantation,* J.S. Thompson et al., eds. Alan Liss, Inc., New York, pp. 169–187.
WM-2 79	7C11	Lansdorp, P.	
WM-2 80	CLBgran7	Geurts van Kessel, A.H.M.	(1984) *J. Immunol.* **133:**1265
WM-2 81	CLBgran10	von dem Borne, A.E.G. Kr.	
WM-2 82	CLBgran11	von dem Borne, A.E.G. Kr.	
WM-2 83	BL-1	Lebacq-Verheyden, A-M.	
WM-2 84	BL-2	Lebacq-Verheyden, A-M.	
WM-2 85	BL-3	Lebacq-Verheyden, A-M.	
WM-2 86	N901	Griffin, J.	(1983) *J. Immunol.* **130:**2947
WM-2 88	Ki-M5	Radzun, H.J.	(1983) *Cell. Immunol.* **82:**174
		Parwaresch, M.R.	(1983) *Blood* **62:**585
		Parwaresch, M.R.	(1983) *J. Immunol.* **131:**2719
		Radzun, H.J.	(In press) *Am. J. Pathol.*

Table 1.2. *Continued*

Code no.	Antibody[a]	First author	Reference
WM-2 89	TS1/18.11	Sanchez-Madrid, F.	(1982) *Proc. Natl. Acad. Sci. U.S.A.* **79:**7489
		Krensky, A.M.	(1983) *J. Immunol.* **131:**611
		Ware, C.F.	(1983) *J. Immunol.* **131:**1182
		Sanchez-Madrid, F.	(1983) *J. Exp. Med.* **158:**1785
		Krensky, A.M.	(1984) *J. Immunol.* **132:**2180
		Anderson, D.C.	(1984) *J. Clin. Invest.* **74:**536
		Springer, T.A.	(In press) *Fed. Proc.*
		Anderson, D.C.	(In press) *Fed. Proc.*
		Kohl, S.	(In press) *J. Immunol.*
WM-2 90	Ki-M3	Radzun, H.J.	(1983) *Cell. Immunol.* **82:**174
		Parwaresch, M.R.	(1983) *Blood* **62:**585
		Parwaresch, M.R.	(1983) *J. Immunol.* **131:**2719
		Radzun, H.J.	(In press) *Am. J. Pathol.*
WM-2 91	28	Hogg, N.	(In press)
WM-2 92	Ki-M6	Radzun, H.J.	(1983) *Cell. Immunol.* **82:**174
		Parwaresch, M.R.	(1983) *Blood* **62:**585
		Parwaresch, M.R.	(1983) *J. Immunol.* **131:**2719
		Radzun, H.J.	(In press) *Am. J. Pathol.*
WM-2 93	Ki-M1	Radzun, H.J.	(1983) *Cell. Immunol.* **82:**174
		Parwaresch, M.R.	(1983) *Blood* **62:**585
		Parwaresch, M.R.	(1983) *J. Immunol.* **131:**2719
		Radzun, H.J.	(In press) *Am. J. Pathol.*
WM-2 94	G035	Thompson, J.	(1984) In: *Advances in immunobiology: Blood cell antigens and bone marrow transplantation*, J.S. Thompson *et al.*, eds. Alan Liss, Inc., New York, pp. 169–187.
WM-2 95	SA-1	Avnstrom, S.	(In press) *Scand. J. Haematol.*
WM-2 96	24	Hogg, N.	(Submitted)
WM-2 97	T10C6	Thompson, J.	
WM-2 98	M5D12	Thompson, J.	(1984) In: *Advances in immunobiology: Blood cell antigens and bone marrow transplantation*, J.S. Thompson *et al.*, eds. Alan Liss, Inc., New York, pp. 169–187.
		Thompson, J.	(1984) *J. Am. Ger. Soc.* **32:**274

Code	Antibody	Author	Reference
WM-2 99	UCHM1	Hogg, N.	(In press) Immunology
WM-2 100	M3C7.2A	Thompson, J.	(1984) In: Advances in immunobiology: Blood cell antigens and bone marrow transplantation, J.S. Thompson et al., eds. Alan Liss, Inc., New York, pp. 169–187.
WM-2 101	PL1	Lebacq-Verheyden, A.-M.	
WM-2 102	CIKM5	Pilkington, G.	(1984) In: Leucocyte typing, A. Bernard et al., eds. Springer-Verlag, Berlin, Heidelberg
WM-2 103	PMN-3	Winchester, R.	
WM-2 104	44	Hogg, N.	
WM-2 105	Leu 7	Abo, T.	(1981) J. Immunol. 127:1024
		McGarry, R.C.	(1983) Nature 306:376
		Kubagawa, H.	(1983) Fed. Proc. 42:1219
		Abo, T.	(1982) J. Exp. Med. 155:321
		Lanier, L.L.	(1983) Immunol. Rev. 74:143
		Lanier, L.L.	(1983) J. Immunol. 131:1789
WM-2 106	Leu M-2	Raff, H.V.	(1980) J. Exp. Med. 152:581
		Hausman, P.B.	(1980) J. Immunol. 125:1374
		Gonwa, T.	Unpublished data
		Gonwa, T.	(In press) J. Immunol.
WM-2 107	CIMT	Pilkington, G.	(Submitted) Clin. Exp. Immunol.
WM-2 110	LMA-1	Tai, P.C.	(Submitted) Clin. Exp. Immunol.
WM-2 111	LMA-3	Tai, P.C.	(Submitted) Clin. Exp. Immunol.
WM-2 112	LMA-2	Tai, P.C.	
WM-2 113	PMN-1	Winchester, R.	
WM-2 114	CIM	Pilkington, G.	
WM-2 115	CAMAL-1	Malcolm, A.	(1984) Exp. Hematol. 12:539
		Logan, P.	(1984) Diagn. Immunol. 2:86
WM-2 116	HG-2	Tai, P.C.	(Submitted) Clin. Exp. Immunol.
WM-2 117	HG-3	Tai, P.C.	(Submitted) Clin. Exp. Immunol.
WM-2 118	PAM-1	Biondi, A.	(1984) J. Immunol. 132:1

a Antibodies CLBGran5 and TS1/22 were withdrawn because of possible contamination with other antibodies; antibodies Leu 112 and Leu M-1 were not analyzed because amounts submitted were insufficient.

Table 1.3. Characteristics of myeloid antibody panel.

Code no.	Lab no.	Antibody	Binds SPA	Fixes C'	Ig subclass
WM-2 1	74	NKH2	No	Yes	IgM
WM-2 2	74	NKH1A	No	Yes	IgM
WM-2 3	80	82H3	No	No	IgG1
WM-2 4	80	81H5	ND[a]	ND	IgG
WM-2 5	80	82H5	No	Yes	IgM
WM-2 6	80	86H1	ND	Yes	IgG2
WM-2 7	66	3BB10	ND	ND	ND
WM-2 8	69	CRIS-6	Yes	Yes	IgG
WM-2 9	69	94-3D1	ND	No	ND
WM-2 10	69	EDU3	Yes	Yes	IgG2a
WM-2 11	69	JOAN-1	Yes	Yes	IgG
WM-2 12	77	Mo5	Yes	Yes	IgG2a
WM-2 13	29	1G10	No	Yes	IgM
WM-2 14	29	5F1	No	Yes	IgM
WM-2 15	29	T5A7	No	Yes	IgM
WM-2 16	29	L1B2	Yes	Yes	IgG2a
WM-2 17	29	L4F3	No	Yes	IgM
WM-2 18	46	TM.2.26	ND	Yes	IgG2b
WM-2 19	88	HLC5	Yes	Yes	IgM
WM-2 20	53	BW242/408	(weak)	Yes	IgG3
WM-2 21	53	BW252/104	(weak)	ND	IgG3
WM-2 22	53	BW243/41	No	Yes	IgM
WM-2 23	15	CLBery2	ND	ND	IgM
WM-2 24	27	MHM31	ND	ND	IgG
WM-2 25	116	VIMC6	ND	Yes	IgM
WM-2 26	116	VIM12	Yes	No	IgG1
WM-2 27	116	VIM8	No	Yes	IgM
WM-2 28	116	VIM2	No	Yes	IgM
WM-2 29	116	VIM3	ND	ND	ND
WM-2 30	116	VIM10	No	Yes	IgM
WM-2 31	116	VIM13	ND	Yes	IgM
WM-2 32	21	E2A7	ND	Yes	IgM
WM-2 33	21	G7C5	ND	Yes	IgM
WM-2 34	21	M5E2	ND	Yes	IgG2a
WM-2 35	27	AN51	Yes	No	IgG1
WM-2 36	26	M101	No	ND	IgG1
WM-2 37	23	MA5	No	No	IgG1
WM-2 38	23	JD2	Yes	Yes	IgG1
WM-2 39	23	GB3	Yes	No	IgG1
WM-2 40	23	MB1	No	Yes	IgG1
WM-2 41	23	CC1.7	Yes	No	IgG1
WM-2 42	35	D12	Yes	Yes	IgG2a
WM-2 43	33	NHL30.5	ND	No	IgG
WM-2 44	27	J.15	No	Yes	IgM
WM-2 45	10	TA-60B	Yes	No	IgG1
WM-2 46	70	SHCL3	Yes	ND	IgG2b
WM-2 47	30	SJ-1D1	No	No	IgG1
WM-2 48	12	MY-9	ND	Yes	IgG2b
WM-2 49	21	G7C10	ND	Yes	IgM
WM-2 50	23	KD3	No	No	IgG1
WM-2 51	67	BL-5	ND	ND	ND
WM-2 52	19	HG-1	ND	ND	ND
WM-2 53	22	UCHL1	ND	No	IgG2a
WM-2 54	21	G022b	ND	Yes	IgM
WM-2 55	27	MHM23	Yes	No	IgG1
WM-2 56	27	MHM24	ND	ND	ND
WM-2 57	15	CLBFcRgran1	Yes	ND	IgG2
WM-2 58	109	VEP13	No	Yes	IgM

Table 1.3. *Continued*

Code no.	Lab no.	Antibody	Binds SPA	Fixes C'	Ig subclass
WM-2 59	53	BW209/2	ND	ND	ND
WM-2 60	53	BW227/19	No	Yes	IgG1
WM-2 61	25	29	ND	ND	IgM
WM-2 62	42	MoU26	Yes	ND	IgG2a
WM-2 63	42	MoS1	Yes	ND	IgG2a
WM-2 64	42	R17	Yes	ND	IgG2a
WM-2 65	42	MoU48	Yes	ND	IgG2a
WM-2 66	42	MoU28	Yes	ND	IgG2a
WM-2 67	42	MoP9	Yes	ND	IgG2b
WM-2 68	42	MoP15	No	ND	IgG1
WM-2 69	42	MoS39	Yes	ND	IgG2a
WM-2 70	81	GA-1	ND	ND	IgG1
WM-2 71	95	Ki-M4b	ND	ND	ND
WM-2 73	51	CLBLFA1/1	ND	ND	IgG1
WM-2 74	39	CIPAN	at pH 8	ND	IgG1
WM-2 75	79	60.3	Yes	ND	IgG2a
WM-2 76	79	60.1	Yes	ND	IgG2a
WM-2 77	21	G9F9	ND	Yes	IgM
WM-2 78	21	M3D11	ND	Yes	IgM
WM-2 79	83	7C11	ND	ND	IgM
WM-2 80	15	CLBgran7	ND	ND	IgM
WM-2 81	15	CLBgran10	ND	ND	IgG1
WM-2 82	15	CLBgran11	ND	ND	IgG2
WM-2 83	67	BL-1	No	ND	IgG
WM-2 84	67	BL-2	ND	ND	ND
WM-2 85	67	BL-3	ND	ND	ND
WM-2 86	12	N901	ND	No	IgG1
WM-2 88	95	Ki-M51	ND	ND	IgG2a
WM-2 89	90	TS1/18.11	Yes	ND	IgG1
WM-2 90	95	Ki-M3	ND	ND	IgM
WM-2 91	25	28	ND	ND	IgM
WM-2 92	95	Ki-M6	ND	ND	IgG2a
WM-2 93	95	Ki-M1	ND	ND	IgM
WM-2 94	21	G035	ND	Yes	IgM
WM-2 95	84	SA-1	Yes	Yes	IgG
WM-2 96	25	24	ND	ND	IgG1
WM-2 97	21	T10C6	ND	Yes	IgM
WM-2 98	21	M5D12	ND	Yes	IgM
WM-2 99	22	UCHM1	Yes	Yes	IgG2a
WM-2 100	21	M3C7.2A	ND	Yes	IgM
WM-2 101	67	PL1	ND	ND	(Rat)
WM-2 102	39	CIKM5	at pH 8	No	IgG1
WM-2 103	42	PMN-3	Yes	ND	IgG2a
WM-2 104	25	44	ND	ND	IgG1
WM-2 105	47	Leu 7	No	Yes	IgM
WM-2 106	47	Leu M-2	No	Yes	IgM
WM-2 107	39	CIMT	ND	ND	IgG1
WM-2 110	19	LMA-1	ND	ND	ND
WM-2 111	19	LMA-3	ND	ND	ND
WM-2 112	19	LMA-2	ND	ND	ND
WM-2 113	42	PMN-1	Yes	ND	IgG2a
WM-2 114	39	CIM	at pH 8	No	IgG1
WM-2 115	118	CAMAL-1	Yes	No	IgG1
WM-2 116	19	HG-2	ND	ND	ND
WM-2 117	19	HG-3	ND	ND	ND
WM-2 118	76	PAM-1	(weak)	ND	IgG1

[a] ND: Not determined.

Table 1.4. Participating laboratories.

	Lab no.	Name
Protocol I		
	12	J. Griffin
	15	A.E.G. Kr. von dem Borne
	21	J. Thompson
	39	G. Pilkington
	42	R. Winchester
	53	R. Kurrle
	77	R. Todd
Protocol II		
	14	G. Cattoretti
	15	A.E.G. Kr. von dem Borne
	21	J. Thompson
	29	I. Bernstein
	30	J. Mirro
	39	G. Pilkington
	42	R. Winchester
	46	H. Maeda
	67	A-M. Lebacq-Verheyden
	77	R. Todd
	80	P. Mannoni
	92	P. Wernet
	116	W. Knapp

In this chapter we have compiled and summarized the data obtained from immunofluorescence studies using normal cells (protocol I) and using malignant cells and cell lines (protocol II). The participants who provided the data for protocols I and II are listed in Table 1.4, and the number of samples analyzed for each cell type in these protocols are presented in Table 1.5. In addition, we have also summarized certain of the biochemical data, particularly those provided by Tetteroo *et al.*, Springer *et al.*, and Symington *et al.*, which are presented in greater detail in Chapters 2, 3, and 4 of this volume. In the following sections in this chapter we detail the statistical methods used to form preliminary clusters, based on results of protocols I and II, followed by a description of the preliminary groupings formed by this analysis. Afterwards, newly defined CD and CDw groupings based on the distribution of antibody reactions with specific cell types as well as biochemical data are described. The remainder of the laboratory studies based on protocols III–VI are detailed in subsequent chapters in this volume.

Statistical Analysis

The data analysis proceeded in three phases: (1) patterns of variation among replicate observations within labs were analyzed, (2) patterns of variation between labs were explored, and (3) distance matrices

Table 1.5. Cell types analyzed.

Target cell code		No. of samples	No. of laboratories
Protocol I			
01	Granulocytes	17	7
02	Monocytes	12	5
03	T lymphocytes	8	3
04	Non-T lymphocytes	8	3
20	Red blood cells	11	4
21	Platelets	5	3
Protocol II			
06	HL-60	12	12
07	KG-1	9	9
08	U 937	12	12
09	K562 leukemia	9	9
12	ANLL M1	11	6
13	ANLL M2	13	7
14	ANLL M3	5	5
15	ANLL M4	9	7
16	ANLL M5	6	6
17	ANLL M6	0	0
18	Fresh acute T leukemic cells	7	7
19	Fresh acute non-T leukemic cells	11	8
22P[a]	T leukemic cell lines	7	7
30P[b]	B leukemic cell lines	5	5

[a] 22P included studies of the HSB-2 (1 laboratory), MOLT-4 (2), Jurkat (2), and CEM (2) cell lines.
[b] 30P included studies of the Daudi (3 laboratories) and NALM-6 (2 laboratories) cell lines.

among the antibodies were calculated and a cluster analysis was performed.

In the first phase, variation among replicate observations was examined by level of response. Because it was found that the variation was much less for those target cell/antibody combinations with responses near zero or 100 than for those combinations with intermediate-level responses, the data were transformed to the arcsin-square root scale in an attempt to stabilize the variance. The rest of the analysis was conducted on this transformed scale. Variation among replicates between labs was also examined. From this examination, data from one lab were eliminated from the rest of the analysis due to poor repeatability. The data were then reduced to a single response for each lab/target cell/antibody combination by averaging over any replicate observations. Also, if fewer than three labs contributed data on any target cell, that target cell was eliminated from further consideration.

The second phase of the analysis proceeded by computing for each target cell/antibody combination, a median response over labs. The distribution of deviations from these median responses was displayed in stem and leaf diagrams for each lab/target cell combination and examined visually. From this procedure it was determined that the results from one lab were not at all congruent with the results from other labs and so data from this lab were excluded from the rest of the analysis. Also, a total of 62 lab/target cell/antibody combinations (out of approximately 15,000) were identified as gross outliers by this procedure and excluded.

Distances between antibodies were calculated for each lab/target cell combination as the absolute difference between levels of response and then these were averaged over labs. A total distance between antibodies was then calculated by averaging over target cells. Cluster analysis based on this total distance was then performed by the hierarchical, complete linkage clustering algorithm. Calculations were performed by the computer program BMDP1M. The calculations of the distance matrix and the clustering method used correspond exactly to those used in the First Workshop (except for the transformation applied to the data described above). Clusters were defined from the tree by using the within-replicate variance as an informal guide as to which groups of antibodies would be significantly different if a formal statistical test were applied.

Results

A summary of the results of the cluster analysis based on the reactivity of the antibodies with all of the cells in protocols I and II is shown in Fig. 1.1. This analysis revealed 28 distinct groupings of antibodies or individual antibodies which failed to cluster. Based on these studies two new CD groupings, CD15 (X-hapten) and CD16 (Fc receptors on granulocytes), and two new CDw groupings, CDw17 (lactosylceramide) and CDw18 (LFA-1), were designated, and antibodies were added to CD11 (C3bi receptor) and CDw14 [monocyte antigen(s)].

These 28 preliminary groups are listed in Table 1.6. The mean percent reactivity of each antibody cluster with cells is shown in Table 1.7(A) and with malignant cell lines and fresh malignant cells in Table 1.7(B) and 1.7(C), respectively. Also listed is the range of reactivity of each antibody cluster with each cell type. A wide range of reactivity denotes heterogeneity of antibody reactivity with the particular target cell. The reactivity of individual antibodies that did not group in clusters is presented in Table 1.8. Below we briefly discuss each of the antibody groupings listed in Table 1.6 that consists of four or more antibodies, as well as one cluster of two antibodies in which both displayed the same biochemical specificity. Following this, the groupings that were sufficiently characterized to be designated as a CD or as a provisional CDw grouping are discussed.

Group 1

The antibodies in this group displayed little or no reactivity with the normal and malignant cell panels, except for two antibodies reactive with

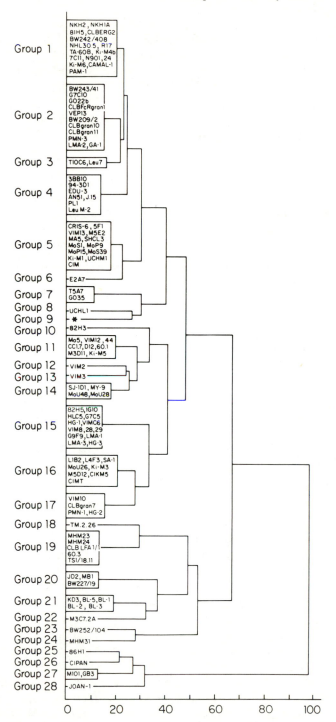

Fig. 1.1. Cluster tree of all cells from the myeloid protocol.

Table 1.6. Preliminary serologically defined groups.

Group 1:	NKH2, NKH1A, 81H5, BW242/408, CLBery2, NHL30.5, TA-60B, R17, Ki-M4b, 7C11, N901, Ki-M6, 24, CAMAL-1, PAM-1
Group 2:	BW243/41, G7C10, G022b, CLBFcRgran1, VEP13, BW209/2, GA-1, CLB-gran10, CLBgran11, PMN-3, LMA-2
Group 3:	T10C6, Leu7
Group 4:	3BB10, 94-3D1, EDU3, AN51, J.15, PL1, Leu M-2
Group 5:	CRIS-6, 5F1, VIM13, M5E2, MA5, SHCL3, MoS1, MoP9, MoP15, MoS39, Ki-M1, UCHM1, CIM
Group 6:	E2A7
Group 7:	T5A7, G035
Group 8:	UCHL1
Group 9:	^{a}
Group 10:	82H3
Group 11:	Mo5, VIM12, CC1.7, D12, 60.1, M3D11, Ki-M5, 44
Group 12:	VIM2
Group 13:	VIM3
Group 14:	SJ-1D1, MY-9, MoU48, MoU28
Group 15:	92H5, 1G10, HLC5, VIMC6, VIM8, G7C5, HG-1, 29, G9F9, 28, LMA-1, LMA-3, HG-3
Group 16:	L1B2, L4F3, MoU26, Ki-M3, SA-1, M5D12, CIKM5, CIMT
Group 17:	VIM10, CLBgran7, PMN-1, HG-2
Group 18:	TM.2.26
Group 19:	MHM23, MHM24, CLB-LFA1/1, 60.3, TS1/18.11
Group 20:	JD2, MB1, BW227/19
Group 21:	KD3, BL-5, BL-1, BL-2, BL-3
Group 22:	M3C7.2A
Group 23:	BW252/104
Group 24:	MHM31
Group 25:	86H1
Group 26:	CIPAN
Group 27:	M101, GB3
Group 28:	JOAN-1

^{a} Group 9 contained the antibody CLB gran 5 which was withdrawn.

red blood cells and six reactive with platelets. Many of the antibodies in this group are known to react with cell types that were not sufficiently enriched in the test cell populations to allow detection of positive reactions. For example, NKH2, NKH1A, and N901 which are known to react with NK cells did not react with sufficient numbers of cells in the populations tested to yield a positive signal. Similarly, the reactivity of TA-60b which reacts with activated T cells was not identified.

Two antibodies reacted with approximately 50% of red blood cells. One of these, CLB ery 2, reacted only with red blood cells and is known to react with globoside. The second erythrocyte-reactive antibody, 7C11, also reacted with platelets. This group also included antibodies that were not expected to react with normal myeloid cells, e.g., CAMAL-1. Nonetheless, it is conceivable that, in some instances, reactivity was not detected because of loss of antibody affinity or titer.

Table 1.7(A). Preliminary groupings of myeloid workshop antibodies (Groups of two or more antibodies): Reactions with normal cells.[a]

Preliminary group	No. Abs	Granulocyte	Monocyte	T-lymph	B-lymph	RBC	Platelets
1	15	6.5	7.5	2.6	5.7	8.1	17.6
		(3.1–12.3)	(1.7–20.3)	(0.3–6.3)	(1.9–16.0)	(0.3–55.3)	(1.8–63.6)
2	11	79.0	12.0	5.2	16.1	0.8	11.6
		(53.7–94.0)	(2.6–23.1)	(1.2–13.0)	(3.0–30.9)	(0.4–1.7)	(1.8–28.9)
3	2	3.6	6.8	42.4	33.1	1.3	15.4
		(2.1–5.0)	(3.7–9.8)	(11.4–73.4)	(22.5–43.7)	(0.9–1.8)	(2.5–28.3)
4	7	11.8	32.4	2.0	12.1	1.5	77.0
		(3.2–23.2)	(18.0–46.5)	(1.1–4.5)	(5.5–20.0)	(1.0–2.4)	(44.4–89.3)
5	13	10.6	75.7	2.5	10.9	0.7	22.0
		(3.3–24.5)	(31.3–89.5)	(1.1–6.1)	(7.4–17.5)	(0.3–1.6)	(1.1–88.9)
7	2	93.4	52.4	5.5	29.3	3.3	80.8
		(93.0–93.8)	(44.4–60.4)	(2.4–8.6)	(15.6–43.0)	(2.8–3.8)	(78.6–83.0)
11	8	85.6	84.1	3.9	18.7	0.7	17.8
		(63.0–93.4)	(72.6–95.8)	(1.5–6.2)	(9.7–36.5)	(0.4–1.4)	(2.0–27.5)
14	4	43.0	54.6	3.3	10.7	0.6	2.3
		(6.2–57.6)	(41.3–69.3)	(1.3–8.3)	(8.3–14.8)	(0.3–1.3)	(1.8–2.5)
15	13	86.4	11.9	3.6	6.8	1.0	9.7
		(76.9–93.2)	(3.8–24.3)	(1.8–9.8)	(4.8–9.3)	(0.3–3.6)	(1.4–27.5)
16	8	12.1	19.6	3.6	6.8	0.5	15.3
		(3.4–28.2)	(4.6–32.7)	(0.8–6.1)	(1.8–14.2)	(0.3–0.7)	(2.0–28.4)
17	4	84.2	6.1	1.9	5.0	5.9	21.7
		(75.4–91.8)	(2.7–9.7)	(1.3–2.6)	(3.4–6.5)	(0.6–19.7)	(1.8–51.9)
19	5	85.5	89.4	79.0	62.2	1.5	30.8
		(63.6–94.8)	(80.0–97.6)	(67.6–89.1)	(55.0–71.0)	(0.6–3.2)	(28.8–38.8)
20	3	11.8	65.3	3.8	32.4	0.2	21.0
		(4.4–18.7)	(52.2–83.4)	(1.9–6.5)	(32.0–33.0)	(0.0–0.3)	(1.9–33.8)
21	5	9.2	15.2	36.2	59.0	0.8	24.1
		(3.8–22.8)	(10.1–22.9)	(27.5–48.0)	(46.1–74.1)	(0.5–1.1)	(4.0–29.6)
27	2	85.9	96.2	92.5	82.7	1.3	29.3
		(77.1–94.6)	(95.2–97.2)	(90.7–94.4)	(76.4–89.0)	(1.2–1.4)	(28.9–29.6)

[a] % Mean reactivity by indirect immunofluorescence (range of reactivity).

Group 2

This group contained a number of antibodies most of which were shown to react with Fc receptors on granulocytic cells. Thus, these antibodies were included in a cluster designated as CD16 which is discussed in greater detail at the end of this chapter.

Group 4

This group included antibodies which reacted mainly with platelets. This group is clearly heterogeneous. Certain antibodies in this group are known to detect platelet glycoproteins IIb/IIIa (EDU3, J.15), while others are known to detect glycoprotein Ib (AN51).

Table 1.7(B). Preliminary groupings of myeloid workshop antibodies: Leukemic cell lines.[a]

Pre-limi-nary group	No. Abs	HL-60	KG-1	U 937	K-562	T-leukemia line	B-leukemia line
1	15	4.8	8.1	12.6	5.0	7.0	4.3
		(2.1–14.0)	(1.7–26.1)	(2.0–34.2)	(1.9–13.0)	(2.0–20.4)	(0.8–22.3)
2	11	8.9	8.6	13.6	4.8	3.7	3.8
		(2.5–21.3)	(2.9–21.6)	(8.9–22.6)	(2.6–8.6)	(2.1–6.7)	(1.0–11.0)
3	2	6.7	15.9	9.1	3.0	19.3	10.3
		(2.5–10.9)	(11.1–20.7)	(2.9–15.2)	(2.6–3.4)	(8.9–29.7)	(2.8–18.0)
4	7	4.0	7.8	12.6	9.9	5.3	3.5
		(2.2–11.6)	(2.3–14.7)	(3.3–29.4)	(3.3–19.3)	(1.6–12.0)	(1.0–9.0)
5	13	6.4	7.3	18.9	5.4	2.4	2.1
		(2.7–14.6)	(2.7–23.0)	(4.8–44.5)	(2.5–15.0)	(1.1–4.6)	(0.8–7.8)
7	2	59.7	19.0	22.1	8.9	24.3	42.6
		(57.1–62.2)	(12.1–25.9)	(21.1–23.0)	(6.5–11.4)	(16.7–31.9)	(33.3–52.0)
11	8	10.8	39.1	51.2	6.7	5.8	3.1
		(3.6–17.8)	(17.4–59.6)	(33.7–74.4)	(2.4–11.1)	(1.7–17.3)	(1.3–13.3)
14	4	67.7	58.1	84.1	6.4	2.9	1.3
		(56.2–80.1)	(42.4–71.1)	(78.0–87.5)	(2.8–15.3)	(1.4–5.3)	(0.8–2.0)
15	13	82.1	39.8	79.5	62.9	30.2	3.2
		(64.3–92.5)	(26.3–54.3)	(69.6–90.4)	(51.6–78.5)	(19.3–37.0)	(0.8–21.3)
16	8	26.0	36.1	51.5	18.3	7.5	8.6
		(11.3–47.9)	(11.9–55.1)	(31.0–75.8)	(2.8–55.3)	(2.1–26.3)	(1.0–29.3)
17	4	57.3	21.2	41.5	26.6	13.9	1.5
		(38.9–69.4)	(17.9–25.9)	(31.6–64.4)	(13.5–40.5)	(6.0–21.3)	(0.8–2.0)
19	5	56.8	78.3	84.8	7.2	72.0	15.0
		(47.8–66.7)	(69.9–83.9)	(75.8–91.8)	(1.6–12.6)	(61.1–78.7)	(2.3–31.5)
20	3	2.7	53.7	36.0	11.5	4.4	94.5
		(1.6–4.1)	(52.9–54.2)	(15.8–47.7)	(3.3–27.3)	(2.7–7.0)	(90.0–98.0)
21	5	3.1	24.9	69.7	9.4	11.4	43.6
		(2.3–4.2)	(20.4–32.9)	(62.8–84.4)	(3.0–12.3)	(2.9–30.3)	(36.5–57.5)
27	2	58.9	94.4	93.1	43.3	76.9	51.3
		(41.5–76.3)	(93.3–95.4)	(92.0–94.1)	(39.5–47.1)	(62.6–91.3)	(44.0–58.5)

[a] % Mean reactivity by indirect immunofluorescence (range of reactivity).

Group 5

The properties of this heterogeneous group of monocyte-reactive antibodies overlap with those of antibodies previously designated as CDw14 and are discussed as additions to that provisional group along with the other CD groupings derived from this Workshop at the end of this chapter.

Group 7

This cluster of two antibodies detects a myelomonocytic-associated glycolipid identified as lactosylceramide. This group has been designated as a provisional cluster, CDw17 (see end of chapter).

Table 1.7(C). Preliminary groupings of myeloid workshop antibodies: Fresh leukemic cells.[a]

Preliminary group	No. Abs	M1	M2	M3	M4	M5	T-ALL	Non-T ALL
1	15	10.9 (2.1–27.1)	9.6 (1.7–32.4)	5.1 (0.6–14.8)	11.8 (2.4–30.7)	19.3 (1.2–38.2)	5.9 (1.7–17.9)	11.8 (1.4–30.6)
2	11	10.8 (5.6–18.1)	7.3 (2.3–10.9)	5.0 (0.6–11.2)	13.5 (5.0–24.1)	13.9 (7.7–18.7)	3.8 (2.4–7.9)	16.5 (5.7–30.5)
3	2	13.2 (2.7–23.9)	2.8 (2.0–3.6)	3.0 (2.2–3.8)	5.9 (5.3–6.5)	11.2 (1.6–20.8)	3.1 (2.3–3.9)	9.9 (9.9–9.9)
4	7	9.0 (3.9–17.5)	6.8 (2.6–12.2)	5.0 (1.4–11.3)	10.2 (3.0–20.1)	22.3 (8.8–30.3)	4.2 (2.4–8.3)	9.3 (5.9–12.8)
5	13	20.0 (8.5–29.1)	11.4 (5.6–25.6)	5.9 (1.6–17.0)	30.7 (22.8–39.0)	42.1 (28.7–48.7)	5.1 (2.4–13.3)	8.4 (2.4–17.4)
7	2	27.2 (25.8–28.5)	25.8 (24.9–26.7)	15.6 (14.6–16.6)	32.5 (26.7–38.3)	46.8 (46.2–47.5)	41.8 (41.1–42.4)	24.7 (22.3–27.1)
11	8	30.7 (23.4–36.8)	23.4 (17.9–36.2)	17.7 (8.0–27.0)	43.4 (37.7–50.4)	54.3 (47.3–59.5)	9.3 (3.3–19.7)	19.7 (12.7–37.6)
14	4	19.3 (12.3–30.8)	57.5 (40.3–68.9)	45.1 (28.4–51.5)	47.7 (43.8–51.8)	58.0 (53.2–63.7)	5.7 (3.0–7.8)	15.2 (10.6–18.6)
15	13	23.2 (18.3–28.7)	17.2 (12.0–24.9)	9.3 (6.2–19.4)	36.1 (27.3–44.6)	34.1 (23.8–43.3)	4.6 (2.6–11.1)	10.5 (2.1–22.7)
16	8	18.9 (6.7–29.9)	17.4 (4.0–31.8)	16.1 (3.0–40.0)	28.3 (13.1–39.4)	28.7 (23.5–35.2)	3.9 (1.9–7.3)	12.1 (6.8–20.4)
17	4	10.2 (6.4–13.9)	11.5 (6.5–17.5)	5.7 (2.0–12.0)	17.6 (14.4–22.9)	16.4 (5.8–27.0)	3.0 (2.0–4.7)	6.2 (1.3–10.3)
19	5	55.6 (48.9–63.4)	61.6 (56.3–68.0)	39.1 (28.5–52.4)	64.7 (58.5–70.9)	74.8 (70.0–80.8)	71.9 (67.0–78.6)	37.8 (33.2–45.1)
20	3	54.8 (47.6–68.3)	41.3 (38.8–44.3)	22.3 (18.6–28.4)	39.2 (34.8–44.3)	69.7 (57.2–78.5)	5.1 (4.7–5.4)	78.1 (77.5–78.5)
21	5	41.9 (33.6–63.1)	37.7 (30.7–57.8)	51.5 (47.4–66.2)	24.1 (19.9–34.3)	45.8 (40.7–62.7)	5.6 (3.7–10.6)	49.6 (43.4–70.0)
27	2	89.7 (87.2–92.1)	92.2 (89.6–94.8)	84.3 (78.6–90.0)	86.9 (86.6–87.3)	85.7 (80.8–90.5)	87.1 (86.9–87.4)	78.6 (73.0–84.3)

[a] % Mean reactivity by indirect immunofluorescence (range of reactivity).

Table 1.8. Reactivity of individual antibodies.[a]

Preliminary group	Normal cells					
	Granulocyte	Monocyte	T-Lymph	B-Lymph	RBC	Platelets
6	8.2	58.3	7.0	16.8	1.1	28.1
8	84.7	66.5	38.0	14.2	0.8	28.2
10	7.7	54.0	1.8	11.2	0.5	17.6
12	95.8	75.9	1.3	9.8	0.5	26.8
13	29.5	85.8	2.4	14.5	1.0	28.1
18	60.3	83.8	18.0	27.5	1.4	59.0
22	25.6	80.1	74.8	42.5	0.8	4.8
23	22.8	48.0	2.8	13.5	3.0	27.4
24	3.2	2.9	0.8	3.4	0.9	26.6
25	13.9	84.5	80.8	76.8	25.7	42.9
26	82.2	94.6	89.4	79.1	75.7	80.9
28	55.2	82.0	83.9	79.3	2.8	57.3

Preliminary group	Leukemic cell lines					
	HL-60	KG-1	U 937	K-562	T-Line	B-Line
6	6.3	29.6	46.8	3.9	19.9	7.3
8	25.1	40.1	14.5	40.0	39.4	14.0
10	25.8	39.7	68.6	62.8	34.9	27.8
12	87.7	40.7	93.1	54.8	28.4	3.3
13	62.4	60.6	92.4	15.0	15.7	40.0
18	23.8	48.7	80.5	6.9	53.9	5.0
22	23.7	58.7	51.0	3.6	7.0	1.5
23	84.5	91.3	85.3	95.6	96.9	99.0
24	75.0	59.1	51.1	69.6	73.7	90.8
25	73.4	79.3	93.9	82.6	97.1	38.5
26	95.8	94.9	91.7	87.5	95.6	96.5
28	91.6	47.1	91.2	8.5	49.7	51.8

Preliminary group	Fresh leukemia cells						
	M1	M2	M3	M4	M5	T-ALL	Non-T ALL
6	34.22	40.81	26.6	42.9	43.7	9.3	19.8
8	23.5	22.1	13.8	49.4	34.8	80.9	15.2
10	47.2	57.6	65.0	50.2	68.2	22.0	21.6
12	47.2	45.0	54.2	58.1	60.3	6.0	15.4
13	39.1	43.5	60.4	56.3	49.5	16.1	37.3
18	60.7	68.2	26.6	74.3	68.5	40.4	44.2
22	50.3	44.5	69.4	53.6	60.3	14.3	24.0
23	44.4	44.3	53.2	56.4	61.5	62.4	52.3
24	20.1	30.2	20.6	12.8	11.7	44.6	21.8
25	90.8	83.3	94.8	83.5	95.2	96.6	87.6
26	91.0	92.8	96.8	91.3	92.0	87.6	90.1
28	84.1	81.3	71.0	69.3	76.2	57.3	86.2

[a] % Mean reactivity by indirect immunofluorescence.

Group 11

Each of the antibodies in this group appears to identify the C3bi receptor, except Mo5. This antigen has previously been defined as CD11 and the antibodies with this specificity are discussed at the end of this chapter with the new CD groupings.

Group 14

This group consists of four antibodies which react with 40–70% of granulocytes and with approximately half of the monocytes, except MY-9 which reacted with only 6% of granulocytes. They also reacted with fresh ANLL cells, particularly M2 to M6, as well as the ANLL cell lines tested except for K562 cells. Two of these antibodies, MoU48 and MoU28, were said to react with a 180–190-Kd molecule. Of interest, MY-9 appears to recognize the same or similar determinant as L1B2 and L4F3 based on competition inhibition experiments (R. Andrews and J. Griffin, personal communications). The reason that these latter two antibodies clustered in group 16 is not known, but it may have resulted from differences in affinity, epitope specificity, or titer.

Group 15

This group of antibodies, except for VIM8, identifies a carbohydrate determinant, the X-hapten, expressed by most granulocytes. It is now designated CD15 and is detailed at the end of the chapter.

Group 16

This group includes eight antibodies which showed little or no reactivity with normal cells, but did recognize fresh ANLL cells and ANLL cell lines. As mentioned above, two of the antibodies in this group, L1B2 and L4F3, may recognize the same glycoprotein as MY-9 (in group 14 above). Also of note is that CIMT was identified by the contributor and in biochemical studies to react with the LFA-1 heterodimers. The failure of this antibody to react appropriately is thought to be due to loss of activity as this antibody was provided in lyophilized rather than ascites form and thus may have been altered during or following the reconstitution process.

Group 17

This group of four granulocyte-reactive antibodies contained two antibodies which reacted with the X-hapten (HG-2 and CLB gran 7) in studies by Symington *et al.* (this volume, Chapter 3) and have therefore been included in CD15.

Group 19

These antibodies, except MHM24, appear to recognize a common determinant on a group of heterodimers described as the LFA family. This antigen grouping has been designated CDw18 and is detailed at the end of the chapter.

Group 21

This group of four antibodies mainly reacted with normal lymphocytes, but also reacted with ANLL cells, non-T ALL cells, and both lymphocytic and nonlymphocytic leukemic cell lines. Of note is the common reactivity of these antibodies with NK cells, which is further discussed in subsequent chapters in this volume.

Clusters Defined by Second Myeloid Workshop

Alterations in Previously Defined Clusters

CD11 (Mo5, VIM12, CC1.7, D12, 60.1, M3D11, Ki-M5, 44, Mo1, B2.12, M522)

This group includes antibodies which have been shown to react with a 170/95-Kd heterodimer identified as the C3bi receptor (CR3) and previously defined as CD11. These antibodies react with the majority of granulocytes and monocytes, and with subsets of lymphoid cells that were not fully characterized. They also react with fresh ANLL cells as well as ANLL cell lines, in particular KG1 and U 937. Within this grouping VIM12, CC1.7, D12, 60.1, M3D11, Ki-M5, and 44 have been shown to identify this receptor. One antibody which clustered with these antibodies (Group 11), Mo5, does not recognize this receptor and was therefore deleted from this group. This cluster also includes antibodies Mo1, B2.12, and M522 from the first Workshop.

CDw14 (CRIS-6, 5F1, VIM13, M5E2, MA5, SCHL3, MoS1, MoP9, MoP15, MoS39, Ki-M1, UCHM1, CIM, FMC17, Mo2, 20.3, MY4, TM18)

The antibodies in the preliminary group 5 included a number of antibodies previously included in CDw14 by the First Workshop (5F1, MoP9, MoP15, and MoS39). Consequently, the present grouping of monocyte-reactive antibodies was added to the previously defined and heterogenous CDw14 provisional working grouping. Antibodies in the present workshop mainly react with monocytes, with each antibody recognizing between 75 and 90% of monocytes except Ki-M1 and CIM which recognized 55 and 31% of monocytes, respectively. One antibody, 5F1, recognized 80% of monocytes and 89% of platelets, and has been reported to recognize an 85-Kd molecule. Thus, 5F1 is distinct from the other monocyte-reactive antibodies. The heterogeneity of antibodies is further suggested by their reactivity with malignant cells, e.g., U 937 where 6 of the 13 antibodies from the present workshop reacted with >20% of the cells. Two antibodies, Ki-M1 and SCHL3, have been identified to react with a 150/95-Kd heterodimer thought to represent one of the LFA family (see this volume, Chapter 4). Thus, CDw14 is clearly a heterogeneous cluster of antibodies that can be separated based on the strength of their reaction with monocytes and platelets, as well as on the basis of biochemical data.

CD15 (82H5, 1G10, HLC5, VIMC6, VIM8, G7C5, HG-1, 29, G9F9, 28, LMA-1, LMA-3, HG-3, HG-2, CLB gran 7, FMC10, FMC12, B4.3, VIMD5)

This group of granulocyte-reactive antibodies will include antibodies from group 15 in the present workshop as well as those from the previous workshop in CDw15 known to react with the X-hapten carbohydrate determinant. In the present workshop, this group consisted of 13 antibodies strongly reactive with granulocytes and weakly and variably reactive with a minor subset of monocytes. They also reacted with ANLL cells, in particular M4 and M5, although little reactivity with M3 was observed, and ANLL cell lines including HL60, KG1, U 937, and K562. Some reactions with T-ALL cell lines were also observed although reactions with fresh T-ALL cells were not noted.

Each of the antibodies within group 15 of this workshop, except VIM8 which has therefore been excluded, were shown to react with the X-hapten, also referred to as lacto-N-fucose-pentaosyl III. In addition, two antibodies in group 17 of this workshop, HG-2 and CLB gran 7, also reacted with this determinant. The reason that the latter two antibodies did not cluster in this group may have been because of differences in affinity. Nonetheless, each of these X-hapten-reactive antibodies have been incorporated into CD15. Group 15 includes one antibody, 1G10, included in CDw15 in the previous workshop, and other antibodies in CDw15 including VIMD5, B4.3, and FMC10, have been shown to recognize this same carbohydrate determinant.

There are known subtle differences between the reactivities of these antibodies. This includes differences in their reactivity with subsets of monocytes as well as known reported differences in their reactivity with hematopoietic stem cells. Presumably these differences are due either to affinity or to the way in which the antibody recognizes the X-hapten. For further discussion of antibodies of this specificity, see Chapter 3.

New Cluster Groupings

CD16 (BW243/41, CLB FcR gran 1, BW209/2, CLB gran 11, VEP13, G022b)

This group includes antibodies which recognize the Fc receptor on granulocytes. Of the antibodies in this group, BW243/41, CLB FcR gran 1,

BW209/2, and CLB gran 11 were shown biochemically to react with Fc receptors on granulocytes. In addition, Rumpold *et al.* (this volume, Chapter 12) used competition inhibition studies to demonstrate that antibodies VEP13 and G022b also identify the Fc receptor. Thus, these antibodies that reacted with the Fc receptor received a formal cluster differentiation designation. Studies by Springer *et al.* (see this volume, Chapter 4) and Tetteroo *et al.* (see this volume, Chapter 2) have shown some heterogeneity in the molecular weights of the proteins precipitated by these antibodies. This is presumably due to microheterogeneity of Fc receptors. Other antibodies which clustered in this group (group 2) and were found not to react with the Fc receptor have not been included in CD16 (G7C10, GA-1, CLB gran 10, PMN-3, and LMA-2).

New Provisional Cluster Designations

CDw17 (T5A7, G035)

This group contains two antibodies T5A7 and G035 that reacted with >90% of the granulocytes, 40–60% of the monocytes, and platelets. They also detected fresh ANLL and ALL cells, and both myelocytic and lymphocytic leukemic cell lines. Thus, these antibodies recognize a glycolipid moiety, lactosylceramide, as shown by Symington *et al.* (see this volume, Chapter 3). Because of the common biochemical target, these antibodies were designated as a cluster differentiation group, but were given a provisional CDw classification pending verification of their biochemical specificity by a second laboratory.

CDw18 (MHM23, CLB-LFA-1/1, 60.3, TS1/18.11)

This group of antibodies appears to identify the group of heterodimers thought to have a common beta chain but different alpha chains, and designated as the LFA (leukocyte function antigen) family. According to Springer *et al.* (see this volume, Chapter 4), this family includes the C3bi receptor (M_r 170/95) found mainly on granulocytes and monocytes, LFA-1 (M_r 180/95) on lymphocytes and monocytes, and at least a third group (M_r 150/95) on monocytes. Antigens of the LFA family are therefore found on monocytes, granulocytes, and lymphocytes, and antibodies which recognize this family precipitate chains of 180, 170, 150, and 95 Kd.

The antibodies which recognize all members of the LFA family included MHM23, CLB-LFA-1/1, 60.3, and TS1/18.11. These antibodies precipitate at least three dimers, including the 180/95-Kd LFA-1 molecule, the 170/95-Kd C3bi molecule (CD11), and a 150/95-Kd heterodimer. Presumably these antibodies recognize a beta chain common to each of these groupings, although this family may turn out to be more heterogeneous. Excluded from this group is MHM24 which appears to recognize the alpha chain on the LFA-1 molecule. CIMT (group 16) is also thought

Table 1.9. New CD groupings determined by the Second International Workshop.

	Second International Workshop	First International Workshop
Previously defined clusters		
CD11:	Mo5, VIM12, CC1.7, D12, 60.1, Ki-M5,44, M3D11	Mo1, B2.12, M522
CDw14:	CRIS-6, 5F1, VIM13, M5E2, MA5, SHCL3, MoS1, MoP9, MoP15, MoS39, Ki-M1, UCHM1, CIM	5F1, MoP9, MoP15, MoS39, FMC17, Mo2, 20.3, MoS1, MY4, TM18
New clusters		
CD15:	82H5, 1G10, HLC5, VIMC6, VIM8, G7C5, HG-1, 29, G9F9, 28, LMA-1, LMA-3, HG-2, HG-3, CLBgran7	FMC10, B4.3, VIMD5, 1G10, FMC12
CD16:	BW243/41, CLBFcRgran1, BW209/2, CLBgran11, VEP13, G022b	
New provisional clusters		
CDw17:	T5A7, G035	
CDw18:	MHM23, CLB-LFA1/1, 60.3, TS1/18.11	

to recognize this chain. Other antibodies thought to identify the individual members within this family also include the anti-C3bi antibodies in CD11 and possibly antibodies SHCL3 and Ki-M1 which recognize a third group (M_r 150/95) in this family. These latter two antibodies have been provisionally included in CDw14.

Conclusion

The new CD groupings determined by the present workshop are summarized in Table 1.9. The experience gained in this workshop has also focused attention on the need to emphasize biochemical analyses of these antigens including studies of glycolipid as well as glycoprotein moieties. Because of the broad expression of myeloid antigens by cells of one or more lineages within the hematopoietic system, there is also the need to further focus on particular cells, e.g., granulocytes and/or monocytes, or functional properties, e.g., stem cells. It is expected that the Third International Workshop will build on the results of the present workshop and further our understanding of the importance of myeloid-associated antigens during normal and malignant hematopoiesis.

CHAPTER 2

Serological, Biochemical, and Cytogenetic Studies with the Granulocyte Monoclonal Antibodies of the "M Protocol"

P.A.T. Tetteroo, F.J. Visser, M.J.E. Bos,
A.H.M. Geurts van Kessel, J.F. Tromp, and
A.E.G. Kr. von dem Borne

In the First International Workshop on Leucocyte Differentiation Antigens, five clusters of myeloid monoclonal antibodies (CD11–CD15) were defined on the basis of their serological reactivity. Compared to the antigens recognized by T cell and non-T/non-B cell mAbs, less attention was paid to the biochemical characterization of the antigens recognized by the myeloid antibodies.

Meanwhile, it has been shown that many of the antibodies of the CD15w cluster (granulocyte specific) are directed against the carbohydrate sequence 3-α-fucosyl-N-acetyllactosamine (X-hapten, lacto-N-fucopentaose III or SSEA-1) (1,2) present on sphingolipids (1) and glycoproteins (1,3). Moreover, it has been found that on various types of cells this structure is present in a sialated form, which makes them unreactive with the antibodies (3,4).

The availability of myeloid human–mouse somatic cell hybrids (5) has allowed a study of the chromosomal localization of the genes that code for some myeloid differentiation antigens. Using such hybrids it was shown that the gene involved in the expression of the 3-α-fucosyl-N-acetyllactosamine structure is located on the long arm of chromosome 11 (5,6). Upon screening a large panel of mAbs with a known myelocytic, monocytic, or myelo-monocytic specificity, we found that 20 mAbs exhibited similar reactivity patterns with the hybrid clones (6).

The aim of the present study was to characterize the granulocyte antigens recognized by mAbs of the M protocol of the Second Workshop. Fifty-eight antibodies of this protocol reacted with human granulocytes. These 58 mAbs, together with three antibodies of the First Workshop [CD15w (G.u) B 13.9, CD15w (G.u) B4.3, and CD11 (M.G.u) B2.12], were selected for immunoprecipitation studies with ^{125}I-labeled granulocytes. The antigens were analyzed by SDS–polyacrylamide gel electrophoresis and autoradiography. The effect of sialidase treatment of leukemic cells on the reactivity of these antibodies was also tested. In order to detect a possible role of these antigens for granulocyte function, the inhibitory effect of the mAbs on adherence of IgG-sensitized erythrocytes was tested, as well as the effect of chemo-attractant activation of PMN on the number of antigens expressed on the cell membrane. Cytogenetic studies were performed by testing the reactivity of the antibodies with four myeloid cell hybrids, in which different human chromosomes segregate. The findings are presented in relation to the data obtained from serological investigation of these antibodies.

Materials and Methods

Immunofluorescence

The binding of the mAbs to granulocytes, to sialidase-treated leukemic cells, and to somatic cell hybrids was tested on paraformaldehyde(1%)-fixed cells by indirect immunofluorescence with FITC-conjugated goat anti–mouse Ig (Gm17-01-F, CLB). Fluorescence intensity was quantified by flow cytometry (cytofluorograph FC 200, Ortho) and was expressed in arbitrary units (μG), defined as the fluorescence intensity of 10,000 FITC molecules per cell.

Sialidase Treatment

The leukemic cells of a patient with AML (M2) and of a patient with ALL (cALL$^+$) were enriched by centrifugation over Ficoll–Isopaque (d = 1.077). The cells were incubated with neuraminidase (EC 3.2.18, Vibrio cholerae, 0.1 U/ml, Behringwerke AG, Marburg, W. Germany) for 30 min at 37°C and then washed twice with phosphate-buffered saline (PBS).

Immunoprecipitation

Before radiolabeling with ^{125}I, the cells were incubated with the protease inhibitor diisopropyl fluorophosphate (DFP, 2 mM) in PBS for 5 min at 4°C. The granulocytes (2×10^7) were ^{125}I surface-labeled with Iodogen. The cells were lysed at 4°C for 1 hr in a 0.01 M Tris-HCl buffer, pH 7.8, containing 1% (v/v) NP40, 0.15 M NaCl, 1 mM PMSF, 5 mM Na$_2$ EDTA, and 0.02 mg/ml trypsin inhibitor. The lysates were centrifuged at 13,000 $\times$ g for 15 min and frozen at −20°C. Immunoprecipitation was performed with preformed immune complexes of mAb and goat anti–mouse Ig, GAM (CLB, Gm 17-01-P02 from our Institute). Twenty μl of ascites were added to 100 μl of GAM and incubated for 16 hr at 4°C. The mAb–GAM complexes were washed three times. Thawed lysate was precleared three times with a preformed complex of normal mouse Ig and GAM, each time for 1 hr at 4°C. Precleared lysates were then incubated for 16 hr with mAb–GAM complexes at 4°C. The precipitate was centrifuged on a discontinuous sucrose gradient. Antigens were eluted by boiling for 5 min in 40 μl of SDS sample buffer, containing 0.15 M Tris-HCl, 6% SDS, and 15% β-mercaptoethanol. This material was subjected to electrophoresis on 10% or 12.5% polyacrylamide (w/v) slab gels. ^{14}C-Methylated proteins (Amersham and NEN) were used as molecular weight standards. For autoradiography, Kodak X-Omat RP X-ray film was used in combination with Cronec intensifier screens.

Blocking of EA-Rosette Formation

Human OR_2R_2 red cells (E) were sensitized with human IgG anti-Rh-D (EA). Two hundred μl of PMNs (2×10^6/ml), sensitized with mAbs, and 50 μl of EA (1.3×10^8/ml) were mixed, spun (10 min, 200 g), and incubated as a pellet for 30 min at room temperature. The percentage of cells binding more than 3 red cells was scored.

Activation of PMNs by FMLP

PMNs (1×10^7/ml) were incubated for 10 min at 37°C in incubation medium (138 mM NaCl, 2.7 mM KCl, 8.1 mM Na_2HPO_4, 0.6 mM $CaCl_2$, 1.0 mM $MgCl_2$, 5.5 mM glucose, 5 g/liter human albumin, pH 7.2) with either 1 μM formyl-methionyl-leucyl-phenylalanine (fMLP) plus 10 μg cytochalasin B per ml or with an equal volume (1 μl) of solvent (DMSO). After incubation, cells were washed with PBS and used for immunofluorescence.

Human–Mouse Myeloid Cell Hybrids

The four selected hybrid clones (Wegli) were originally obtained (6) after fusion of the murine myeloid cell line WEHI-TG with peripheral leukocytes from a CML patient carrying a complex Philadelphia (Ph[1]) translocation: t(9;11;22) (q34;q12,q11). Characterization and chromosome analysis of reinitiated cultures of these hybrids were carried out as described before (5).

Results

Serologically Defined Groups of Granulocyte-Reactive Antibodies

Fifty-eight mAbs of the M protocol were selected for further studies because these mAbs reacted with granulocytes. On the basis of their reactivity with other peripheral blood cells 53 of the mAbs could be distributed among seven groups: (a) granulocyte specific (groups I and II),

(b) reactive with granulocytes and monocytes (III), (c) reactive with granulocytes, monocytes, and a subpopulation of lymphocytes [large granular lymphocytes (LGL)?] (IV), (d) reactive with granulocytes, monocytes, and lymphocytes (V), (e) reactive with granulocytes and a subpopulation of lymphocytes (LGL?) (VI), and (f) reactive with granulocytes, monocytes, and platelets (VIII) (Table 2.1). The other five mAbs had a unique reaction pattern (Table 2.2).

Twenty-three antibodies including CD15w (G.u) B4.3 and CD15w (G.u) B13.9 were granulocyte specific (groups I and II). Because of their differences in reactivity with the immature cells of the cell lines HL60, K562, U937, HSB, and KG-1, these 21 mAbs were divided into 2 groups: group I reacted with precursor cells and mature cells, whereas group II reacted only with mature cells (Table 2.3). With the exception of M81, group I antibodies gave similar reactivities with the cell lines. M82 (a mAb of group II) reacted with the granulocytes of two of the three tested donors

Table 2.1. Serologically defined groups of granulocyte-reactive antibodies of the myeloid protocol.

Group	Workshop cluster	Reactivity	Workshop antibodies	Range of mean fluorescence per granulocyte (μG)
I	CD15w	Granulocytes and precursors	a. M5, M13, M19, M25, M80, M91, M117, M52[a], M110[a], M33, M61, M77, M111, M116, M27, CD15w (G.u) B4.3	42–166
			b. M81	14
II	CD15w	Granulocytes	a. M70, M103, M112, CD15w (G.u) B13.9	12–18
			b. M49, M82	50–89
III	CD12w	Granulocytes and monocytes	a. M12, M78, M66, M65	10–17
			b. M28	71
IV	CD11	Granulocytes, monocytes, sub-population lymphocytes	M26, M41, M42, M46, M47, M53, M76, M88, M104, CD11 (M.G.u) B2.12	11–41
V		Granulocytes monocytes, lymphocytes	M36, M38, M39, M55, M56, M72, M73, M75, M89, M100	12–37
VI		Granulocytes and subpopulation lymphocytes	M22, M54, M57, M58, M59	61–94
VII		Granulocytes, monocytes, and platelets	M15, M94	65–87

[a] <15% monocytes positive

Table 2.2. Non-grouped granulocyte-reactive antibodies.

Workshop antibodies	Reactivity	Mean fluorescence intensity per granulocyte (μG)
M11	Granulocytes, monocytes, lymphocytes, and platelets	9
M18	Granulocytes, monocytes, platelets, and a subpopulation of lymphocytes	9
M30	Granulocytes, platelets, and a subpopulation of lymphocytes	36
M102	Granulocytes, platelets	15
M74	Granulocytes, monocytes, lymphocytes, platelets, red cells	20

and therefore seems to react with a polymorphic structure (this volume, Chapter 9).

The binding of these antibodies with granulocytes was quantified by measuring the fluorescence by flow microfluorimetry. The mean fluorescence intensity per cell for each mAb was very similar within the serologically defined groups, except for groups I, II, and III (Table 2.1) in which, therefore, two subgroups (Ia, Ib, IIa, IIb, IIIa, IIIb) were defined. Strong reactions (>40 μg) were found for the groups Ia, IIb, IIIb, VI, and VII (Table 2.1).

Effect of Sialidase Treatment of Leukemic Cells

Neoplastic cells from two patients (AML, M2 and ALL, cALL[+]) were exposed to neuraminidase and subsequently tested with all 58 mAbs and measured by flow microfluorimetry. All group Ia mAbs, except M27, reacted after enzyme treatment with these cells (Fig. 2.1, Table 2.4). The reactivity of mAbs from group VII increased after sialidase treatment. No changes were observed with the other antibodies.

Table 2.3. Reactivity of the granulocyte-specific antibodies with cell lines.

Group	Workshop antibodies	Range of percent positive cells					
		HL60	KG-1	K562	U937	HSB	SB
I	a. M5, M13, M19, M25, M80, M91, M117, M52, M110, M33, M61, M77, M111, M116, M27	80–95	0–5	6–46	38–88	42–89	0
	b. M81	2	37	0	0	0	5
II	M70, M103, M112, M49, M82	0–2	0–5	0–2	0	0	0

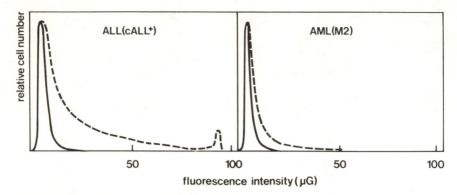

Fig. 2.1. Effect of neuraminidase treatment of the leukemic cells of a patient with ALL (positive for cALL antigen) and of a patient with AML (M2) on the reactivity of CD15w (G.u) B4.3. Reactivity of untreated cells (-) and the reactivity of neuraminidase-treated cells (---) are shown.

Immunochemistry

Immunoprecipitation studies with preformed immune complexes and gel analysis were performed with the 58 mAbs, to characterize the molecular nature of the antigens and to identify similarities between the mAbs within the serologically defined groups (Table 2.5).

Table 2.4. Effect of sialidase treatment of cells on the reactivity of mAbs.

Workshop antibodies	Group	cALL Native (% Pos.)	cALL Neuraminidase (% Pos.)	M2 Native (% Pos.)	M2 Neuraminidase (% Pos.)
M15		4	82	17	57
M13		4	65	12	47
M19		6	59	11	42
M25		6	52	8	36
M80		7	51	8	18
M91		6	71	12	56
M117		6	67	9	58
M52	Ia	5	74	11	51
M110		6	75	9	48
M33		5	66	10	51
M61		6	57	10	30
M77		5	62	7	38
M111		5	74	9	43
M116		6	61	5	23
CD15w (G.u) B4.3		6	66	12	35
M27		4	6	6	16
M15	VII	21	89	54	80
M94	VII	20	80	50	72

Table 2.5. Immunoprecipitation results of the granulocyte-specific antibodies (groups I and II).

Group	Antigen mol.wt. ^{125}I-labeled PMNs (Kd)[a]	Donor	Workshop antibodies	Antigen mol.wt. provided by the Workshop participants (Kd)
Ia	260/220/160/135/95		M5, M19, M80, M117,	—
			M110, M52, M116	—
	64/60	1	M13	85
			M25	150/105
			CD15w (G.u) B4.3	150/105
	260/220/150/90/64	3	M91	155/85
	165/105	2	M27	175
	—	2	M33, M61, M77, M111	100, 165, 129, —
Ib	170	3	M81	170
IIa	160/68–80	2	M70	—
	80–90	2	CD15w (G.u) B13.9	—
	—	2	M103, M112	—
IIb	60–42	2	M82	72–50
	—	2	M49	43.7

[a] The underlined figures indicate the more intense bands.

In group Ia (reacting with granulocytes and precursors) 9 of the 15 mAbs precipitated a series of polypeptides (260, 220, 160, 135, 95, 64, and 60 Kd) similar to the polypeptides obtained with the anti-3-α-fucosyl-N-acetyllactosamine antibody CD15w (G.u) B4.3 (Fig. 2.2). The 160, 135, and 95-Kd bands were more intense than the other bands. With M91 we obtained a 150 and 90-Kd band as the most intense. M27 precipitated two polypeptides of 165 and 105 Kd. With M33, M61, M77, and M111 (Table 2.5) no detectable polypeptides were obtained. mAb M81 precipitated a single polypeptide of 170 Kd (group Ib). In group II, the antibodies reacting with mature antigens, three different antigens were precipitated. M70 bound 160/68–80 Kd, M82 bound 42–60 Kd, and CD15w (G.u) B13.9 bound 80–90 Kd.

The results of the immunoprecipitation and gel analysis of the antibodies also reacting with other leukocytes are summarized in Table 2.6. Figure 2.3 shows the results for some mAbs of groups IV and V. In group VI (reacting with granulocytes and some lymphocytes) mAbs M22 and M57 precipitated a 72–42-Kd protein, earlier described as the Fc receptor of granulocytes (7). Figure 2.4 shows the results for all group VI antibodies and the group IIb antibodies. In Table 2.7 the immunoprecipitation results for the non-grouped antibodies are summarized.

Blocking of EA-Rosette Formation

mAbs of group IIb (M49 and M82) and of group VI recognize Fc receptor-like molecules (Fig. 2.4) and were therefore tested for their ability to

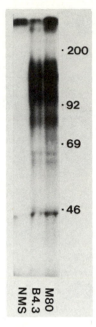

Fig. 2.2. Autoradiogram of the SDS–gel analysis of the polypeptides precipitated by CD15w (G.u) B4.3 and CLB gran 7 (M80) (10% gel).

Table 2.6. Immunoprecipitation results of the granulocyte-reactive leukocyte antibodies of groups III–VI.

Group	Antigen mol.wt. ^{125}I-labeled PMN (Kd)[a]	Donor	Workshop antibodies	Antigen mol.wt. provided by the Workshop participants (Kd)
III	75	2	M65	24
	—	2	M12, M78, M66, M28	94, —, 24, —
IV	—	3	M41, M47, M88	80/160, 100, —
	41/32	3	M42	—
	130/80	3	CD11 (M.G.u) B2.12	—
	$\overline{130/90}$, 130/90	3	M46, M26	150/90, 155/94
	$\overline{150}$/130	3	M104	180
	$\underline{150}$/130/110/90	3	M76	110/95
	$\overline{190}$/170/140–100	2	M53	—
V	190/170/140–100	2	M36	—
	130	3	M55	180/95
	160	3	M56	180/95
	160/130/90/80	3	M72	177/95
	130/$\overline{90}$	3	M73, M89	180/155/95
	$\overline{130}$/40/32	3	M75	150/130/95
	—	2	M38, M39, M100	—
VI	72–42	2	M22, M57	—, 72–50
	60–42	2	M54, M59	37.8, —
	50	2	M58	—

[a] The underlined figures indicate the more intense bands.

Table 2.7. Immunoprecipitation results of group VII and non-grouped antibodies.

Group	Antigen mol.wt. ^{125}I-labeled PMN (Kd)	Donor	Workshop antibodies	Antigen mol.wt. provided by the Workshop participants (Kd)
VII	135	1	M94	—
	—	2	M15	—
	290/53/52/50	1	M11	—
	46–50	2	M18	120
	—	2	M30	165
	41/32	3	M102	—
	160/85–60	2	M74	—

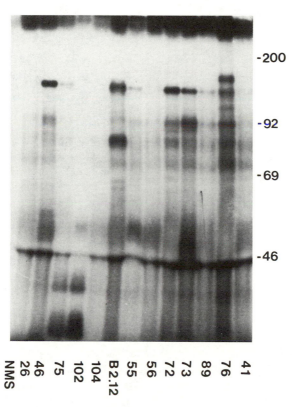

Fig. 2.3. Autoradiogram of the SDS–gel analysis of the mAbs of group IV (CD11): VIM12 (M26), SHCL3(M46), 44 (M104), CD11 (M.G.u) B2.12, 60.1 (M76), CC1.7 (M41), and of group V: 60.3 (M75), MHM23 (M55), MHM24 (M56), TS1/22 (M72), CLB LFA-1/1 (M73), TS1/18.11 (M89), CIKM5 (M102).

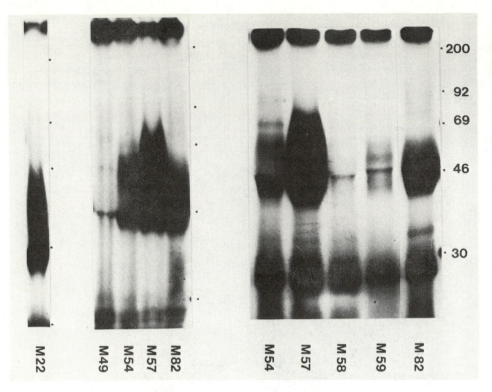

Fig. 2.4. Autoradiogram of the SDS–gel analysis of the mAbs of group IIb: G7C10 (M49) and CLB gran 11 (M82), and the mAbs of group VI: BW 243/41 (M22), G022b (M54), CLB FcR gran 1 (M57), VEP13 (M58), BW209/2 (M59) (10% and 12.5% gels). PMNs were used from a CLB gran 11 (M82) positive donor.

inhibit the binding of IgG-sensitized erythrocytes to PMNs. Because M82 recognizes a polymorphic structure, we tested this antibody with the cells from a donor positive for M82 (donor 1) and with cells from an M82-negative donor (donor 2). The PMNs were sensitized with 1 : 250 diluted ascites fluid. Figure 2.5 shows that all seven antibodies were able to decrease EA-rosette formation. However, differences in blocking efficiency of the various antibodies were obtained per cell donor.

Effect of Granulocyte Activation

The binding of the 58 mAbs was quantified on resting PMNs and on FMLP-activated PMNs. The differences in reactivity per mAb were expressed as the mean fluorescence intensity per cell after activation divided by the mean fluorescence intensity per cell of unstimulated PMNs. Table 2.8 summarizes the results. A large increase of binding after activation was found with M81 (the mean fluorescence increased by a factor of 2.1), M70 (a factor of 2.2), and CD15w (G.u) B13.9 (a factor of 3.5) (Fig.

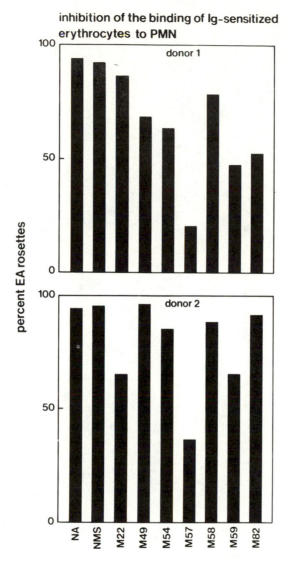

Fig. 2.5. Inhibition of the binding of Ig-sensitized erythrocytes to PMNs by mAbs of group IIb: G7C10 (M49) and CLB gran 11 (M82), and the mAbs of group VI: BW 243/41 (M22) G022b (M54), CLB FcR gran 1 (M57), VEP13 (M58), BW209/2 (M59). 1 : 250 diluted ascites were used to sensitize the PMNs.

2.6). A decrease of binding was obtained with all antibodies of groups VI, IIb, and VII (Fig. 2.6).

Cytogenetic Studies

Four myeloid human–mouse hybrid cell lines as well as the parent mouse myelo-monocytic cell line WEHI-TG were tested for reactivity with the

Table 2.8. Granulocyte-reactive mAbs of the M protocol grouped in relation to the effect of activation of granulocytes with the chemo-attractant fMLP on the reactivity of the antibodies.

Group	<0.8	0.8–0.9	0.9–1.3	1.3–1.5	1.5–2	>2
			Activation factor[a]			
I		M110, 33, 61, 77	M5, 13, 19, 25, 80, 117, 52, 116, 111	M91, B4.3	M27	M81
II	M49, 82				M112	
III			M66, 28	M12	M78	M70, B13.9
IV			M46, 47, 53		M26, 41, 42, 76, 88, 104 B2.12	
V			M38, 39, 55, 56, 72, 73, 75, 100	M89	M36	
VI	M22, 54, 57, 58, 59					
VII	M15, 94					
non-grouped			M11, 18, 74, 102	M30		

[a] Activation factor defined as the fluorescence intensity of the mAb with granulocytes after activation with fMLP divided by fluorescence intensity with nonactivated granulocytes.

differences of reactivity of MoAb induced by fMLP–activation

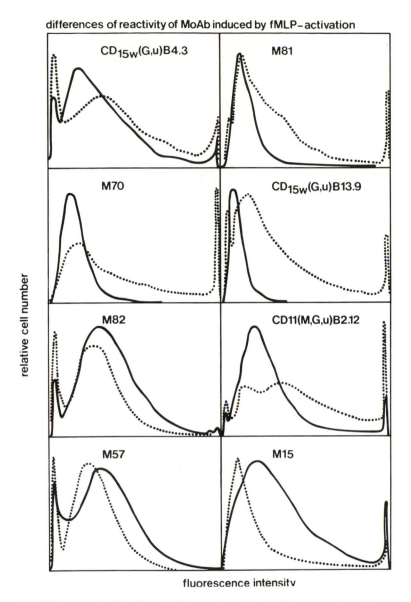

Fig. 2.6. Differences in binding of mAbs induced by fMLP-activation. Nonactivated PMN (—) and activated PMN (---).

58 mAbs. Only a part of the whole human genome was covered by these four lines. We selected these four because they are discriminative for chromosome 11.

Twenty-five of the 58 mAbs reacted with one or more of the hybrid cell lines (Table 2.9). The granulocyte-specific mAbs from group Ia, except

Table 2.9. Reactivity of the mAbs with myeloid human–mouse hybrid cell lines.[a]

Group	Workshop antibodies	WEHI-TG	WEGLI-9	WEGLI-11	WEGLI-16 85% Chr 11	WEGLI-18 50% Chr 11
Ia	CD15w (G.u) B4.3	0	0	0	55	41
	M5, M13, M19,					
	M25, M80, M91,					
	M117, M52, M110,					
	M33, M61, M77,					
	M111, M116	0	0	0	69–45	59–34
	M27	5	12	10	63	53
IIIb	M28	26	11	14	65	61
IV	CD15w (M.G.u) B12.2	28	ND[b]	8	ND	7
	M47	<1	0	24	24	33
	M53	0	ND	7	ND	30
V	M55, M72	0	20–24	0	23–26	8
	M56	0	0	0	2	8
	M73, M89	0	25–17	15–8	39	26–31
	M75	13	41	13	46	28
—	M30	0	0	0	26	36

[a] Values are percentage positive cells.
[b] ND: Not determined.

M27, showed the same binding pattern as CD15w (G.u) B4.3. We already showed that the gene(s) responsible for the expression of the 3-α-fucosyl-*N*-acetyllactosamine (FAL) antigen is located on chromosome 11. The results obtained with the four Wegli cell lines, the co-segregation of the reactivity with the presence of chromosome 11, suggest that the antigens recognized by group Ia are similar or related.

Because of the small number (four) of lines used we could not, as yet, localize the genes for the other antigens. Also a number of human chromosomes were absent from all four lines and some of these antigens may, therefore, not be expressed on any of them. A complicating factor is that some of the antibodies cross-reacted to some extent with the mouse parent line (M27, M28, B2.12, M75).

Discussion

The goal of these Workshops is to compare submitted antibodies and to classify them into groups recognizing similar antigens. This should provide a basis for standardization and should facilitate the future characterization of new reagents.

This approach resulted in the definition of 15 clusters of antibodies after the First Workshop: 8 clusters of T cell antibodies, 2 clusters of non-T/non-B antibodies, and 5 clusters of myeloid antibodies. But, in contrast to the antigen recognized by the T and non-T/non-B clusters, the antigens recognized by the myeloid antibodies were only characterized serologically. Therefore, we studied in this Second Workshop the myeloid antibodies, reactive with granulocytes, in more detail in order to group them and to characterize the antigens involved. We tested (a) their reactivity with peripheral blood cells and cell-line cells, (b) the density distribution of the recognized antigens, (c) the immunoprecipitation pattern of their target antigens from ^{125}I-labeled granulocytes, (d) the effect of PMN activation on their reactivity, and (e) their reactivity with human–mouse cell hybrids.

Fifty-eight of the 116 antibodies included in the M protocol reacted with granulocytes and were assigned to one of seven different specificity groups, based on their reaction pattern with peripheral blood- and cell-line cells. Immunoprecipitation analysis, however, showed that some of these serologically defined groups II, III, IV, V were heterogeneous, because they contain antibodies which react with different antigens.

The results obtained in this study will be discussed in more detail in relation to known antigens.

Carbohydrate Antigens

3-α-Fucosyl-N-acetyllactosamine (FAL)

In group I a series of 15 mAbs (Ia) behaved rather homogeneously in all tests. They reacted strongly with granulocytes and they were all of the IgM type. Antibody CD15w (G.u) B4.3 belongs also to this group. This mAb as well as 92H5 (M5), 1G10 (M13) (1), HLC5 (M19) (8) and VIMC6 (M25) (2) react with 3-α-fucosyl-N-acetyllactosamine (FAL) (3). We already showed that this structure is associated with a series of granulocyte membrane proteins (3,5,9) with those of 105 and 150 Kd as most pronounced polypeptides. When ^{3}H-labeled or ^{125}I-labeled HL60 cells are used this structure is found on two proteins of 200 and 240 Kd (1,3). In this study we found three intense bands of 95, 135, and 160 Kd for nine of the 15 antibodies, two intense bands (150 and 90 Kd) for one, and no bands for five of the 15. However, lysates from three different donors were used. The producers of some of these antibodies also obtained different values for the mol. wt. of the antigens (Table 2.5), although no information was given about their target cells. With exception of VIM2 (M27), all group Ia mAbs reacted with myeloid and cALL$^+$ lymphatic leukemic cells after removal of sialic acid. Also these 14 mAbs reacted with human–mouse hybrid cell lines which contained human chromosome 11. All these results suggest that the latter 14 mAbs may be directed against 3-α-fucosyl-N-acetyllactosamine.

Lactosylceramide

mAb T5A7 (M15), a member of group VII, reacts with lactosylceramide (10). Together with G035 (M94) it reacts strongly with granulocytes but also with monocytes and platelets, and for both antibodies the reactivity increased after neuraminidase treatment of leukemic cells. Thus, lactosylceramide, like FAL, becomes more accessible to the antibodies (both IgM) after removal of sialic acid. In contrast to T5A7, G035 precipitated a 135-Kd polypeptide; however, different donors were used for both antibodies.

Functional Antigens

Fc Receptor

Two mAbs of the granulocyte-restricted group II (IIb) as well as the mAbs of group VI (reactive with granulocyte and large granular lymphocytes) reacted strongly with granulocytes. Six of the seven mAbs precipitated proteins with a mol. wt. around 50 Kd. BW243/41 (M22) and CLB FcR gran1 (M57) gave a broad band (42–72 Kd). Both mAbs blocked EA-rosette formation with granulocytes. G022b (M54), BW209/2 (M59), and CLB gran 11 (M82) gave a smaller band (42–60 Kd) and also blocked the EA-rosette formation. VEP13 (M58) precipitated a 50-Kd band and only partly blocked EA-rosette formation. These results indicate that the mAbs of both groups react with the receptor for the Fc portion of IgG (FcR). mAbs (3G8 and B73.1) against this receptor have been described by others (7,11), but only the antigen recognized by antibody 3G8 was characterized using granulocytes. In this study we demonstrate that the SDS analyses of the antigens precipitated by the anti-FcR show differences in the recognized antigens. The broad band suggests microheterogeneity in glycosylation of the FcR. The different mol. wt. determinants could then be explained by reactivity of the antibodies with only part of the FcRs. Perussia *et al.* (12) compared five mAbs against the FcR including G022b and VEP13 and demonstrated that these antibodies detected two distinct epitopes on FcR, one by B73.1 and one by 3G8, G022b, VEP13.

One of the anti-FcRs, CLB gran 11 (M82) reacts with a polymorphic structure, similar to the allotypic structure NA1 (this volume, Chapter 9) only present on granulocytes. B73.1 also reacts with a polymorphic structure on PMNs, but this mAb also reacts with NK cells (11).

Of interest is the finding that activation of PMNs causes a decrease in expression of the FcR (Table 2.8).

C3bi Receptor

Group IV (CD11-like) was detected by nine antibodies reacting with granulocytes, monocytes, and large granular lymphocytes. Also mAb CD11 (M.G.u) B2.12 is a member of this group. The antibodies reacted weakly and moderately with granulocytes.

mAb Mo-1, studied in the First Workshop and a member of the CD11 cluster, identifies the C3bi receptor (13). More antibodies directed against this receptor have been described (14). However, different mol. wts. for this receptor were found (13,14). Thus, the conclusion that some of the mAbs of group IV react with the C3bi receptor cannot be based on their estimated mol. wt.

LFA-1 Antigen

Antibodies directed against leukocyte function-associated antigen-1 (LFA-1), react with granulocytes, lymphocytes, and monocytes, like the antibodies of group V. The five mAbs, MHM23 (M55), MHM24 (M56), TS1/22 (M72), CLB LFA-1/1 (M73), and TS1/18.11 (M89), known to react with LFA-1 as well as 60.3 (M75) directed against gp 95-150 were incorporated into this group. The LFA-1 antigen was characterized on granulocytes in only one study (15). In our study only CLB LFA-1/1 and TS1/18.11 gave the same antigen pattern (130/90 Kd), whereas for the other antibodies unique mol. wts. were obtained. The results obtained for TS1/22 (M72) and TS1/18.11 (M89) were different from the results obtained by Sanchez-Marid *et al.* (15). The techniques used were different: (a) preformed immune complex method versus Staph A precipitation and (b) the percentage of polyacrylamide (10% versus 7%). The results could also be explained by polymorphism of the LFA-1 antigen.

All anti-LFA-1 antibodies reacted with some of the human–mouse hybrid cell lines. Further studies may allow us to locate the genes coding for this structure.

Summary

Seven different specificity groups were defined by the serological features of 58 granulocyte-reactive mAbs of the M protocol. The antigens recognized by these groups of antibodies were characterized by SDS–polyacrylamide gel electrophoresis. This analysis revealed that some groups were homogeneous (groups I, VI, VII) while other groups were heterogeneous (groups II, III, IV, V). Initial studies to assign the genes coding for the different antigens to specific human chromosomes were performed by testing the reactivity of these 58 antibodies with four myeloid human–mouse cell hybrids. Antibodies directed against 3-α-fucosyl-N-acetyllactosamine and leukocyte function-associated antigen-1 (LFA-1) were positive with some of these cell hybrids.

Acknowledgments. This work was supported by the "Koningin Wilhelmina Fonds," The Netherlands Cancer Foundation, under grant no. CLB 81-2.

References

1. Urdal, D., T.A. Brentall, I.D. Bernstein, and S.I. Hakarmoni. 1983. A granulocyte reactive monoclonal antibody, 1G10, identifies the Gal β 1-4 (Fuc α 1–3) Glc NAc (X determinant) expressed in HL60 on both glycolipid and glycoprotein molecules. *Blood* **62**:1022.
2. Knapp, W., O. Majdic, and P. Bettelheim. 1984. Immunocytology: Myeloid leukemias. In: *Immunohaematology,* C.P. Engelfriet, J.J. Loghem, and A.E.G. Kr. von dem Borne, eds. Elsevier Biomedical Press, Amsterdam, p. 323.
3. Tetteroo, P.A.T., A. Mulder, P.M. Landsdorp, H. Zola, D.A. Baker, F.J. Visser, and A.E.G. Kr. von dem Borne. 1984. Myeloid-associated antigen 3-α-fucosyl-N-acetyl-lactosamine (FAL); location on various granulocyte-membrane glycoproteins and masking upon monocytic differentiation. *Eur. J. Immunol.*, **14**:1089.
4. Tetteroo, P.A.T., M.B. van 't Veer, J.F. Tromp, and A.E.G. Kr. von dem Borne. 1984. Detection of the granulocyte-specific antigen 3-fucosyl-N-acetyl-lactosamine on leukemic cells after neuraminidase treatment. *Int. J. Cancer* **33**:355.
5. Geurts van Kessel, A.H.M., P.A.T. Tetteroo, A.E.G. Kr. von dem Borne, A. Hagemeyer, and D. Bootsma. 1983. Expression of human myeloid-associated surface antigens in human–mouse myeloid cell hybrids. *Proc. Natl. Acad. Sci. U.S.A.* **80**:3748.
6. Geurts van Kessel, A.H.M., P.A.T. Tetteroo, T. van Agthoven, R. Paulussen, J. van Dongen, A. Hagemeyer, and A.E.G. Kr. von dem Borne. 1984. Localization of human myeloid-associated surface antigen detected by a panel of 20 monoclonal antibodies to the q12-q ter region of chromosome 11. *J. Immunol.*, **133**:1265.
7. Fleit, H.B., S.A. Wright, and J.C. Unkeless. 1982. Human neutrophil Fc receptor distribution and structure. *Proc. Natl. Acad. Sci. U.S.A.* **79**:3275.
8. Girardet, C., S. Ladisch, D. Heumann, J.P. Mach, and S. Carrel. 1983. Identification by a monoclonal antibody of a glycolipid highly expressed by cells from the human myeloid lineage. *Int. J. Cancer* **32**:177.
9. Tetteroo, P.A.T., P.M. Lansdorp, A.H.M. Geurts van Kessel, A. Hagemeyer, and A.E.G. Kr. von dem Borne. 1984. Serological and biochemical characterization of human myeloid-associated antigens and their expression in human–mouse cell hybrids. In: *Leucocyte typing,* A. Bernard, L. Boumsell, J. Dausset, C. Milstein, and S.F. Schlossman, eds. Springer-Verlag, Berlin, Heidelberg, pp. 419–423.
10. Symington, F.W., I.D. Bernstein, and S.I. Hakomori. 1984. Monoclonal antibody specific for lactosyl ceramide. *J. Biol. Chem.* **259**:6008.
11. Perussia, B., D. Acuto, C. Terhorst, J. Gaust, R. Lazarus, V. Fanning, and G. Trinchieri. 1983. Human Natural killer cells analyzed by B73.1, a monoclonal antibody blocking Fc receptor functions. II. Studies of B73.1 antibody–antigen interaction of the lymphocyte membrane. *J. Immunol.* **130**:2142.
12. Perussia, B., G. Trinchieri, A. Jackson, N.L. Warner, J. Faust, H. Rumpold, D. Kraft, and L.L. Lanier. 1984. The Fc receptor for IgG on human natural killer cells: phenotypic, functional, and comparative studies with monoclonal antibodies. *J. Immunol.* **133**:180.

13. Arnaout, M.A., R.F. Todd, N. Dana, J. Melamed, S.F. Schlossmann, and H. R. Colten. 1983. Inhibition of phagocytosis of complement C3- or Immunoglobulin G-ocated particles and of C3bi binding by monoclonal antibodies to a monocyte-granulocyte membrane glycoprotein (Mo-1). *J. Clin. Invest.* **72:**171.

14. Wright, S.D., P.E. Rao, W.C. van Voorhis, L.S. Craigmyle, K. Iida, M.A. Talle, E.F. Westberg, G. Goldstein, and S.C. Silverstein. 1983. Identification of the C3bi receptor of human monocytes and macrophages by using monoclonal antibodies. *Proc. Natl. Acad. Sci. U.S.A.* **80:**5699.

15. Sanchez-Marid, F., J.A. Nagy, E. Robbins, P. Simon, and T.A. Springer. A human leukocyte differentiation antigen family with distinct α subunits and a common β subunit: the lymphocyte function-associated antigen (LFA-1), the C3bi complement receptor (OKMl/Macl) and the p 150,95 molecule. *J. Exp. Med.* **158:**1785.

CHAPTER 3

Glycolipid Specificities of Anti-Hematopoietic Cell Antibodies

Frank W. Symington, Brian E. McMaster,
Sen-itiroh Hakomori, and Irwin D. Bernstein

Introduction

Lineage-specific and stage-dependent oligosaccharides are expressed both by mature blood elements and their precursors, and by leukemias believed to represent developmentally arrested populations of hematopoietic cells (1,2). The reasons why changes in cell surface carbohydrate expression accompany differentiation and distinguish mature hematopoietic subsets are obscure, although several roles for oligosaccharides of cell surface glycoconjugates in cell–cell interaction, growth control, and oncogenesis have been considered (3,4). We have utilized monoclonal antibodies to probe cell surface oligosaccharide antigenicity during hematopoiesis in order to map patterns of carbohydrate expression which may further our understanding of these processes.

In the present work we screened 116 Workshop hybridoma ascitic fluids for possible anti-carbohydrate reactivity. With the expectation that many antibodies would show specificity for the Gal$\beta1\rightarrow4$(Fuc$\alpha1\rightarrow3$) GlcNAc$\beta1\rightarrow$R (X-hapten) structure detected on certain myeloid cells (5,6), we first screened on glycolipids expressing this determinant. Subsequent screens of samples reactive with HL-60 myeloid leukemia cells were performed on a panel of granulocyte glycolipids. Using solid-phase assays to detect antibody binding to glycolipids, we detected strong anti-carbohydrate reactivity in 17 of the submitted samples. Antibodies reacting with at least two distinct myeloid-associated antigens and one erythroid-specific structure were identified.

Materials and Methods

Glycolipids

Glycolipids were prepared from organic extracts of human erythrocyte fractions, tumors, and the HL-60 promyelocytic cell line as previously described (7). Rat kidney globoside (globo-*iso*-tetraosylceramide, Cytolipin R) was donated by the late Dr. Kawanami. Fractionation of acetylated and nonacetylated glycolipid mixtures into purified components was achieved by preparative thin layer chromatography (TLC) or by high-pressure liquid chromatography. Chemical characterization of the major glycolipid species was by exoglycosidase or acid hydrolysis, methylation analysis, and gas chromatography–mass spectroscopy (8). Predicted glycolipid antigenicities were verified using defined monoclonal antibodies. Glycolipid concentrations were visually estimated by orcinol staining of high-performance thin layer chromatographs.

Antibodies

Control antibodies T_5A_7 [anti-lactosylceramide, (9)], 1B2 [anti-*N*-acetyl-lactosamine, (10)], 1G10 [anti-X-hapten, (6)], and ID4 [anti-Fuc$\alpha1\rightarrow$2Gal$\beta1\rightarrow$4(Fuc$\alpha1\rightarrow$3)GlcNAc$\beta1\rightarrow$3R, Y-hapten, (11)] were used as culture supernate or ascitic fluid. For certain tests, ID4 was purified by boric acid precipitation and radioiodinated using Iodogen according to instructions supplied by the manufacturer (Pierce Chemical Co., Rockford, IL).

Antibody Binding Assays

Binding of test antibodies to films of glycolipid, phosphatidylcholine, and cholesterol adsorbed to poly(vinyl chloride) plate wells was assessed as previously described (12) using rabbit anti–mouse IgG + M and ^{125}I-protein A. Initial tests employed 1/400 dilutions of ascitic fluids. Antibody binding to glycolipids chromatographed on silica gel plates was measured using a slight modification of the autoradiographic procedure described by Magnani *et al.* (13).

Results

Panel samples were screened for reactivity with the glycolipid, lacto-*N*-fucopentaosyl III ceramide, which contains the X-hapten. Positive samples (generally at least 5-fold greater binding to glycolipid coated wells than to control wells) were retested on this and the related glycolipid, difucosyl-Y_2 (Table 3.1). Samples 5, 13, 19, 25, 33, 52, 61, 77, 80, 91, 110, 111, 116, and 117 appeared to contain high-titered reactivity for both of

Table 3.1. Selected glycolipid structures.

LNFIIICer (III^{3}Fuc nLc$_4$)	Galβ1→4GlcNAcβ1→3Galβ1→4Glcβ1→1Cer 3 ↑ Fucα1
Y$_2$ (V^{3}Fuc nLc$_6$)	Galβ1→4GlcNAcβ1→3Galβ1→4GlcNAcβ1→3Galβ1→4Glcβ1→1Cer 3 ↑ Fucα1
difucosyl-Y$_2$ (III^{3}V^{3}Fuc$_2$ nLc$_6$)	Galβ1→4GlcNAcβ1→3Galβ1→4GlcNAcβ1→3Galβ1→4Glcβ1→1Cer 3$$3 ↑$$↑ Fucα1$Fuc\alpha$1
typical Y-active glycolipid (III^{3}IV^{2}Fuc$_2$ nLc$_4$)	Galβ1→4GlcNAcβ1→3Galβ1→4Glcβ1→1Cer 2$$3 ↑$$↑ Fucα1$$Fucα1
GlcCer	Glcβ1→1Cer
GalCer	Galβ1→1Cer
LacCer	Galβ1→4Glcβ1→1Cer
Lc$_3$	GlcNAcβ1→3Galβ1→4Glcβ1→1Cer
nLc$_4$	Galβ1→4GlcNAcβ1→3Galβ1→4Glcβ1→1Cer
nLc$_6$	Galβ1→4GlcNAcβ1→3Galβ1→4GlcNAcβ1→3Galβ1→4Glcβ1→1Cer
Gb$_3$	Galα1→4Galβ1→4Glcβ1→1Cer
Gb$_4$, globoside	GalNAcβ1→3Galα1→4Galβ1→4Glcβ1→1Cer
Gb$_{4b}$, Cytolipin R	GalNAcβ1→3Galα1→3Galβ1→4Glcβ1→1Cer
Gb$_5$, Forssman	GalNAcα1→3GalNAcβ1→3Galα1→4Galβ1→4Glcβ1→1Cer
X$_2$^{a}	GalNAcβ1→3Galβ1→4GlcNAcβ1→3Galβ1→4Glcβ1→1Cer
GL5^b	Galβ1→3GalNAcβ1→3Galα1→4Galβ1→4Glcβ1→1Cer

^{a} Ref. 14.
^{b} Ref. 15.

these structures. The possibility that these antibodies could cross-react with Y-hapten [Fucα1→2Galβ1→4(Fucα2→3)GlcNAcβ1→R] was difficult to assess directly, because difucosyl-Y$_2$ and Y-active structures can co-purify. That such a cross-reaction should be of very low avidity is suggested by the fact that none of these antibodies could inhibit the binding of ^{125}I-ID4 (anti-Y hapten) to its glycolipid antigen (data not shown). Finally, these 14 samples bound to difucosyl-Y$_2$ with higher titer and plateau values than to the shorter, monofucosylated pentaosylceramide.

Antibodies which lacked X-hapten reactivity but which reacted with HL-60 promyelocytic cells were screened on a panel of other neutral glycolipid antigens thought to be expressed by these cells, including Glc-Cer, GalCer, LacCer, Lc$_3$, nLc$_4$, nLc$_6$, and a Type 2 H-active heptaosyl-ceramide with Fucα1→2Galβ1→4GlcNAcβ1→3R terminal structure. In this and subsequent assays involving titrated glycolipids and ascitic fluids, Workshop samples 15 and 94 specifically bound LacCer but did not appreciably bind other tested glycolipids (Table 3.2). Of the two antibodies, 15 could detect about ten times less plate-adsorbed LacCer than could 94,

Table 3.2. Binding specificity of workshop antibodies for LacCer.

Glycolipid[a]	Antibody		
	15	94	1B2
LacCer	11,650[b]	6550	ND[c]
nLc₄	870	1090	8290
GlcCer	760	1520	ND
Control	450	440	480

[a] Approximately 200 ng of the indicated glycolipid and carrier lipids (or carrier lipids alone, control) were reacted with 1/400 dilutions of the indicated ascitic fluids.
[b] Bound antibody (cpm) was detected using rabbit anti–mouse Ig and [125I]protein A.
[c] Not determined.

suggesting differences in the concentrations or avidities of antibody in the ascitic fluid samples.

All of the antibodies not reactive with X-hapten or LacCer (93 samples) were tested on sialosyl-nLc₄ (sialosylα2→3 and α2→6 species), sialosyl-nLc₆, GM₃, GD₃, GM₂, GM₁, GD₁ₐ, GD₁ᵦ, Gg₃, Gg₄, Gb₃, Gb₄, Gb₅, and a tumor glycolipid extract rich in sialosylated-X-hapten-containing structures. Five ascitic fluids, 9, 55, 56, 59, and 60, appeared to specifically react with GM₃ (SAα2→3Galβ1→4Glcβ1→1Cer), albeit at low titer and with a high background (data not shown). The failure of these five samples to mediate complement-dependent lysis of GM₃-liposomes suggested that the nature of their apparent reactions with plate-bound GM₃ was nonim-

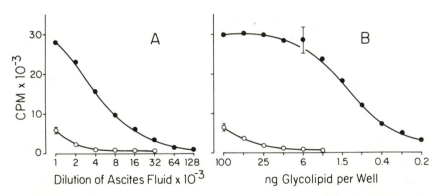

Fig. 3.1. Role of internal Galα1→4Gal structure in the epitope recognized by antibody 23. (A) Dilutions of 23 were reacted with 200 ng of Gb₄ (●) and Gb₄ᵦ (○) adsorbed to plastic wells. (B) Serial dilutions of Gb₄ (●) and Gb₄ᵦ (○) were reacted with antibody 23 diluted 1/1000. Results were obtained as described for Table 3.2. Means of duplicate determinations are given (±SD where they exceed 10% of the mean).

munologic. Alternatively, these samples could have contained low-avidity contaminating non-hybridoma antibody (16).

Most interestingly, these tests also uncovered sensitive, high-titered binding specificity of antibody 23 for Gb_4, the major human erythrocyte glycolipid, but not for a variety of related glycolipids (Gb_3, Gb_5, X_2, or GL5; see Table 3.1). Results of parallel titrations of this antibody on Gb_4 and its positional isomer, Gb_{4b} (Cytolipin R), indicated a prominent role for internal Galα1→4Gal structure in the epitope recognized by this sample (Fig. 3.1). In addition, detection by 23 of Gb_4 on thin layer plates was at least 10-fold more sensitive than detection of Cytolipin R (data not shown).

Discussion

Originally identified in human adenocarcinoma and later on early mouse embryos (17,18), the antigen detected by 14 of the panel reagents (Table 3.3) is now known to be specifically expressed by human polymorphonuclear neutrophils (PMN) and some monocytes. The epitope involves the trisaccharide Galβ1→4(Fucα1→3)GlcNAcβ1→R (X-hapten) and is found on oligosaccharide chains of both glycolipids and glycoproteins of these cells (6,19). Several laboratories have described monoclonal reagents with X-hapten specificity, designated CDw15 by the First International Workshop on Human Leucocyte Differentiation Antigens (20). The antigen appears to be quite immunogenic in the mouse (5,6,21). Although some monoclonal antibodies can clearly discriminate between mono-, di-, and tri-α1→3 fucosylated glycolipids (22), all of the relevant panel reagents reacted with both mono- and difucosylated structures. The possibility that some of the panel could recognize internal and terminal X-hapten on difucosyl-Y_2 as opposed to terminal hapten alone was not explored, although evidence for the existence of these fine specificities has been obtained with other X-reactive antibodies (F. Symington, unpublished data). While its function is unknown, this structure may be involved in margination, marrow egress, phagocytosis, or resistance to bacterial hydrolases.

Table 3.3. Summary of workshop antibody glycolipid reactivities.

Antibodies	Nominal structure(s) recognized
5, 13, 19, 25, 33, 52, 61, 77, 80, 91, 110, 111, 116, 117	Galβ1→4GlcNAcβ1→3Galβ1→4Glcβ1→1Cer 3 ↑ Fucα1
	(and other glycolipids containing X-hapten)
15, 94	Galβ1→4Glcβ1→1Cer
23	GalNAcβ1→3Galα1→4Galβ1→4Glcβ1→1Cer

In addition to their content of glycoconjugates reactive with anti-X reagents, human PMN contain abundant lactosylceramide, a simple glycolipid whose cell surface antigenic expression by hematocytes seems restricted to PMN and subpopulations of monocytes and activated T cells (23). Neutrophils contain prodigious amounts of LacCer, and it is, quantitatively speaking, a glycolipid characteristic of myeloid cells (24). A monoclonal antibody generated from mice immunized with myelogenous leukemia cells was previously shown to bind LacCer (9), suggesting that the existence of two Workshop antibodies with this specificity may reflect these antigenic and quantitative aspects of myeloid LacCer expression. The increase in LacCer content associated with granulocyte differentiation was correctly inferred by Hildebrand *et al.* (24); however, the requirement for increased LacCer in mature neutrophils is unexplained. The availability of sensitive, specific monoclonal probes to this glycolipid may prove essential to addressing this issue.

Globoside (Gb_4) is the major human erythrocyte glycolipid, and it is the major neutral glycolipid of various tissues and organs (25–27). Its chemical structure was characterized as having an unusual internal α-galactosyl residue, and this tetraosylceramide is the blood group P antigen (28,29). Still, only polyclonal human and rabbit antisera specific for globoside have been available for analyzing the expression of this unique marker. We have shown here that one of the Workshop panel antibodies reacts almost exclusively with this glycolipid among those tested. The ability of this reagent to detect picomole amounts of Gb_4 in our assay system suggests its potential for extending our knowledge of globoside synthesis, distribution, and function among human erythrocytes and their progenitors.

We have outlined results of solid-phase glycolipid binding assays conducted with 116 Workshop ascitic fluids. Our results (summarized in Table 3.3) indicate that antibodies against lineage-related carbohydrate antigens represent a significant fraction of the submitted panel. It may be expected that future selective immunizations and screenings of hybridomas with purified carbohydrate antigens will further broaden the spectrum of antibodies available for dissecting normal and pathologic hematopoiesis and immunoregulation.

Acknowledgments. We are grateful to E. Nudelman and Drs. Y. Okada and E. Holmes for their gifts of purified glycolipids and to D. Hedges for technical assistance. This work was supported by NCI postdoctoral fellowship grant CA07461 to F. Symington and USPHS grant AM HL 31232.

References

1. Symington, F.W., and S. Hakomori. 1985. Glycosphingolipids of immune cells. *Lymphokines,* Vol. 12. E. Pick, ed. Academic Press, New York.
2. Klock, J.C., B.A. Macher, and W.M.F. Lee. 1981. Complex carbohydrates

as differentiation markers in malignant blood cells: glycolipids in the human leukemias. *Blood Cells* **7:**247.

3. Critchley, D.R. 1979. Glycolipids as membrane receptors important in growth regulation. In: *Surfaces of normal and malignant cells,* R.O. Hynes, ed. John Wiley & Sons, New York, pp. 63–101.

4. Hakomori, S. 1981. Glycosphingolipids in cellular interaction, differentiation, and oncogenesis. *Annu. Rev. Biochem.* **50:**733.

5. Gooi, H.C., S.J. Thorpe, E.F. Hounsell, H. Rumpold, D. Kraft, O. Forster, and T. Feizi. 1983. Marker of peripheral blood granulocytes of man recognized by two monoclonal antibodies VEP8 and VEP9 involves the trisaccharide 3-fucosyl-*N*-acetyllactosamine. *Eur. J. Immunol.* **13:**306.

6. Urdal, D.L., T.A. Brentnall, I.D. Bernstein, and S. Hakomori. 1983. A granulocyte reactive monoclonal antibody, 1G10, identifies the Galβ1-4(Fucα1-3) GlcNAc (X determinant) expressed in HL-60 cells on both glycolipid and glycoprotein molecules. *Blood* **62:**1022.

7. Kannagi, R., E. Nudelman, S.B. Levery, and S. Hakomori. 1982. A series of human erythrocyte glycosphingolipids reacting to monoclonal antibody directed to a developmentally regulated antigen, SSEA-1. *J. Biol. Chem.* **257:**14865.

8. Hakomori, S. 1983. Chemistry of glycosphingolipids. In: *Sphingolipid biochemistry.* J.N. Kanfer and S. Hakomori, eds. Plenum Press, New York, pp. 1–165.

9. Symington, F.W., I.D. Bernstein, and S. Hakomori. 1984. Monoclonal antibody specific for lactosylceramide. *J. Biol. Chem.* **259:**6008.

10. Young, W.W., Jr., J. Portoukalian, and S. Hakomori. 1982. Two monoclonal anticarbohydrate antibodies directed to glycosphingolipids with a lacto-*N*-glycosyl Type II chain. *J. Biol. Chem.* **256:**10967.

11. Abe K., J.M. McKibbin, and S. Hakomori. 1983. The monoclonal antibody directed to difucosylated Type 2 chain (Fucα1$\rightarrow$2Galβ1$\rightarrow$4[Fucα1$\rightarrow$3] GlcNAc; Y determinant. *J. Biol. Chem.* **258:**11793.

12. Kannagi, R., R. Stroup, N.A. Cochran, D.L. Urdal, W.W. Young, and S. Hakomori. 1983. Factors affecting expression of glycolipid tumor antigens: influence of ceramide composition and coexisting glycolipid on the antigenicity of gangliotriosylceramide in murine lymphoma cells. *Cancer Res.* **43:**4997.

13. Magnani, J.L., D.F. Smith, and V. Ginsburg. 1980. Detection of gangliosides that bind cholera toxin: direct binding of ^{125}I-labeled toxin to thin-layer chromatograms. *Anal. Biochem.* **109:**399.

14. Kannagi, R., S.B. Levery, F. Ishigami, S. Hakomori, L.H. Shevinsky, B.B. Knowles, and D. Solter. 1983. New globoseries glycosphingolipids in human teratocarcinoma reactive with the monoclonal antibody directed to a developmentally regulated antigen, stage-specific embryonic antigen 3. *J. Biol. Chem.* **258:**8934.

15. Kannagi, R., M.N. Fukuda, and S. Hakomori. 1982. A new glycolipid antigen isolated from human erythrocyte membranes reacting with antibodies directed to globo-*N*-tetraosylceramide (globoside). *J. Biol. Chem.* **257:**4438.

16. Gooi, H.C., and T. Feizi. 1982. Natural antibodies as contaminants of hybridoma products. *Biochem. Biophys. Res. Commun.* **106:**539.

17. Yang, H-J., and S. Hakomori. 1971. A sphingolipid having a novel type of ceramide and "lacto-*N*-fucopentaose III." *J. Biol. Chem.* **246:**1192.

18. Solter, D., and B.B. Knowles. 1978. Monoclonal antibody defining a stage-specific mouse embryonic antigen (SSEA-1). *Proc. Natl. Acad. Sci. U.S.A.* **75**:5565.
19. Gooi, H.C., T. Feizi, A. Kapadia, B.B. Knowles, D. Solter, and M.J. Evans. 1981. Stage-specific embryonic antigen involves $\alpha 1 \rightarrow 3$ fucosylated type 2 blood group chains. *Nature* **292**:156.
20. Bernard, A., L. Boumsell, J. Dausset, C. Milstein, and S.F. Schlossman, eds. 1984. *Leucocyte typing.* Springer-Verlag, Berlin, Heidelberg.
21. Huang, L.C., C.I. Civin, J.L. Magnani, J.H. Shaper, and V. Ginsburg. 1983. My-1, the human myeloid-specific antigen detected by mouse monoclonal antibodies, is a sugar sequence found in lacto-*N*-fucopentaose III. *Blood* **61**:1020.
22. Fukushi, Y., S. Hakomori, E. Nudelman, and N. Cochran. 1984. Novel fucolipids accumulating in human adenocarcinoma. II. Selective isolation of hybridoma antibodies that differentially recognize mono-, di-, and trifucosylated Type 2 chain. *J. Biol. Chem.* **259**:4681.
23. Andrews, R.G., B. Torok-Storb, and I.D. Bernstein. 1983. Myeloid-associated differentiation antigens on stem cells and their progeny identified by monoclonal antibodies. *Blood* **62**:124.
24. Hildebrand, J., P. Stryckmans, and P. Stoffyn. 1971. Neutral glycolipids in leukemic and nonleukemic leukocytes. *J. Lipid. Res.* **12**:361.
25. Klenk, E., and K. Lauenstein. 1951. Uber die zuckerhaltigen Lipoide der Formbestandteile des menschlichen Blutes. *Z. Physiol. Chem.* **288**:220.
26. Yamakawa, T., and S. Suzuki. 1952. The chemistry of the lipids of posthemolytic residue or stroma of erythrocytes. III. Globoside, the sugar-containing lipid of human blood stroma. *J. Biochem. (Tokyo)* **39**:393.
27. Siddiqui, B., S. Kawanami, Y-T. Li, and S. Hakomori. 1972. Structure of globoside from various sources: uniqueness of rat kidney globoside. *J. Lipid Res.* **13**:657.
28. Hakomori, S., B. Siddiqui, Y-T. Li, S-C. Li, and C.G. Hellerqvist. 1971. Anomeric structures of "globoside" and ceramide trihexoside of human erythrocytes and "BHK" fibroblasts. *J. Biol. Chem.* **246**:2271.
29. Marcus, D.M., M. Naiki, and S.K. Kundu. 1976. Abnormalities in the glycosphingolipid content of human P^k and p erythrocytes. *Proc. Natl. Acad. Sci. U.S.A.* **73**:3263.

CHAPTER 4

Antibodies Specific for the Mac-1, LFA-1, p150,95 Glycoproteins or Their Family, or for Other Granulocyte Proteins, in the 2nd International Workshop on Human Leukocyte Differentiation Antigens

Timothy A. Springer and Donald C. Anderson

Introduction

A family of functionally important leukocyte surface glycoproteins which share a common β subunit of $M_r = 95,000$ has recently been defined in humans and mice (1,2). These glycoproteins, the lymphocyte function-associated 1 (LFA-1), macrophage 1 (Mac-1), and p150,95 molecules each contain a different α subunit noncovalently associated with the common β subunit in an $\alpha_1\beta_1$ structure. Monoclonal antibodies specific for the LFA-1 and Mac-1 molecules have allowed definition of their cell distributions and functions.

The LFA-1 molecule is expressed on B and T lymphocytes, NK cells, monocytes, and granulocytes. mAbs to LFA-1 block cytolytic T lympho-cyte-mediated killing, natural killing, and T helper cell responses (3,4). Anti-LFA-1 blocks the adhesion of cytolytic T lymphocytes to target cells, and it has been proposed that LFA-1 is an adhesion protein of wide importance in interactions of leukocytes with other cells (5,6).

The Mac-1 molecule, identical to OKMI and Mol, is expressed on granulocytes, monocytes, and natural killer cells, and is absent on lymphocytes (reviewed in Ref. 1). In response to chemo-attractants and secretagogues, Mac-1 surface expression on granulocytes is increased or "up-regulated" 5-fold (7)]. Mac-1 is identical to the complement receptor type 3 on myeloid cells, which mediates adherence to and phagocytosis of particles bearing iC3b, a cleaved, hemolytically inactive fragment of the third component of complement (8–10).

A genetic deficiency of this glycoprotein family has been found in certain patients with recurring bacterial infections (7). Deficiency is inherited as an autosomal, recessive trait. Over 19 patients have been identified worldwide within the last year (reviewed in Ref. 11). All patients lack the α subunits of Mac-1 and LFA-1 and the common β subunit on leukocyte surfaces. A latent, intracellular pool of Mac-1 is also deficient. Biosynthesis experiments have suggested that β subunit synthesis is defective in patient cells, and that normal α chain precursors are present, but are not transported to the cell surface, due to a lack of association with the β subunit (7).

Monoclonal antibodies specific for LFA-1, specific for Mac-1, or cross-reacting between LFA-1, Mac-1, and p150,95 have previously been obtained. Studies on antigen preparations in which the α and β subunits had been dissociated by brief exposure to high pH showed that these mAbs were specific for the LFA-1 α (αL), the Mac-1 α (αM), or the common β subunit, respectively (1). Two-dimensional isoelectric focusing SDS–PAGE and peptide mapping have confirmed that these molecules contain different α subunits and identical β subunits (1,12–14).

The third member of the family, p150,95, named after the $M_r \times 10^{-3}$ of its subunits, has been defined biochemically after isolation from myeloid cells with anti-β mAb (1). The p150,95 α subunit, designated αX, was found to be antigenically distinct from αM and αL, since it did not react with mAbs defining multiple epitopes on the latter subunits. It was predicted that it should be possible to obtain mAbs to unique epitopes of αX, which would be useful for characterization of the structure and function of p150,95.

In this study, the 118 mAbs submitted in the myeloid panel of the Second International Conference on Human Leukocyte Differentiation Antigens were tested for reactivity with members of the Mac-1, LFA-1, p150,95 family. It was of interest to compare mAbs from different labs, which have been the subject of several publications (1,3,4,7,11,15–22). Furthermore, the following questions were addressed: (1) Which mAbs reacted with specific members of the family and which cross-reacted with all three members? Did any show unusual specificities, such as cross-reaction between only two members of the family? (2) Could mAbs specific for the p150,95 molecule be identified in the panel? (3) Which members of the family were up-regulated on myeloid cell surfaces by chemo-attractants? (4) Which mAbs in the myeloid panel were negative on Mac-1-, LFA-1-deficient patients? How specific was the deficiency to the Mac-1, LFA-1, p150,95 glycoprotein family?

In the course of these studies, information was also obtained on molecules distinct from Mac-1, LFA-1, and p150,95. This is presented as an appendix.

Methods

Lactoperoxidase-catalyzed ^{125}I-labeling of granulocytes, immunoprecipitation and SDS–PAGE, labeling of cells with mAbs and FITC anti–mouse IgG followed by analysis on an Epics V flow cytometer, and Mac-1-, LFA-1-deficient patients have been previously described (1,7,11).

Results and Discussion

Identification of mAbs to Mac-1, LFA-1, and p150,95

Granulocytes were used to screen for mAbs to Mac-1, LFA-1, or p150,95, since they express all three molecules (1). Granulocytes were stimulated with f-Met-Leu-Phe to increase Mac-1 and p150,95 expression, labeled with ^{125}I, and subjected to immunoprecipitation with each workshop mAb. Immunoprecipitates of positive mAbs were then compared side-by-side in further experiments with the previously characterized TS1/22 anti-αL, OKM1 (or OKM10) anti-αM, and TS1/18 anti-β MAb (1) (Fig. 4.1). The LFA-1 αL and β subunits of M_r = 177,000 and 95,000, respectively, were precipitated by TS1/22,* MHM24, 25.3.1, and CIMT (Fig. 4.1, lanes 1, 2, 11, and 13). The Mac-1 αM and β subunits of 165,000 and 95,000 M_r, respectively, were precipitated by VIM12, CC1.7, OKM10, D12, M3D11, Ki-M5, and 44 (Fig. 4.1, lanes 14–22). The workshop mAb 60.1 appeared to precipitate Mac-1 and, more weakly, other members of the family (Fig. 4.1, lane 5), but further testing with authentic 60.1 showed precipitation of only Mac-1 (not shown). The MHM23, TS1/18, 60.3 (Fig. 4.1, lanes 3, 6–8, and 12) and CLB-54 mAb (not shown) precipitated the αL, αM, and αX subunits and the common β subunit.

Interestingly, the SHCL3 and Ki-M1 mAb appeared to specifically precipitate the αX and β subunits of p150,95 [Fig. 4.1, lanes 9 and 10, and see below, Fig. 4.5(A), lane 1]. The αX and β subunits precipitated by SHCL3 were identical to those precipitated by the TS1/18 anti-β mAb in isoelectric focusing and in coprecipitation experiments (T. Springer and L. Miller, unpublished). Thus, SHCL3 clearly is specific for p150,95, the third member of the glycoprotein family which had previously been detected by precipitation with anti-β mAb (1). Because of the identity of the β subunits among Mac-1, LFA-1, and p150,95, the specificity of SHCL3 suggests reactivity with the p150,95 αX subunit; however, this remains to be experimentally verified.

---

* Since M72, the Workshop TS1/22, appeared contaminated with an anti-β MAb (Fig. 4.1, lane 4), authentic TS1/22 was used in these studies.

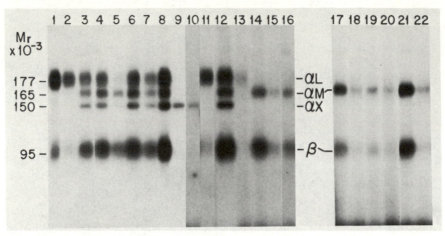

Fig. 4.1. Precipitation from ^{125}I-labeled granulocytes by mAbs specific for Mac-1, LFA-1, p150,95, or their family. ^{125}I-labeled f-Met-Leu-Phe-stimulated granulocyte lysates were immunoprecipitated with TS1/22–Sepharose CL-4B (lane 1); M56, MHM24 (lane 2); M55, MHM23 (lane 3); M72, TS1/22 (lane 4); M76, 60.1 (lane 5); M89, TS1/18.11 (lane 6); TS1/18.11 (lane 7); TS1/18.11–Sepharose CL-4B (lane 8); M46, SHCL3 (lane 9); M93, Ki-M1 (lane 10); 25.3.1 (lane 11); M75, 60.3 (lane 12); M107, CIMT (lane 13); M26, VIM12 (lane 14); M41, CC1.7 (lane 15); OKM10 (lane 16); M26, VIM12 (lane 17); M41, CC1.7 (lane 18); M42, D12 (lane 19); M78, M3D11 (lane 20); M88, Ki-M5 (lane 21); M104, 44 (lane 22). Immunoprecipitates were subjected to SDS–7.5% PAGE and autoradiography for 4 days (lanes 1–9, 17–22) or 20 days (lanes 10–16). Precipitates in lanes 2–7 and 9–16 were formed with 2–4-μl Workshop mAb, 60 μl rabbit anti–mouse IgG, and 50 μl *S. aureus* bacteria; those in lanes 17–22 were formed with 1-μl Workshop mAb and a mixture of rabbit anti–mouse IgG (60 μl) and sheep anti–mouse MOPC 104E IgM myeloma (5 μl) and no *S. aureus*.

The distribution of these antigens was examined by immunofluorescent flow cytometry. The staining pattern on mononuclear cells was characteristic for each type of antibody specificity. Representative immunofluorescence histograms are shown for Mac-1, LFA-1, p150,95-cross-reactive mAbs [Fig. 4.2(a)–(c)], anti-Mac-1 mAbs [Fig. 4.2(d)–(f)], anti-LFA-1 mAbs [Fig. 4.2(g)–(i)], and anti-p150,95 mAbs [Fig. 4.2(j) and (k)]. Anti-β mAbs [Fig. 4.2(a)–(c)] and anti-LFA-1 mAbs [Fig. 4.2, (g)–(i)] stained all mononuclear cells, and revealed differences among subpopulations in the quantity of antigen expressed, as previously described (3). Weaker staining by CIMT than other anti-LFA-1 mAbs (Fig. 2i) was due to its lower titer, as revealed by staining at different CIMT mAb concentrations (not shown). Anti-Mac-1 and anti-p150,95 mAbs stained a subpopulation of mononuclear cells [Fig. 4.2(d)–(f), (j) and (k)]. Scatter gating showed this

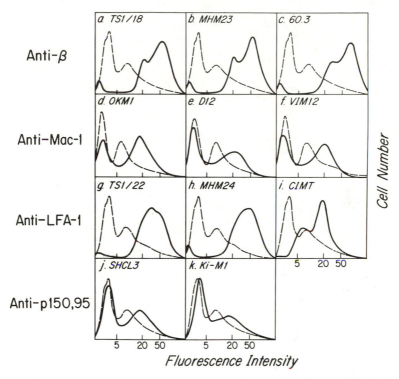

Fig. 4.2. Fluorescent staining of mononuclear cells by antibodies specific for Mac-1, LFA-1, p150,95, and their family. Mononuclear cells prepared by Ficoll–Hypaque centrifugation (2.5×10^7/ml in 50 μl) were labeled with an equal volume of 1/200 Workshop mAb (solid curves) or P3X63 IgG1 control mAb (dashed curves), then with FITC anti–mouse IgG, and subjected to immunofluorescence flow cytometry. Fluorescent unit = 1/20 the intensity of 2% bright beads (Coulter).

positive subpopulation corresponded to monocytes, and the vast majority of lymphocytes were negative (data not shown, and see below, Fig. 4.3). The intensity of staining of granulocytes was also characteristic for each antibody specificity; staining by anti-Mac-1 and anti-family mAbs was stronger than staining by anti-LFA-1 and anti-p150,95 mAbs (Table 4.1).

The mAbs to the Mac-1, LFA-1, p150,95 glycoproteins could be placed into four distinct groups based on specificity for one of the three members of the family or reactivity with the entire family (Table 4.1). The specificities suggest reaction with individual α subunits or the common β subunit. However, these subunit specificities remain speculative unless confirmed by reaction with isolated subunits. Subunit specificity of these antibodies has thus far been directly demonstrated only for the OKM1 anti-Mac-1 α mAb, the TS1/22 anti-LFA-1 α mAb, and the TS1/18 anti-β mAb (1).

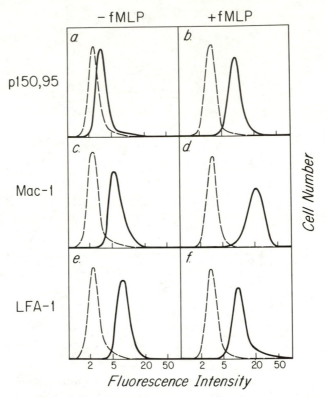

Fig. 4.3. Chemo-attractant stimulates increased expression of p150,95 and Mac-1 but not LFA-1 on monocytes. Mononuclear cells were incubated ½ hr at 37°C with 10^{-8} M f-Met-Leu-Phe, or held at 4°C as indicated. Cells were stained at 4°C with specific (solid curves) or control (dashed curves) mAb, followed by FITC anti–mouse IgG, and subjected to immunofluorescence flow cytometry. Fluorescence of monocytes was determined by gating on 90° and forward-angle light scatter to exclude lymphocytes.

Up-regulation of Mac-1 and p150,95 Surface Expression

Previous work has shown that Mac-1 but not LFA-1 surface expression is increased when granulocytes are stimulated with the chemo-attractant and secretagogue f-Met-Leu-Phe (7). All Workshop anti-Mac-1 mAbs detected such "up-regulation," with an average increase of 3.7-fold, while LFA-1 was not up-regulated. The β subunit showed an intermediate increase (data not shown).

It was further studied whether p150,95 was up-regulated on granulocytes, and whether the expression of these molecules on monocytes was altered by f-Met-Leu-Phe stimulation. p150,95 was strikingly increased an average of 4.4-fold when granulocytes were f-Met-Leu-Phe-stimulated, as shown with the two different anti-p150,95 mAbs. The granulocyte fluores-

Table 4.1. Myeloid workshop anti-Mac-1, LFA-1, p150,95 antibodies.

Speci-ficity	Work-shop no.	Local name	Precip-itation	Lymph/monocytes[a]	Gran. fluor. inten.[b]	Deficient on patients[c]	Subclass
Mac-1	—	OKM1	Mac-1	−/+	130	+	$\gamma 2$
Mac-1	M26	V1M12	Mac-1	−/+	104	+	$\gamma 1$
Mac-1?[d]	M41	CC1.7	Mac-1	+/+	106	+	$\gamma 1$
Mac-1	M42	D12	Mac-1	−/+	114	+	$\gamma 2a$
Mac-1	M76	60.1	Mac-1	−/+	128	+	$\gamma 2a$
Mac-1	M78	M3D11	Mac-1	−/+	60	+	μ
Mac-1	M88	Ki-M5	Mac-1	−/+	90	+	$\gamma 2a$
Mac-1	M104	44	Mac-1	−/+	102	+	$\gamma 1$
LFA-1	—	TS1/22	LFA-1	+/+	15	+	$\gamma 1$
LFA-1	M56	MHM24	LFA-1	+/+	14	+	ND
LFA-1	M107	C1MT	LFA-1	+/+	4	+	$\gamma 1$
LFA-1	—	25.3.1	LFA-1	ND/ND[e]	ND	ND	—
p150,95	M46	SHCL3	p150,95	−/+	13	+	$\gamma 2b$
p150,95	M93	Ki-M1	p150,95	−/+	8	+	μ
Family	M55	MHM23	Family	+/+	122	+	$\gamma 1$
Family	M73	CLB-54	Family	+/+	93	+	$\gamma 1$
Family	M75	60.3	Family	+/+	172	+	$\gamma 2a$
Family	M89	TS1/18	Family	+/+	102	+	$\gamma 1$

[a] By immunofluorescence, as described in captions to Figs. 4.2 and 4.3.
[b] Fluorescence intensity (relative to TS1/18 anti-β = 100) on f-Met-Leu-Phe-stimulated granulocytes. Background fluorescence has been subtracted.
[c] By immunofluorescence, as described in caption to Fig. 4.4.
[d] This mAb was specific for Mac-1 by immunoprecipitation, but an unexpected reactivity with lymphocytes was found in immunofluorescence. Since this could be due to an accidental contamination, and no reactivity with LFA-1 was seen, this mAb has been designated a tentative anti-Mac-1.
[e] ND: Not done.

cence intensities shown in Table 4.1 are for f-Met-Leu-Phe-treated cells; on unstimulated granulocytes p150,95 is the most weakly expressed member of the family.

Stimulation of monocytes with f-Met-Leu-Phe resulted in dramatic 6.5-fold and 5.0-fold increases in surface expression of p150,95 and Mac-1, respectively [Fig. 4.3, (a)–(d)]. In contrast, LFA-1 expression on monocytes [Fig. 4.3(e) and (f)] and on lymphocytes (not shown) was unaffected. These large increases in surface expression occurred within ½ hr, suggesting a latent store of Mac-1 and p150,95 in granulocytes and monocytes. In contrast, LFA-1 was not up-regulated. Since the β subunits of these molecules are identical, the α chains appear to control subcellular targetting to the plasma membrane or to intracellular storage sites.

Studies on Mac-1, LFA-1-Deficient Patients

As a second means of identifying antibodies to the Mac-1, LFA-1 family, each mAb in the myeloid panel was tested with granulocytes from a Mac-

1, LFA-1-deficient patient in an [^{125}I]anti–mouse IgG indirect binding assay (23). mAbs giving lower binding to patient than to normal cells were retested by immunofluorescent flow cytometry (Table 4.1; representative histograms are shown in Fig. 4.4). All eight mAbs to Mac-1, three mAbs to LFA-1, two mAbs to p150,95, and four mAbs to the family were negative on f-Met-Leu-Phe-stimulated patient granulocytes. The deficiency in patient cells was also demonstrated by precipitation with anti-Mac-1, anti-LFA-1, anti-p150,95, and anti-β subunit mAbs [Fig. 4.5(B) compared to 4.5(A)]. The results with the putative anti-p150,95 αX mAb extend previous findings that all three α subunits and the common β subunit are deficient on the cell surface (7). The primary defect in patient cells appears to be in the β subunit; α precursor is made but is neither processed nor transported to the cell surface or to latent stores in the absence of association with the β subunit (7).

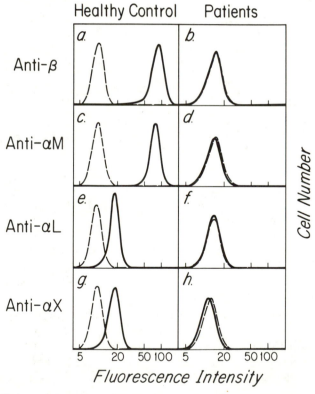

Fig. 4.4. Deficiency of α and β subunits on granulocytes of a patient with recurring infections. Granulocytes of a patient with severe deficiency [patient 1 (11)] or of a healthy adult control were incubated ½ hr at 37°C with 10^{-8} M f-Met-Leu-Phe, then stained with specific mAb (solid curves) or control mAb (dashed curves) and subjected to immunofluorescence flow cytometry as described for Fig. 4.2.

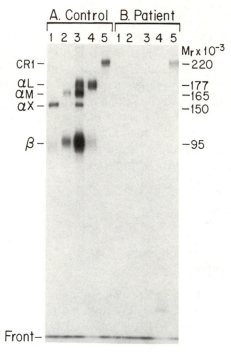

Fig. 4.5. Immunoprecipitation of p150,95, Mac-1, and LFA-1 from normal and deficient granulocytes. Triton X-100 lysates of ^{125}I-labeled healthy adult control (A) or severely deficient patient (B) f-Met-Leu-Phe-stimulated granulocytes were immunoprecipitated with anti-p150,95 SHCL3 mAb (lane 1); anti-Mac-1 α mAb OKM10 (lane 2); anti-β mAb TS1/18 (lane 3); anti-LFA-1 α mAb TS1/22 (lane 4); or anti-CR1 mAb 44D as positive control (lane 5). Immunoprecipitates were subjected to SDS–7% PAGE and autoradiography.

None of the antigens defined by the 102 non-Mac-1, LFA-1 mAbs were deficient. M103 antigen showed lower than normal expression on patients' cells, but was clearly detectable by immunoprecipitation (not shown). The complement receptor type 1 (CR1) was also normally expressed on patient cells [Fig. 4.5(A) and (B), lane 5]. Thus, the deficiency is highly specific to the Mac-1, LFA-1 glycoprotein family.

The idea that LFA-1, Mac-1, and p150,95 function as leukocyte adhesion molecules has received support from functional studies on deficient patient cells. Patient cells show defects in natural, cytolytic T-lymphocyte-mediated, and antibody-dependent cell-mediated killing (6,20,22), and in the complement receptor type 3 (19,21). This is consistent with blocking of these functions on normal cells by anti-LFA-1 and anti-Mac-1 mAbs, respectively. Furthermore, patient granulocytes and monocytes show multiple adhesion-dependent defects, including adherence to surfaces, spreading, chemotaxis, and aggregation (11,21). The finding here that the entire Mac-1, LFA-1 family, including p150,95, is specifically deficient on patient cells, is of great importance in understanding these multiple defects in adhesion-dependent functions. Stimulated adherence, a phenomenon in which granulocytes and monocytes exhibit increased adherence and aggregate in response to chemo-attractants such as f-Met-Leu-Phe, is also defective in the patients (21). Since other surface receptors are up-regulated normally upon stimulation of patient cells (7), these functional deficiencies may be related to a lack of Mac-1 and p150,95 up-regulation. These deficiencies *in vitro* appear to explain the inability of patient neutrophils *in vivo* to extravasate and migrate into inflammatory sites (11).

Appendix

Non-Mac-1, LFA-1 mAbs

Data were also obtained on mAbs which immunoprecipitated non-LFA-1, Mac-1 antigens. Thirteen mAbs precipitated a spectrum of proteins at $M_r = 220,000$, 180,000, 155,000, 130,000, and 98,000 (Fig. 4.6, lanes 1–10, 13–15, and Table 4.2). These mAbs are specific for Galβ1→4(Fuc-α1→3)GlcNAcβ1→R (X-hapten) (this volume, Chapters 2 and 3), which thus appears to be linked to the above polypeptide chains. The X-hapten is also found on glycolipids, which would not be labeled with ^{125}I and hence would not be visualized in these experiments.

The M103 mAb precipitated two chains with mobility identical to the $M_r = 220,000$ and 130,000 chains precipitated by the anti-hapten X mAbs

Table 4.2. Molecular properties of non-LFA-1, Mac-1 antigens on granulocytes.

Workshop no.	Polypeptide (M.W. $\times$ 10^{-3})	Fluorescence intensity[a] −fMLP	Fluorescence intensity[a] +fMLP	Up-reg-ulation (-fold)
M5, M13, M19, M25, M33, M52, M61, M77, M80, M91, M110, M111, M117	220, 180, 155, 130, 98	442	653	1.5
M103	220, 130	ND[b]	10	
M87	68	80	158	2.0
M115	65	1.5	1.4	—
M22, M57, M59, M82	62, 54 (diffuse)	270	384	1.4
M3	50, 25	ND	ND	
M102	27	ND	ND	

[a] Relative to TS1/18 anti-β = 100 on f-Met-Leu-Phe-stimulated granulocytes. Cells were treated with 10^{-8} M f-Met-Leu-Phe for ½ hr at 37°C or held at 4°C as described for Fig. 4.3. Fluorescence data are for the underlined MAb.
[b] ND: Not done.

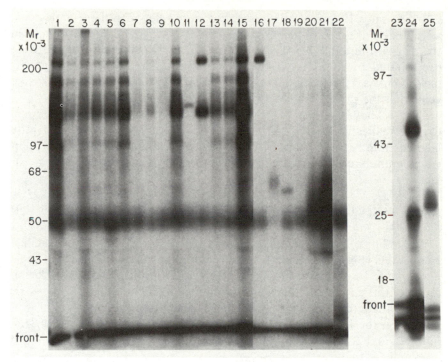

Fig. 4.6. Non-Mac-1, LFA-1 proteins precipitated from ^{125}I-labeled granulocytes. ^{125}I-labeled f-Met-Leu-Phe-stimulated granulocyte lysates were immunoprecipitated with myeloid Workshop antibodies M5, 92H5 (lane 1); M13, 1G10 (lane 2); M19, HLC5 (lane 3); M25, VIMC6 (lane 4); M33, G7C5 (lane 5); M52, HG-1 (lane 6); M61, 29 (lane 7); M77, G9F9 (lane 8); M80, CLB gran 7 (lane 9); M91, 28 (lane 10); M93, Ki-M1 (lane 11); M103, PMN-3 (lane 12); M110, LMA-1 (lane 13); M111, LMA-3 (lane 14); M117, HG-3 (lane 15); anti-CR1, 44D (lane 16); M87, CLB gran 5 (lane 17); M115, CAMAL-1 (lane 18); irrelevant control mAb (lane 19); M22, BW243/41 (lane 20); M57, CLB FcR gran 1 (lane 21); M102, CIKM5 (lane 22); M2, nonprecipitating mAb (lane 23); M3, 82H3 (lane 24); M102, CIKM5 (lane 25). Precipitates were electrophoresed on SDS gels of 7.5% polyacrylamide (lanes 1–22) or 11% polyacrylamide (lanes 23–25). The positions of marker proteins and the buffer front are marked.

(Fig. 4.6, lane 12). Other bands precipitated by anti-hapten X mAbs were not present, showing M103 has a different specificity. Furthermore, the fluorescence intensity (Table 4.2) shows there are many fewer M103 antigen than X-hapten sites on granulocytes. It is possible that mAb M103 defines a protein determinant on a subset of X-hapten-bearing polypeptide chains. Under nonreducing conditions, more of the 220,000 and less of the 130,000 M_r chains were seen (not shown). The 130,000-M_r chain may therefore be a monomer or a fragment of the 220,000-M_r chain.

The CR1 molecule (Fig. 4.6, lane 16) has a mobility similar to that of the $M_r = 220,000$ chains precipitated by anti–X-hapten and M103 mAb (cf.

Fig. 4.6, lanes 12 and 13). To test for identities between these molecules, advantage was taken of the previously described genetic polymorphism in the size of the CR1 (24). Both CR1 bands of 240,000 and 220,000 were precipitated by anti-CR1 from a heterozygote, but M103 and anti–X-hapten mAbs only precipitated a 220,000 M_r band (not shown). This suggests that the molecules defined by these mAbs are distinct.

Two mAbs, M87 and M115, precipitated bands of M_r = 68,000 and 65,000, respectively (Fig. 4.6, lanes 17 and 18). M115 gave very little labeling of intact cells (Table 4.2), suggesting it might define an intracellular or granule constituent which was released or accessible during ^{125}I-labeling.

The M22, M57, M59, and M82 mAbs precipitated diffuse bands of M_r = 62,000 and 54,000 (Fig. 4.6, lanes 20, 21) similar to the neutrophil FcR (25).

The M3 mAb precipitated two chains of 50,000 and 25,000 M_r, reminiscent of IgG (Fig. 4.6, lane 24). Under nonreducing conditions, the chains appeared to be associated together into a M_r = 150,000 complex. Precipitation of this antigen by M3 was not inhibited by human IgG. It is possible that M3 recognizes an Fc receptor associated with IgG.

Finally, the M102 mAb precipitated a chain of M_r = 27,000 (Fig. 4.6, lanes 22 and 25).

Acknowledgments. This work was supported by Council for Tobacco Research grant 1307, and NIH grants CA 31798 and AI 19031. The excellent technical assistance of Scott Thompson is acknowledged.

References

1. Sanchez-Madrid, F., J. Nagy, E. Robbins, P. Simon, and T.A. Springer. 1983. A human leukocyte differentiation antigen family with distinct alpha subunits and a common beta subunit: the lymphocyte-function associated antigen (LFA-1), the C3bi complement receptor (OKM1/Mac-1), and the p150,95 molecule. *J. Exp. Med.* **158**:1785.
2. Sanchez-Madrid, F., P. Simon, S. Thompson, and T.A. Springer. 1983. Mapping of antigenic and functional epitopes on the alpha and beta subunits of two related glycoproteins involved in cell interactions, LFA-1 and Mac-1. *J. Exp. Med.* **158**:586.
3. Krensky, A.M., F. Sanchez-Madrid, E. Robbins, J. Nagy, T.A. Springer, and S.J. Burakoff. 1983. The functional significance, distribution, and structure of LFA-1, LFA-2, and LFA-3: cell surface antigens associated with CTL–target interactions. *J. Immunol.* **131**:611.
4. Hildreth, J.E.K., F.M. Gotch, P.D.K. Hildreth, and A.J. McMichael. 1983. A human lymphocyte-associated antigen involved in cell-mediated lympholysis. *Eur. J. Immunol.* **13**:202.
5. Springer, T.A., D. Davignon, M.K. Ho, K. Kürzinger, E. Martz, and F. Sanchez-Madrid. 1982. LFA-1 and Lyt-2,3, molecules associated with T lym-

phocyte-mediated killing; and Mac-1, an LFA-1 homologue associated with complement receptor function. *Immunol. Rev.* **68**:111.

6. Springer, T.A., R. Rothlein, D.C. Anderson, S.J. Burakoff, and A.M. Krensky. 1985. The function of LFA-1 in cell-mediated killing and adhesion: Studies on heritable LFA-1, Mac-1 deficiency and on lymphoid cell self-aggregation. In: *Mechanisms of cell-mediated cytotoxicity II,* Plenum Press, New York, p. 311.

7. Springer, T.A., W.S. Thompson, L.J. Miller, F.C. Schmalstieg, and D.C. Anderson. 1984. Inherited deficiency of the Mac-1, LFA-1, p150,95 glycoprotein family and its molecular basis. *J. Exp. Med.,* **160**:1901.

8. Beller, D.I., T.A. Springer, and R.D. Schreiber. 1982. Anti-Mac-1 selectively inhibits the mouse and human type three complement receptor. *J. Exp. Med.* **156**:1000–1009.

9. Wright, S.D., P.E. Rao, W.C. Van Voorhis, L.S. Craigmyle, K. Iida, M.A. Talle, E.F. Westberg, G. Goldstein, and S.C. Silverstein. 1983. Identification of the C3bi receptor of human monocytes and macrophages with monoclonal antibodies. *Proc. Natl. Acad. Sci. U.S.A.* **80**:5699.

10. Arnaout, M.A., R.F. Todd III, N. Dana, J. Melamed, S.F. Schlossman, and H.R. Colten. 1983. Inhibition of phagocytosis of complement C3- or immunoglobulin G-coated particles and of C3bi binding by monoclonal antibodies to a monocyte-granulocyte membrane glycoprotein (Mol). *J. Clin. Invest.* **72**:171.

11. Anderson, D.C., F.C. Schmalstieg, M.J. Finegold, B.J. Hughes, R. Rothlein, L.J. Miller, S. Kohl, M.F. Tosi, R.L. Jacobs, A. Goldman, W.T. Shearer, and T.A. Springer. 1984. Inherited deficiency of the leukocyte Mac-1, LFA-1, glycoprotein family: relationship of clinical expression to severity of molecular and functional abnormalities. *J. Inf. Dis.,* in press.

12. Kürzinger, K., and T.A. Springer. 1982. Purification and structural characterization of LFA-1, a lymphocyte function-associated antigen, and Mac-1, a related macrophage differentiation antigen. *J. Biol. Chem.* **257**:12412.

13. Kürzinger, K., M.K. Ho, and T.A. Springer. 1982. Structural homology of a macrophage differentiation antigen and an antigen involved in T-cell-mediated killing. *Nature* **296**:668.

14. Trowbridge, I.S., and M.B. Omary. 1981. Molecular complexity of leukocyte surface glycoproteins related to the macrophage differentiation antigen Mac-1. *J. Exp. Med.* **154**:1517.

15. Sanchez-Madrid, F., A.M. Krensky, C.F. Ware, E. Robbins, J.L. Strominger, S.J. Burakoff, and T.A. Springer. 1982. Three distinct antigens associated with human T lymphocyte-mediated cytolysis: LFA-1, LFA-2, and LFA-3. *Proc. Natl. Acad. Sci. U.S.A.* **79**:7489.

16. Ware, C.F., F. Sanchez-Madrid, A.M. Krensky, S.J. Burakoff, J.L. Strominger, and T.A. Springer. 1983. Human lymphocyte function associated antigen-1 (LFA-1): identification of multiple antigenic epitopes and their relationship to CTL-mediated cytotoxicity. *J. Immunol.* **131**:1182.

17. Krensky, A.M., E. Robbins, T.A. Springer, and S.J. Burakoff. 1984. LFA-1, LFA-2 and LFA-3 antigens are involved in CTL–target conjugation. *J. Immunol.* **132**:2180.

18. Beatty, P.G., J.A. Ledbetter, P.J. Martin, T.H. Price, and J.A. Hansen. 1983.

Definition of a common leukocyte cell-surface antigen (Lp95-150) associated with diverse cell-mediated immune functions. *J. Immunol.* **131**:2913.

19. Dana, N., R.F. Todd III, J. Pitt, T.A. Springer, and M.A. Arnaout. 1984. Deficiency of a surface membrane glycoprotein (Mol) in man. *J. Clin. Invest.* **73**:153.

20. Beatty, P.G., J.M. Harlan, H. Rosen, J.A. Hansen, H.D. Ochs, T.D. Price, R.F. Taylor, and S.J. Klebanoff. 1984. Absence of monoclonal-antibody-defined protein complex in boy with abnormal leucocyte function. *Lancet* **1984(1)**:535.

21. Anderson, D.C., F.C. Schmalstieg, S. Kohl, M.A. Arnaout, B.J. Hughes, M.F. Tosi, G.J. Buffone, B.R. Brinkley, W.D. Dickey, J.S. Abramson, T.A. Springer, L.A. Boxer, J.M. Hollers, and C.W. Smith. 1984. Abnormalities of polymorphonuclear leukocyte function associated with a heritable deficiency of high molecular weight surface glycoproteins (GP138): Common relationship to diminished cell adherence. *J. Clin. Invest.* **74**:536.

22. Kohl, S., T.A. Springer, F.C. Schmalstieg, L.S. Loo, and D.C. Anderson. 1984. Defective natural killer cytotoxicity and polymorphonuclear leukocyte antibody-dependent cellular cytotoxicity in patients with LFA-1/OKM-1 deficiency. *J. Immunol.*, **133**:2972.

23. Springer, T.A., A. Bhattacharya, J.T. Cardoza, and F. Sanchez-Madrid, 1982. Monoclonal antibodies specific for rat IgG1, IgG2a, and IgG2b subclasses, and kappa chain monotypic and allotypic determinants: Reagents for use with rat monoclonal antibodies. *Hybridoma* **1**:257.

24. Fearon, D.T., and W.W. Wong. 1983. Complement ligand–receptor interactions that mediate biological responses. *Annu. Rev. Immunol.* **1**:243.

25. Fleit, H.B., S.D. Wright, and J.C. Unkeless. 1982. Human neutrophil Fc-gamma receptor distribution and structure. *Proc. Natl. Acad. Sci. U.S.A.* **79**:3275.

CHAPTER 5

Functional Studies with Monoclonal Antibodies Against Function-Associated Leukocyte Antigens

Frank Miedema, Fokke G. Terpstra, and
Cornelius J.M. Melief

Introduction

Lymphocyte function-associated antigen-1 (LFA-1) has initially been described in the mouse system (1,2). Monoclonal antibodies (mAbs) specific for the human counterpart to murine LFA-1 have subsequently been produced in several laboratories (3–6). Anti-LFA-1 mAbs inhibit a variety of T cell functions including cytotoxic T cell (CTL) activity (1–6), natural killer (NK) cell activity (3–6), killer (K) cell activity (6), T cell-dependent immunoglobulin (Ig) synthesis (1), and lectin-induced T cell proliferation (1,2,5). The inhibiting effect of anti-LFA-1 mAbs was shown to be caused by a blockade of cell–cell contact, required for cytotoxic activity (effector–target) or T cell activation (T–macrophage interaction) (2,7).

No effect of anti-LFA-1 on the interleukin-2(IL-2)-dependent proliferation of T cell clones and on T-independent B cell functions (Ig production and proliferation) was demonstrable (1).

Recently a mAb, designated 60.3, was described that was shown to react with a function-associated leukocyte antigen (8). Although 60.3 was shown to bind the same antigen as anti-LFA-1 mAb MHM23 (4) and precipitated two polypeptide chains with a M.W. similar to polypeptides precipitated by mAb TS1/22.1.1 (3), the finding that 60.3 precipitated an additional polypeptide chain led the authors to the conclusion that 60.3 did not detect LFA-1 (8).

This additional chain however, was also observed in immunoprecipitates with anti-LFA-1 mAb CLB LFA-1/1* (6) and TS1/18.11, a mAb that detects a β-chain epitope (3), and most likely is a member of the LFA-1 polypeptide chain family.

---

* The name of AB 43 was changed from CLB 54 to CLB LFA-1/1.

This report deals with the inhibiting effect of anti-LFA-1 mAbs on K and NK cell activity and on T cell-dependent Ig synthesis. We previously demonstrated that IL-2 plays an important role in T-dependent Ig synthesis (9). IL-2 production is an early and necessary step in PWM-induced Ig synthesis as was concluded from the finding that anti-Tac completely inhibited PWM-driven Ig synthesis (9). It was shown that IL-2 itself induces Ig synthesis in a T cell-dependent fashion (9). Anti-LFA-1 mAbs abrogate PWM-induced Ig synthesis, but only weakly inhibit IL-2-induced Ig synthesis (Miedema *et al.*, in preparation). A selected series of mAbs that supposedly recognized LFA-1 or LFA-1-related structures was tested for inhibition of PWM- and IL-2-driven Ig synthesis and K and NK activity.

Materials and Methods

Cell Separations

Healthy donor peripheral blood lymphocytes (PBL) were obtained by leukapheresis. Mononuclear cells were isolated by Ficoll–Isopaque density gradient centrifugation. T and non-T cells were separated by E-rosette sedimentation. For the separation of T cell subsets based on their reactivity with monoclonal antibodies (OKT4 and OKT8) a panning technique was used as described in detail elsewhere (10).

K and NK Cell Assays

Antibody-dependent cellular cytotoxicity (K cell activity) was measured with IgG-sensitized P815 mouse mastocytoma cells as described before (10). NK cell activity was assayed with K562 cells as target cells (10). For inhibition studies with mAbs, effector cells were incubated (30 min, room temperature) with diluted dialyzed ascites prior to the addition of ^{51}Cr-labeled target cells. mAb was present throughout the effector phase of the assays. As a negative control an anti-glycophorin mAb was used (kindly provided by Marjolein Bos). Final ascites dilution was 1:750.

T Cell-Dependent Immunoglobulin Production

PWM- and IL-2-induced polyclonal IgM synthesis was measured in a microculture system as described (10), using Iscove's modified Dulbecco's medium with 20% fetal calf serum and antibiotics. PWM was used at a final concentration of 50 μg/ml and IL-2 (kindly provided by Dr. L.A. Aarden) at a final concentration of 30 U/ml, one Unit of IL-2 being defined as the amount of IL-2 that induces half-maximal proliferation in an IL-2-dependent T cell line. IgM levels in culture supernatants were measured after 7 days by means of ELISA. For inhibition studies with mAbs, serial dilutions of dialyzed ascites were added to cultures containing 40,000 non-T cells and 10,000 T4$^+$ cells. After 7 days IgM production was measured and the percentage of inhibition of IgM synthesis determined. IgM production in control cultures without ascites ranged from 200–1500 ng/well.

Results

Inhibition of T Cell-Dependent Ig Synthesis

The results of inhibition studies with the selected mAbs in PWM-driven Ig synthesis are shown in Table 5.1. mAbs 55, 56, 72, 73, and 76 strongly inhibited Ig synthesis in this system. mAbs 75 and 89 showed a moderate but significant inhibition. mAb 26 did not block Ig production in the PWM system.

IL-2-induced Ig synthesis was only moderately inhibited by mAbs 55, 56, 73, and 76. mAbs 72, 75, and 89 weakly blocked with IL-2-driven system. No inhibition whatsoever was observed with mAb 26 (Table 5.2).

Inhibition of NK and K Cell Activity

The cytolysis blocking effect of mAbs in K and NK activity was tested with two donor PBLs as effector cells. Table 5.3 shows that mAbs 26, 55, and 56 did not inhibit NK cell activity. mAbs 75, 89, 73, and 76 inhibited NK cell activity in that order.

All mAbs except mAb 26 inhibited K cell activity of one of the two effector cells. mAbs 75, 73, and 89 completely blocked K cell activity (Table 5.4).

Table 5.1. Inhibiting effect of mAbs against function-associated leukocyte antigens on PWM-driven Ig synthesis.

		Inhibition (%)					
		1:10,000[a]		1:1000[a]		1:100[a]	
No.	mAb Name	Exp. 1	Exp. 2	Exp. 1	Exp. 2	Exp. 1	Exp. 2
26	VIM12	55	41	46	2	27	3
55	MHM23	93	90	82	75	93	55
56	MHM24	75	58	61	77	87	71
72	TS1/22	96	87	96	87	94	95
73	CLB-1/1	80	87	89	81	89	89
75	60.3	42	24	22	39	93	71
76	60.1	89	86	88	90	92	95
89	TS1/18.11	50	68	63	53	73	50

[a] Final ascites dilution.

Table 5.2. Inhibiting effect of mAbs against function-associated leukocyte antigen on IL-2-induced Ig synthesis.

		Inhibition (%)					
		1 : 10,000[a]		1 : 1000[a]		1 : 100[a]	
No.	Name	Exp. 1	Exp. 2	Exp. 1	Exp. 2	Exp. 1	Exp. 2
26	VIM12	12	−10	25	−53	−7	−117
55	MHM23	59	73	57	55	48	20
56	MHM24	57	47	69	63	55	56
72	TS1/22	11	51	32	29	34	31
73	CLB-1/1	64	60	60	58	41	19
75	60.3	−7	−43	−6	−19	39	51
76	60.1	43	71	19	28	38	66
89	TS1/18.11	56	−11	46	34	34	54

[a] Final ascites dilution.

Discussion

Our results indicate that mAb 26 (VIM12) does not inhibit leukocyte functions. Serological studies show that mAb 26 only reacts with a subpopulation of lymphocytes (this volume, Chapter 2). Together with the functional studies this indicates that mAb 26 probably recognizes Mo1 or a related structure.

mAbs 72, 73, 75, 76, and 89 inhibited PWM-driven Ig synthesis and K and NK activity. mAbs 55 and 56 strongly inhibited PWM-driven Ig synthesis and K cell activity but not NK activity (Table 5.5). Although mAbs 75 (60.3) and 55 (MHM23) were shown to cross-block in binding studies (8), these mAbs differed with respect to their capacity to inhibit leukocyte functions (Table 5.5). This could be due to a difference in affinity in the binding of the mAb to the antigen. Of special interest is the finding that mAb 76 inhibited K cell activity of only one of the two effector cells. Perhaps mAb 76 detects a polymorphic determinant on a human function-associated antigen not expressed on the E2 donor cells.

The data suggest that PWM-driven Ig synthesis is far more sensitive to the inhibiting effect of anti-LFA-1 than IL-2-driven Ig synthesis. PWM-induced Ig synthesis is completely abrogated, but IL-2-induced Ig synthesis is only partially or not at all affected. We have shown that anti-LFA-1 mAb inhibits PWM-induced Ig synthesis by a blockade of PWM-induced IL-2 production, which is cell–cell contact dependent (Miedema et al., in preparation). Only mAbs 55 and 56 inhibited IL-2-driven Ig synthesis to a fairly large extent (60%). Whether this inhibition is the result of interference with the binding of IL-2 to the IL-2 receptor or of cell–cell contact inhibition is unclear. Over a large series of experiments mAb 73 (CLB LFA-1/1) did not reproducibly block IL-2-induced Ig synthesis, but strongly inhibited PWM-driven Ig synthesis (data not shown).

Table 5.3. Inhibiting effect of anti-LFA-1 mAbs on NK cell activity.

E:T ratio	Control ascites[a]		26 (VIM12)		55 (MHM23)		Specific release (%) 56 (MHM24)		73 (CLB-1/1)		75 (60.3)		76 (60.1)		89 (TS1/18.11)	
	E1[b]	E2[b]	E1	E2	E1	E2	E1	E2	E1	E2	E1	E2	E1	E2	E1	E2
0.6	10	4	3	3	6	9	0	1	0	0	3	0	0	3	2	7
1.25	4	6	3	8	7	10	0	5	0	0	0	0	2	1	4	3
2.5	7	12	10	9	7	16	4	11	0	3	1	0	3	5	2	5
5	13	21	7	17	15	21	5	18	7	8	0	0	4	8	4	11
10	14	38	12	23	17	33	12	28	9	15	0	4	7	21	3	12
20	28	37	20	38	20	42	17	41	10	23	0	7	10	32	8	19

[a] Final ascites dilution was 1:750, control ascites was an anti-glycophorin mAb.
[b] E1 and E2 are effector PBLs of two different donors.

Table 5.4. Inhibiting effect of anti-LFA-1 mAbs on K cell activity.

| | Specific release (%) | | | | | | | | | | | | | | | |
| | Control ascites[a] | | 26 (VIM12) | | 55 (MHM23) | | 56 (MHM24) | | 73 (CLB-1/1) | | 75 (60.3) | | 76 (60.1) | | 89 (TS1/18.11) | |
E:T ratio	E1[b]	E2[b]	E1	E2	E1	E2	E1	E2	E1	E2	E1	E2	E1	E2	E1	E2
0.3	12	nt[c]	0	3	3	0	0	0	0	0	0	0	0	0	0	0
0.6	13	nt	0	4	0	3	0	0	0	0	0	2	0	0	0	0
1.25	3	nt	16	3	6	2	0	2	0	4	0	0	0	1	0	0
2.5	3	nt	13	13	12	3	0	7	0	4	0	0	0	5	0	1
5	25	nt	20	12	11	7	7	10	0	6	0	0	0	13	0	3
10	26	nt	18	15	4	11	0	12	0	7	0	2	3	20	0	7

[a] Final ascites dilution was 1:750, control ascites was an anti-glycophorin mAb.
[b] E1 and E2 are effector PBLs of two different donors.
[c] nt = Not tested.

Table 5.5. Summary of blocking studies.

| | Inhibition | | | |
| | T-helper activity | | Cytotoxicity | |
mAb	PWM	IL-2	NK	K
26 (VIM12)	−	−	−	−
55 (MHM23)	+++	+	−	++
56 (MHM24)	++	+	−	++
72 (TS1/22)	+++	+/−	nt^a	nt^a
73 (CLB-1/1)	+++	+	+	+++
75 (60.3)	+	−	++	+++
76 (60.1)	++	+/−	+	++
89 (TS1/18.11)	+	+/−	+	+++

^{a} Not tested because of ascites paucity.

Our data clearly show that not only anti-α chain mAbs, but also anti-β chain mAbs (89; TS1/18.11) (3) inhibited leukocyte functional activities.

In conclusion we showed that all selected mAbs, except 26 (VIM12), detect function-associated antigens on leukocytes. The mAbs differed with respect to their function-inhibiting capacities but they all detected members of the family of LFA-1 related antigens.

Summary

Blocking studies were performed with a selected series of monoclonal antibodies (mAbs) from the myeloid Workshop directed against function-associated antigens on leukocytes. mAbs 55 (MHM23), 56 (MHM24), 72 (TS1/22), 73 (CLB LFA-1/1), 75 (60.3), 76 (60.1), and 89 (TS1/18.11) strongly inhibited pokeweed mitogen-induced immunoglobulin (Ig) synthesis and killer cell activity. mAbs 73, 75, 76, and 89 in addition inhibited natural killer (NK) cell activity. mAb 76 (60.1) inhibited K and NK activity of only one of the two donor effector cells. Perhaps mAb 76 detects a polymorphic determinant on a function-associated antigen. mAb 26 (VIM12) did not inhibit leukocyte functions and does not seem to react with a function-associated antigen.

Acknowledgment. This study was supported by grant CLB 80-2 of the Koningin Wilhelmina Fonds/Netherlands Cancer Foundation.

References

1. Springer, T.A., D. Davignon, M.K. Ho, K. Kürzinger, E. Martz, and F. Sanchez-Madrid. 1982. LFA-1 and Lyt-2,3 molecules associated with T lymphocyte-mediated killing; and Mac-1 and LFA-1 homologue associated with complement receptor function. *Immunol. Rev.* **68**:111.

2. Goldstein, P., C. Gordis, A.M. Schmitt-Verhulst, B. Hayst, A. Pierres, A. van Agthoven, Y. Kaufmann, Z. Eshhar, and M. Pierres. 1982. Lymphoid cell surface interaction structures detected using cytolysis-inhibiting monoclonal antibodies. *Immunol. Rev.* **68**:5.

3. Sanchez-Madrid, F., A.M. Krensky, C.F. Ware, E. Robbins, J.L. Strominger, S.J. Burakoff, and T.A. Springer. 1982. Three distinct antigens associated with human T-lymphocyte mediated cytolysis: LFA-1, LFA-2 and LFA-3. *Proc. Natl. Acad. Sci. U.S.A.* **79**:7489.

4. Hildreth, J.E.K., F.M. Gotch, Ph.D.K. Hildreth, and A.J. Mc.Michael. 1983. A human lymphocyte function associated antigen involved in cell-mediated lympholysis. *Eur. J. Immunol.* **13**:202.

5. Spits, H., G. Keizer, J. Borst, C. Terhorst, A. Hekman, and J.E. de Vries. 1984. Characterization of monoclonal antibodies against cell surface molecules associated with cytotoxic activity of natural and activated killer and cloned CTL lines. *Hybridoma* 2:423.

6. Miedema, F., P.A.T. Tetteroo, W.G. Hesselink, G. Werner, H. Spits, and C.J.M. Melief. 1984. Both Fc receptors and LFA-1 are required for the effector cell function in antibody-dependent cellular cytotoxicity (K-cell activity) mediated by Tγ cells. *Eur. J. Immunol.* **14**:518.

7. Krensky, A.M., E. Robbins, T.A. Springer, and S.J. Burakoff. 1984. LFA-1, LFA-2 and LFA-3-antigens are involved in CTL–target conjugation. J. Immunol. **132**:2180.

8. Beatty, P.G., J.A. Ledbetter, P.J. Martin, Th.H. Prince, and J.A. Hansen. 1983. Definition of a common leukocyte cell-surface antigen (Lp 95-150) associated with diverse cell mediated immune functions. *J. Immunol.* **131**:2913.

9. Miedema, F., J.W. van Oostveen, R.W. Sauerwein, F.G. Terpstra, L.A. Aarden, and C.J.M. Melief. 1985. Induction of immunoglobulin synthesis by Interleukin-2 is T4$^+$8$^-$ cell dependent. A role for IL-2 in the pokeweed-mitogen-driven system. *Eur. J. Immunol.* **15**:107.

10. Rümke, H.C., F. Miedema, I.J.M. ten Berge, F. Terpstra, H.J. van der Reijden, R.J. Van de Griend, H.G. de Bruin, A.E.G. Kr. von dem Borne, J.W. Smit, W.P. Zeijlemaker, C.J.M. Melief. 1982. Functional properties of T-cells in patients with chronic Tγ lymphocytosis and chronic T-cell neoplasia. *J. Immunol.* **129**:419.

CHAPTER 6

Phosphorylation of α,β Subunits of 180/100-Kd Polypeptides (LFA-1) and Related Antigens

Toshiro Hara and Shu Man Fu

Human leukocyte function-associated antigens (LFA) have been defined by their association with human T lymphocyte-mediated cytolysis (1). One of these, LFA-1, is present on lymphocyte, thymocytes, monocytes, and granulocytes. Antibodies to LFA-1 have been shown to interfere with T lymphocyte-mediated cytotoxicity, NK cell-mediated cytolysis, and T cell proliferation to soluble antigens, alloantigens, and mitogens as well as various myeloid cell functions (1–4). OKM1 is a biomolecular structure and it has been identified to be the C3bi receptor of human monocytes and macrophages (5). Recently, LFA-1 and OKM1 were found to have a common β subunit and they belong to a human leukocyte differentiation antigen family (6).

Receptor phosphorylation has been studied intensely and it is thought to be a possible mechanism for transmembrane signaling (7). We have studied phosphorylation of human lymphocyte surface structures with monoclonal antibodies as probes. One of the phosphorylated proteins was found to be LFA-1. In this report, our studies on the phosphorylation of LFA-1 and related antigens are described.

Materials and Methods

Chemicals

12-O-Tetradecanoylphorbol 13-acetate (TPA), phosphoserine, phosphothreonine, and phosphotyrosine were purchased from Sigma (St. Louis, MO).

Monoclonal Antibodies

OKM1 monoclonal antibody was purchased from Ortho Diagnostic Systems (Raritan, NJ). TS1/22 monoclonal antibody against LFA-1 antigen was kindly provided by Dr. T.A. Springer and 60.3 monoclonal antibody against 95-Kd/130-Kd/150-Kd molecules by Drs. J.A. Hansen and P.G. Beatty. 44.1 monoclonal antibody was produced in BC_3F_1 females immunized with activated T cells and was selected by screening the hybridoma supernatants for their ability to block T cell proliferation. Details of the procedures have been described (8).

Cells and Radiolabeling

Peripheral mononuclear cells (PMC) were separated from buffy coats using Ficoll–Hypaque density gradient centrifugation. Monocytes were purified by Percoll gradient centrifugation after Ficoll–Hypaque separation (9). Cells were washed three times with phosphate-free buffer (10 mM Hepes, pH 7.4, 140 mM NaCl, 5.4 mM KCl, 0.4 mM Ca(NO$_3$)$_2$, 0.4 mM MgSO$_4$: PFB). Phosphorylation reaction was performed as described (10) with slight modifications. Cells were suspended in phosphate-free RPMI 1640 with 10% dialyzed fetal calf serum at 3×10^7 cells/ml. After preincubation for 45 min at 37°C, 0.1–0.3 mCi/ml of [^{32}P]orthophosphate (Amersham) was added, and incubations were continued for 3 hr, which is the time necessary to equilibrate the ATP pool (11). Then, the cells were divided into aliquots with or without various factors and incubated for 10 min at 37°C. The reactions were stopped with cold PBS containing 10 mM sodium pyrophosphate, 50 mM NaF, and 2 mM EDTA and cells were washed once more with this buffer.

Cell surface iodination with ^{125}I, immunoprecipitation, and autoradiography were performed as described previously (12). Phosphoamino acids were analyzed according to the method described by Cooper et al. (13). The bands corresponding to β subunits, localized by autoradiography, were excised from the gel and the proteins were eluted from the gel by electrophoresis at 150 V for 20 hr into a dialysis bag containing 10 mM sodium phosphate buffer (pH 7.0) and 0.05% NaDodSO$_4$. The samples were dialysed against 10 mM NaHCO$_3$, lyophilized, and then subjected to acid hydrolysis in 6N HCl for 1 hr at 110°C. After evaporation, the hydrolysates were spotted onto thin layer plates (100 μm, E.M. Laboratories). Two-dimensional electrophoresis was performed at pH 1.9 (88% formic acid : glacial acetic acid : H$_2$O, 50 : 156 : 1794) at 1000 V for 70 min in the first dimension and at pH 3.5 (pyridine : glacial acetic acid : H$_2$O, 10 : 100 : 1890) at 1000 V for 50 min in the second dimension. Phosphoserine, phosphothreosine, and phosphotyrosine were added to the samples and identified by ninhydrin reaction. The radioactive material was identified by autoradiography.

Results

Characteristics of the Monoclonal Antibodies

Monoclonal antibody 44.1 was selected for its ability to block T cell proliferation in allogeneic mixed lymphocyte cultures. This monoclonal antibody was shown to precipitate a 180/100-Kd bimolecular complex (Fig. 6.1, lane 2) from ^{125}I-labeled peripheral blood mononuclear cells. This complex was similar to that precipitated by mAb TS1/22 (LFA-1) kindly provided by Dr. T.A. Springer (Fig. 6.1, lane 1). When the cell lysate was precleared with anti-LFA-1 monoclonal antibody (TS1/22), neither TS1/22 nor 44.1 precipitated any molecule (Fig. 6.1, lanes 3 and 4).

Fig. 6.1. Immunoprecipitation of 180-Kd/100-Kd polypeptides with TS1/22 (LFA-1) and 44.1 monoclonal antibodies from ^{125}I-labeled peripheral mononuclear cells. ^{125}I-labeled peripheral mononuclear cell lysate was first precleared with control irrelevant monoclonal antibody and immunoprecipitated with TS1/22 (lane 1) and 44.1 (lane 2). Immunoprecipitates were analyzed in 9% polyacrylamide gel electrophoresis in the presence of NaDodSO$_4$ under reducing conditions. No bands were seen with either TS1/22 (lane 3) or 44.1 (lane 4) when the cell lysate was precleared with TS1/22.

Table 6.1. Summary of the molecular structures identified by three monoclonal antibodies.

	LFA-1 44.1	60.3	OKM1
χ Subunit	180 Kd	180 Kd	165 Kd
β Subunit	100 Kd	100 Kd	100 Kd
χ Subunit		165 Kd	

Preclearance with 44.1 showed similar results (data not shown). Thus, 44.1 monoclonal antibody and anti-LFA-1 monoclonal antibody recognized identical polypeptides. For the experiments described, 44.1 was used to identify LFA-1. Two other monoclonal antibodies were used. They were OKM1 and 60.3. The molecular structure and staining characteristics of these three antibodies are summarized in Tables 6.1 and 6.2.

Phosphorylation of LFA-1 and Related Antigens

The phosphorylation of LFA-1 antigen was examined by labeling peripheral blood mononuclear cells with ^{32}P and immunoprecipitation with monoclonal antibody 44.1. As shown in Fig. 6.2(A) (lane 1), the α subunit (180 Kd) was highly phosphorylated while the β subunit (100 Kd) showed little phosphorylation. Similar experiments were performed with antibodies 60.3 and OKM1. In the case of 60.3, the α subunit was highly phosphorylated while the β subunit and the χ subunit (165 Kd) did not show significant phosphorylation. In the case of OKM1, both α and β subunits showed little phosphorylation.

Phosphorylation of the β Subunit Induced by Tumor Promoter Phorbol Ester

In the presence of 12-O-tetradecanoylphorbol 13-acetate (TPA), the β subunit of LFA-1 was phosphorylated in a dose-dependent manner (Fig.

Table 6.2. Reactivity of LFA-1, 44.1, 60.3, and OKM1.

	LFA-1, 44.1	60.3	OKM1
Lymphocytes	+	+	−
Monocytes	+	+	+
Granulocytes	+	+	+
Erythrocytes	−	−	−
Platelets	−	−	−
Receptor function	?	?	C3bi receptor

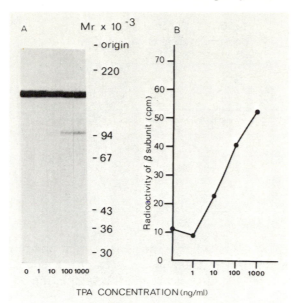

Fig. 6.2. Phosphorylation of LFA-1 in the presence of various concentrations of TPA. (A) Peripheral mononuclear cells were labeled for 3 hr with ^{32}P and incubated in the presence of various concentrations of TPA for 10 min at 37°C. After washing, cells were lysed and immunoprecipitated with 44.1. Immunoprecipitates were analyzed in SDS–polyacrylamide (9%) gel electrophoresis under reducing conditions and autoradiographed. (B) β-Subunit bands were cut from the gel and counted in the scintillation counter.

6.2). The addition of TPA had no appreciable effect on the phosphorylation of the α subunit. Phosphorylation of the β subunit was detected 5 min after TPA addition when the first determination was made. It reached the maximal level 10 min after TPA addition.

Phosphorylation of the β subunits induced by TPA of OKM1 and antigen 60.3 was similarly observed. The addition of TPA had no appreciable effect on the α subunits of these antigens and on the χ subunit of antigen 60.3. The effects of TPA on LFA-1 and related antigens are summarized in Table 6.3.

The phosphorylated β chain of LFA-1 was isolated and subjected to hydrolysis. By two-dimensional thin layer electrophoresis, the phosphorylated amino acid was identified as serine.

Discussion

TPA selectively enhanced the phosphorylation of β subunits (100 Kd) of LFA-1, OKM1, and antigen 60.3 but not that of α subunits. This effect was TPA specific. Other mitogens such as PHA, Con A, and calcium

Table 6.3. Effect of TPA on the phosphorylation of 44.1 (LFA-1), 60.3, and OKM1.

		Phosphorylation	
		\-	+
44.1 (LFA-1)	α	+++	+++
	β	\-	++
OKM1	α	±	±
	β	\-	++
60.3	α	+++	+++
	χ	±	±
	β	\-	++

ionophore A 23187 did not change the phosphorylation of LFA-1 antigen. In addition, preliminary examinations suggest that cyclic AMP- or cyclic GMP-dependent protein kinases are not involved in the phosphorylation of LFA-1 antigen.

Protein kinase C appears to be a receptor protein for phorbol esters (14) and is suggested to have a role in the regulation of receptor functions (15–18). Therefore, it is possible that the phosphorylation of the β subunits of LFA-1 and OKM1 antigens is mediated by protein kinase C.

The selective enhancement of phosphorylation of the β subunits of LFA-1 and related antigens by TPA is of considerable interest. This is analogous to the studies on insulin and somatomedin C receptors (15). TPA-induced phosphorylation of epidermal growth factor receptor, β-adrenergic receptor, and transferrin receptor has also been reported (16–20). Thus, TPA-induced phosphorylation of receptor molecules is a rather common phenomenon.

Phorbol ester has been reported to significantly alter the C3bi receptor function as well as the activity of cytotoxic T lymphocytes and NK cells (9,21–23).

Evidence is accumulating to suggest that some, if not all, pleiotropic actions of tumor promoter phorbol esters are mediated through the action of protein kinase C (18). Therefore, the phosphorylation of LFA-1 and OKM1 antigens may play a role in the effects of TPA on the C3bi receptor and the activity of cytotoxic T lymphocytes and NK cells.

Summary

Monoclonal antibody 44.1 was shown to precipitate a 180/100-Kd bimolecular complex which was identified to be LFA-1. This antibody and two other monoclonal antibodies (OKM1 and 60.3) identified a family of leukocyte differentiation antigens with a common β (100 Kd) subunit. These three monoclonal antibodies were used to probe the phosphorylation of these antigens. In the case of LFA-1 and antigen 60.3, the α subunits were highly phosphorylated. The α subunit of OKM1 showed little phosphorylation. The β subunits of these antigens were not phosphorylated and the χ subunit of antigen 60.3 showed little phosphorylation. Phorbol ester (TPA) induced marked phosphorylation of the β subunits of all three antigens. The phosphorylation of other subunits was not appreciably affected. The phosphorylation was found to be dose dependent and serine was the only amino acid found to be phosphorylated under TPA stimulation. The relation between TPA-induced phosphorylation and the receptor functions of these antigens has been discussed.

Acknowledgment. This work was supported in part by a Public Health Service Grant CA-34546 from the National Institutes of Health.

References

1. Sanchez-Madrid, F., A.M. Krensky, C.F. Ware, E. Robbins, J.L. Strominger, S.J. Burakoff, and T.A. Springer. 1982. Three distinct antigens associated with human T-lymphocyte-mediated cytolysis: LFA-1, LFA-2, and LFA-3. *Proc. Natl. Acad. Sci. U.S.A.* **79**:7489.
2. Hildreth, J.E.K., F.M. Gotch, P.D.K. Hildreth, and A.J. McMichael. 1983. A human lymphocyte-associated antigen involved in cell-mediated lympholysis. *Eur. J. Immunol.* **13**:202.
3. Krensky, A.M., F. Sanchez-Madrid, E. Robbins, J.A. Nagy, T.A. Springer, and S.J. Burakoff. 1983. The functional significance, distribution, and structure of LFA-1, LFA-2, and LFA-3: cell surface antigens associated with CTL–target interactions. *J. Immunol.* **131**:611.
4. Beatty, P.G., J.A. Ledbetter, P.J. Martin, T.H. Price, and J.A. Hansen. 1983. Definition of a common leukocyte cell surface antigen (Lp 95-150) associated with diverse cell mediated immune functions. *J. Immunol.* **131**:2913.
5. Wright, S.D., P.E. Rao, W.C. Van Voorhis, L.S. Craigmyle, K. Iida, M.A. Talle, E.F. Westberg, G. Goldstein, and S.C. Silverstein. 1983. Identification of the C3bi receptor of human monocytes and macrophages by using monoclonal antibodies. *Proc. Natl. Acad. Sci. U.S.A.* **80**:5699.
6. Sanchez-Madrid, F., J.A. Nagy, E. Robbins, P. Simon, and T.A. Springer. 1983. A human leukocyte differentiation antigen family with distinct α-subunits and a common β-subunit: the lymphocyte function associated antigen (LFA-1), the C3bi complement receptor (OKM1/Mac-1), and the p150,95 molecule. *J. Exp. Med.* **158**:1785.
7. Cohen, P. 1982. The role of protein phosphorylation in neural and hormonal control of cellular activity. *Nature* **296**:613.
8. Yes, S.H., F. Gaskin, and S.M. Fu. 1983. Neurofibrillary tangles in senile dementia of the Alzheimer type share an antigenic determinant with intermediate filaments of the vimentin class. *Am. J. Pathol.* **113**:373.
9. Wright, S.D., and S.C. Silverstein. 1982. Tumor-promoting phorbol esters stimulate C3b and C3b′ receptor-mediated phagocytosis in cultured human monocytes. *J. Exp. Med.* **156**:1149.

10. Chaplin, D.D., H.J. Wedner, and C.W. Parker. 1980. Protein phosphorylation in human peripheral blood lymphocytes: mitogen-induced increases in protein phosphorylation in intact lymphocytes. *J. Immunol.* **124**:2390.

11. Johnstone, A.P., J.H. DuBois, and M.J. Crumpton. 1981. Phosphorylated lymphocyte plasma-membrane proteins. *Biochem. J.* **194**:309.

12. Wang, C.Y., A. Al-Katib, C.L. Lane, B. Koziner, and S.M. Fu. 1983. Induction of HLA-DC/DS (Leu 10) antigen expression by human precursor B cell lines. *J. Exp. Med.* **158**:1757.

13. Cooper, J.A., B.M. Sefton, and T. Hunter. 1983. Detection and quantification of phosphotyrosine in proteins. *Methods Enzymol.* **99**:387.

14. Castagna, M., Y. Takai, K. Kaibuchi, K. Sano, U. Kikkawa, and Y. Nishizuka. 1982. Direct activation of calcium-activated, phospholipid-dependent protein kinase by tumor-promoting phorbol esters. *J. Biol. Chem.* **257**:7847.

15. Jacobs, S., N.E. Sahyoun, A.R. Saltiel, and P. Cuatrecasas. 1983. Phorbol esters stimulate the phosphorylation of receptors for insulin and somatomedin C. *Proc. Natl. Acad. Sci. U.S.A.* **80**:6211.

16. May, W.S., S. Jacobs, and P. Cuatrecasas. 1983. Association of phorbol ester-induced hyperphosphorylation and reversible regulation of transferrin membrane receptors in HL60 cells. *Proc. Natl. Acad. Sci. U.S.A.* **81**:2016.

17. Sibley, D.R., P. Nambi, J.R. Peters, and R.J. Lefkowitz. 1984. Phorbol diesters promote β-adrenergic receptor phosphorylation and adenylate cyclase desensitization in duck erythrocytes. *Biochem. Biophys. Res. Commun.* **121**:973.

18. Nishizuka, Y. 1984. The role of protein kinase C in cell surface signal transduction and tumour promotion. *Nature* **308**:693.

19. Iwashita, S., and C.F. Fox. 1984. Epidermal growth factor and potent phorbol tumor promoters induce epidermal growth factor receptor phosphorylation in a similar but distinctively different manner in human epidermoid carcinoma A431 cells. *J. Biol. Chem.* **259**:2559.

20. Moon, S.O., H.C. Palfrey, and A.C. King. 1984. Phorbol esters potentiate tyrosine phosphorylation of epidermal growth factor receptors in A431 membranes by a calcium-independent mechanism. *Proc. Natl. Acad. Sci. U.S.A.* **81**:2298.

21. Goldfarb, R.H. and R.B. Herberman. 1981. Natural killer cell reactivity: regulatory interactions among phorbol ester, interferon, cholera toxin and retinoic acid. *J. Immunol.* **126**:2129.

22. Andreotti, P.E. 1982. Phorbol ester tumor promoter modulation of alloantigen-specific T lymphocyte responses. *J. Immunol.* **129**:92.

23. Orosz, C.G., D.C. Roopenian, and F.H. Bach. 1983. Phorbol ester mediates reversible reduction of cloned T lymphocyte cytolysis. *J. Immunol.* **130**:2499.

CHAPTER 7

Serological, Immunochemical, and Functional Analysis of the Heterogeneity of the Workshop Monoclonal Antibodies Recognizing the LFA-1 Antigen

Neal Flomenberg, Nancy A. Kernan, Bo Dupont, and Robert W. Knowles

The human lymphocyte function-associated antigen, LFA-1, was initially identified by monoclonal antibodies (mAbs) which were screened for their ability to block T cell-mediated cytotoxicity in the absence of complement. The cell surface molecule recognized by these antibodies is expressed on a variety of hematopoietic cells and consists of a heterodimer with subunits of 90 and 160 Kd (1). This molecule is one of a family of structurally related molecules which appear to have a common 90-Kd light chain but distinct heavy chains. Some mAbs are specific for the LFA-1 molecule, while other antibodies also bind to determinants expressed on the other structurally related members of this molecular family. The reactivity of these antibodies with different lymphoid and myeloid cells shows considerable heterogeneity, which reflects the differential expression of these various molecules on different populations of hematopoietic cells (2). In functional studies these antibodies also show heterogeneity in their ability to inhibit cytotoxic T cells (3). The following studies were performed in order to assess the heterogeneity of the Workshop antibodies recognizing the LFA-1 antigen.

Materials and Methods

Isolation of Peripheral Blood Leukocytes

Buffy-coat cells were obtained by sedimentation of erythrocytes in 6% dextran 75 in 0.9% sodium chloride (Abbott Laboratories, North Chicago, IL). Residual erythrocytes were lysed by incubating them at 37°C in Tris–

ammonium chloride for 10 min (0.013 *M* Trisma–HCl, 0.15 *M* NH₄Cl). Granulocytes were recovered from the pellet after Ficoll–Hypaque density gradient centrifugation.

Alloreactive T Lymphocyte Clones

Functional analysis was performed utilizing four CD4-positive allocytotoxic T lymphocyte clones K1D, K1G, K1I, and K1L. K1I recognizes the HLA-DR7 allospecificity, while K1G recognizes HLA-DPw4 (SB4). K1L recognizes a subset of HLA-DQw2 (MB2)-positive cells. K1D recognizes a private specificity expressed by a DP molecule on a limited number of DR7 haplotypes. All clones were derived by limiting dilution and propogated by weekly restimulation and culture in interleukin-2 (IL-2)-containing conditioned medium (4).

Indirect Immunofluorescence Analysis

Cells were washed in phosphate-buffered saline with 1% bovine serum albumin and 0.02% sodium azide. Cells (0.5 × 10⁶ per tube) were incubated with monoclonal antibody for 30 min at 4°C. Cells were rewashed and then incubated with fluoresceinated goal anti–mouse immunoglobulin antibodies (Litton Bionetics, Kensington, MD) for an additional 30 min at 4°C. After extensive washing, cells were examined using an EPICS-C cell sorter (Coulter Electronics, Hialeah, FL).

Immunochemical Analysis

Cells were surfaced labeled with ¹²⁵I using lactoperoxidase and lysed with NP-40. Immunoprecipitation and SDS–gel electrophoresis were performed as described previously (Knowles, this volume) using 1 μl of ascites containing the Workshop monoclonal antibodies.

Functional Analysis

Cell-mediated cytotoxicity assays were performed using Workshop monoclonal antibodies at a final dilution of 1 : 80, 1 : 800, or 1 : 8000 in the test wells. Antibody was preincubated with the effector T cell clones for ½ hr prior to the addition of ⁵¹Cr-labeled target cells. Percent specific cytolysis was measured in a 4-hr chromium release assay. Triplicate determinations were performed for each clone. The percent inhibition by each antibody was calculated as the ratio of the percent specific cytolysis observed in the presence and absence of antibody (4). Antibodies producing 51–100% inhibition were scored as positive (+). Those producing 0–20% inhibition were scored as negative (−), while those producing 21–50% were scored as weak (w).

Results

Selection of Antibodies for This Study

Antibodies in the myeloid and T cell workshops which bound to a T cell line (HSB-2) and a B cell line (EBV-transformed), but failed to bind to an erythroleukemia cell line (K562), were selected as potentially specific for the LFA-1 related antigens. Eight antibodies were therefore selected for further study, including M11, M55, M56, M72, M73, M75, M89, and T60. Biochemical analysis showed that each of the antibodies, except M11, immunoprecipitated the characteristic LFA-1 molecule with subunits of 90 and 160 Kd from a lysate of the T cell line HSB-2. M11 failed to immunoprecipitate detectable surface-labeled molecules from any of the cell lysates examined. It presumably does not recognize the LFA-1 antigen but served as a strongly binding control antibody throughout the biochemical and functional studies.

Serological Analysis: Indirect Immunofluorescence

Serological analysis of peripheral blood leukocytes by indirect immunofluorescence using antibodies detecting LFA-1 revealed two distinct patterns of reactivity. Two of these patterns are presented as two-dimensional histograms of 90° light scatter (a measurement of relative granularity) and log fluorescence intensity (a measurement of relative antibody binding). This allows a simultaneous comparison of the amount of antibody binding to the lymphocyte, monocyte, and granulocyte populations in peripheral blood, as shown in Fig. 7.1 for antibodies M56 and

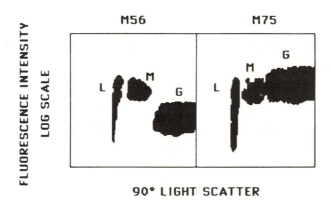

Fig. 7.1. Two-dimensional immunofluorescent analysis of two monoclonal antibodies (M56 and M75) recognizing LFA-1-associated determinants. Ninety-degree light scatter is depicted on the x-axis and log fluorescence intensity on the y-axis. The various populations of peripheral blood leukocytes are indicated: L (lymphocytes), M (monocytes), and G (granulocytes).

M75. For both antibodies, binding to lymphocytes was heterogeneous, while binding to monocytes was uniformly high. In contrast, the amount of antibody bound to the granulocytes differed significantly for the two antibodies. With M56, the granulocyte staining was uniformly weak, at a level comparable to the weakest staining of lymphocytes, but with M75, the granulocyte staining was uniformly bright, at a level comparable to the monocytes and the most intensely stained lymphocytes. This is further demonstrated in Fig. 7.2, with histograms of cell number and log fluorescence intensity. With both antibodies the lymphocyte population shows two peaks of fluorescence intensity, while the granulocyte population shows a single peak which is of low intensity with M56 and high intensity with M75. These patterns were obtained using saturating levels of antibody, and the differential binding to granulocytes cannot be explained by differences in titer of the two antibody preparations. When the other six antibodies were examined in the same way, T60 showed the same pattern as M56, while M55, M72, M73, and M89 showed the M75 pattern of reactivity. In contrast, the pattern using M11 was different from all of the other antibodies. It showed uniformly high binding to all lymphocytes as well as monocytes and granulocytes. This is consistent with the results of the biochemical analysis, suggesting that M11 does not recognize the LFA-1-related molecules.

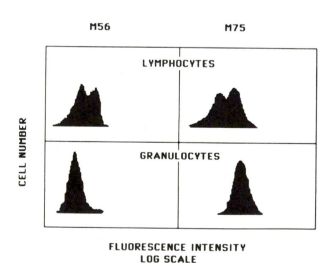

Fig. 7.2. One-dimensional immunofluorescent analysis of two monoclonal antibodies (M56 and M75) recognizing LFA-1-associated determinants on the lymphocyte and granulocyte populations in peripheral blood. Log fluorescence intensity is depicted on the *x*-axis and cell number on the *y*-axis.

Biochemical Analysis: Immunoprecipitation and Molecular Weight Determinations on SDS Gels

Each antibody was used to immunoprecipitate ^{125}I-labeled cell surface antigens from lysates of various hematopoietic cells as described elsewhere. Analysis of EBV-transformed B cell lines, IL-2-expanded PHA-activated peripheral blood T cells, the T cell line, HSB-2, and peripheral blood granulocytes demonstrated that M55, M56, M72, M73, M75, M89, and T60 all recognize LFA-1-related structures expressed on the surface of these cells. M73 immunoprecipitated considerably less of the labeled molecules than the other antibodies, while M11, tested in parallel, failed to immunoprecipitate any detectable antigens. Each immunoprecipitate contained antigens with subunits of 90 Kd and approximately 160 Kd.

The 160-Kd chain immunoprecipitated with M56 and M72 contained considerably more radioactivity relative to the 160-Kd chain found using the other antibodies. This can be seen with the activated T cell lysate shown in Fig. 7.3. This band also showed a somewhat more heterogeneous mobility on these gels. Further structural studies will be needed to determine whether the 160-Kd subunits immunoprecipitated by each of these antibodies are identical.

Another heavy chain of approximately 150 Kd was also found in most immunoprecipitates from these cell lysates. It is most strongly labeled in the granulocyte lysates. This heavy chain was completely missing, however, from the immunoprecipitates using the M56 antibody with the granulocyte lysate, as shown in Fig. 7.3. This chain was also missing from the immunoprecipitates using T60 (data not shown). These two antibodies appear to bind only the LFA-1 molecule on granulocytes and do not bind to the additional structurally related molecules on these cells. This may explain the weaker binding to the granulocytes observed with these two antibodies by immunofluorescence.

Functional Analysis: Blocking Cytotoxic T Cell Clones

Each of the LFA-1 related antibodies was examined for its ability to block the cytolytic activity of four human alloreactive T cell clones. Each of these clones recognizes a distinct HLA-class II polymorphic determinant. The cytotoxic cells were preincubated with the monoclonal antibodies prior to adding target cells and were present throughout the chromium release assay as described in the Materials and Methods section. Most of the antibodies showed significant inhibition of cytotoxicity with at least some of the clones, but heterogeneity was found in the amount of blocking observed, as summarized in Table 7.1.

To examine whether the amount of inhibition was dependent on antibody concentration, four antibodies were examined at various dilutions

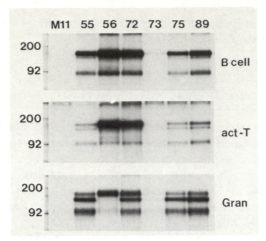

Fig. 7.3. Immunoprecipitation of LFA-1 molecules. Seven Workshop antibodies were used to immunoprecipitate LFA-1 and structurally related molecules from NP-40 lysates of ^{125}I-surface-labeled cells. The molecules immunoprecipitated from an EBV-transformed B cell line (B cell), from IL-2-expanded PHA-activated peripheral blood T cells (act-T), and peripheral blood granulocytes (Gran) with each antibody were separated by SDS–polyacrylamide gel electrophoresis in adjacent wells of 10% slab gels. Molecular weight marker proteins were also run on each gel and the position of two of these markers is indicated on the left side of the figure with their molecular weights ($\times 10^{-3}$).

with two T cell clones, as shown in Fig. 7.4. For M56, M75, and M89 the amount of inhibition varied from 40–80% over a 100-fold range of antibody concentration. This was considerably less than the amount of blocking which can be obtained with some of the T cell-specific antibodies including most of the CD3 and CD2 antibodies examined in the T cell section of this Workshop using the same T cell clones (this series, Volume 1, Chapter 9). Increasing the concentration 100-fold did not significantly

Table 7.1. Monoclonal antibodies recognizing the LFA-1 antigen.

mAb name	mAb no.	Biochem. ID[a]	Blocking activity[a]
MHM23	M55	+	w
MHM24	M56	+v[b]	+/w
TS1/22	M72	+	+
CLB-54	M73	+	+
60.3	M75	+	+
TS1/18.11	M89	+	+
CRIS-3	T60	+v	+/−
JOAN-1	M11	−	−

[a] + = 51–100% inhibition; w = 21–50% inhibition; − = 0–20% inhibition.
[b] V = Variant (only LFA-1 molecules on granulocytes).

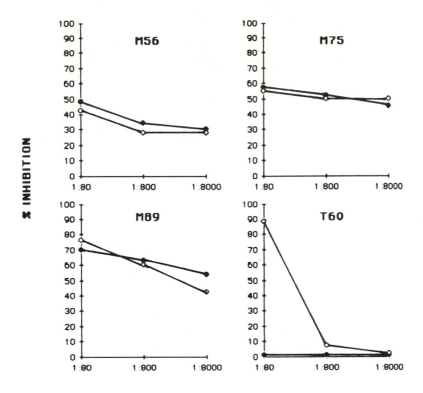

DILUTION OF ASCITES

Fig. 7.4. The ability of monoclonal antibodies binding to the LFA-1 antigen to interfere with cell-mediated cytotoxicity when preincubated with effector cells. The percentage inhibition in cytotoxic activity, shown on the y-axis, is calculated as the ratio of cytotoxic activity observed in the presence of antibody compared to the activity in wells receiving no antibody or an irrelevant antibody. The dilution of ascites refers to the final concentration in the test wells.

increase the blocking effect of these antibodies, suggesting that complete blocking could not obtained, even when these cell surface antigens were saturated with antibody under these conditions.

In contrast, T 60 was completely ineffective in blocking one of the clones, K1G, while inhibiting other clones, including clone K1I, but only at the highest concentration tested, as shown in Fig. 7.4. The ability of T60 to inhibit clone K1I was lost, however, when the antibody was diluted ten times further, at concentrations where the other antibodies were still effective.

The M11 antibody was completely ineffective in blocking any of the T cell clones tested, providing further evidence that it does not react with the LFA-1-related antigens. It did, however, serve as a negative control antibody for the blocking studies while at the same time binding strongly to both the cytotoxic T cell clones and the target B cell lines.

Discussion

The LFA-1 antigen is one of several structurally related molecules expressed on various hematopoietic cells. Monoclonal antibodies have been produced which recognize several distinct epitopes carried on these molecules, some unique to the LFA-1 antigen (1), some unique to the Mac-1, Mol antigen (5,6), some unique to the p150,90 molecule (this volume, Chapter 4), and others which recognize common determinants expressed on all the members of this molecular family (2). Unlike the analysis of T cell-specific antigens which are only expressed on a limited population of lymphocytes, antibodies to the LFA-1 antigens must be studied on a wider spectrum of hematopoietic cells in order to observe the more subtle differences in the expression of these molecules. The antibodies included in the present study were selected to react with lymphocytes so that they would include the antibodies to the LFA-1 antigen, previously reported to inhibit T cell function. Therefore antibodies such as Mac-1 and Mol, which do not react with T cell lines or B cell lines, were not included. All of the antibodies selected for further study were either specific for the LFA-1 antigen or reacted with common determinants shared with the LFA-1 antigen. It was only through analysis of the granulocyte population of peripheral blood leukocytes that these antibodies could be distinguished serologically and biochemically.

The two antibodies which were demonstrated to react only with the LFA-1 antigen, M56 and T60, did not show a consistent pattern of blocking the cytolytic activity of the T cell clones examined. Both antibodies were effective in immunoprecipitating LFA-1 molecules but showed differential and rather weak blocking of some clones which were strongly blocked by most of the antibodies recognizing common determinants expressed on additional related molecules on granulocytes. Although these other antibodies, M55, M72, M73, M75, and M89, must recognize epitopes distinct from those recognized by M56 and T60, they still showed effective blocking of most of the T cell clones. These antibodies must recognize epitopes which are carried on both of the molecules identified immunochemically on granulocytes, those containing the 160-Kd subunit and those containing the related 150-Kd subunit. One of the antibodies, M55, showed somewhat weaker blocking, suggesting that functional heterogeneity may also exist within this group of antibodies as well.

Since each of the antibodies binding to the LFA-1 antigen on T cells were found to inhibit at least some of the T cell clones studied, there is little doubt that this molecule plays a role in T cell function. However, given the extensive heterogeneity observed in this study, caution should be observed before drawing firm conclusions from the functional analysis of individual antibodies studied only with lymphocyte populations from peripheral blood.

Summary

The present study illustrates the heterogeneity found with the antibodies recognizing the LFA-1 antigen. This has been demonstrated serologically, immunochemically, and functionally. Analysis of peripheral blood granulocytes was required in order to distinguish the antibodies which only recognized the LFA-1 antigen from those which recognize all of the structurally related molecules expressed on these cells. The LFA-1 specific antibodies reacted weakly with granulocytes and only immunoprecipitated the LFA-1 molecules with 160-Kd heavy chains from these cells. In functional studies the LFA-1-specific antibodies showed considerable heterogeneity in their ability to block cytotoxic alloreactive T cell clones. In contrast, the antibodies recognizing determinants common to all of the molecules structurally related to the LFA-1 antigens were, as a group, more effective in blocking the T cell clones. These results confirm that the LFA-1 antigen is clearly important to T cell function, but antibodies recognizing distinct epitopes have distinct functional properties and appear to be more heterogeneous functionally than antibodies recognizing the T cell-specific antigens CD2, CD3, and CD4 on these clones.

Acknowledgments. Expert technical assistance was provided by Donna Williams, Jackie Chin-Louie, Debbie Mosheif, Carol Bodenheimer, Lisa Juliano, and Michael Moon. This work was supported by grants from the U.S. Public Health Service, CA-22507, CA-08748, CA-19267, CA-23766, CA-33050, and a grant from the Xoma Corporation.

References

1. Sanchez-Madrid, F., A.M. Krensky, C.F. Ware, E. Robbins, J.L. Strominger, S.J. Burakoff, and T.A. Springer. 1982. Three distinct antigens associated with human T-lymphocyte-mediated cytolysis: LFA-1, LFA-2, and LFA-3. *Proc. Natl. Acad. Sci. U.S.A.* **79**:7489.
2. Sanchez-Madrid, F., J.A. Nagy, E. Robbins, P. Simon, and T.A. Springer. 1983. A human leukocyte differentiation antigen family with distinct alpha-subunits and a common beta-subunit: The lymphocyte function-associated antigen (LFA-1), the C3bi complement receptor (OKM1/Mac-1), and the p150,95 molecule. *J. Exp. Med.* **158**:1785.
3. Ware, C.F., F. Sanchez-Madrid, A.M. Krensky, S.J. Burakoff, J.L. Strominger, and T.A. Springer. 1983. Human lymphocyte function associated antigen-1 (LFA-1): identification of multiple antigenic epitopes and their relationship to CTL-mediated cytotoxicity. *J. Immunol.* **131**:1182.

4. Flomenberg, N., K. Naito, E. Duffy, R.W. Knowles, R.L. Evans, and B. Dupont. 1983. Allocytotoxic T cell clones: both Leu 2+3− and Leu 2−3+ T cells recognize class I histocompatibility antigens. *Eur. J. Immunol.* **13**:905.
5. Ault, K.A., and T.A. Springer. 1981. Cross reaction of a rat anti-mouse phago-cyte-specific monoclonal antibody (anti-Mac-1) with human monocytes and natural killer cells. *J. Immunol.* **126**:359.
6. Todd, R.F., III, A. van Agthoven, S.F. Schlossman, and C. Terhorst. 1982. Structural analysis of differentiation antigens Mo1 and Mo2 on human mono-cytes. *Hybridoma* **1**:329.

CHAPTER 8

Monoclonal Antibodies That Identify Mo1 and LFA-1, Two Human Leukocyte Membrane Glycoproteins: A Review

Robert F. Todd III and M. Amin Arnaout

Monoclonal antibody technology has facilitated the study of plasma membrane determinants expressed by human leukocytes. In the lymphoid system, monoclonal reagents have been used as specific probes for the identification and characterization (functional and structural) of plasma membrane receptors that participate in antigen recognition, proliferation, and target cell cytotoxicity (1–3). Likewise, in the myeloid lineage, other monoclonal antibodies identify receptor structures involved in ligand binding [Fc (4–6) and complement receptors (7,8)], phagocytosis (9), antigen presentation (10), and migration (11). Recently, several laboratories have reported the development of a series of monoclonal reagents that bind to a pair of structurally similar leukocyte glycoproteins Mo1 and LFA-1 (Table 8.1). Antibody blocking studies have suggested the importance of these structures in functions that include binding of C3bi-opsonized particles, phagocytosis of C3 and IgG-coated particles, phagocyte adhesion and spreading to substrates, neutrophil aggregation, neutrophil chemotaxis, lymphocyte proliferation and cytotoxic effector capacity, and NK activity (Table 8.2). In this report, we shall review the convergent observations made by several groups of investigators which have led to an understanding of the structural and functional characteristics of the Mo1 and LFA-1 glycoproteins.

The Mo1 Glycoprotein Expressed by Human Monocytes, Neutrophils, and Null (NK) Cells

In 1979, Springer and his colleagues developed a series of rat monoclonal antibodies specific for murine leukocyte antigens (12). Among these, Mac-1 (M1/70) identified a noncovalently associated two-subunit structure (gp190,105) expressed by murine peritoneal macrophages and neutro-

Table 8.1. Monoclonal antibodies that identify Mol and LFA-1 glycoproteins.

Monoclonal antibody[a]	Antigen M.W. determination(s)[b]	Cellular expression[c]				References
		Mono	PMN	NK null	Lymph	
Anti-Mac-1 (M1/70) [IgG2b]	190, 105 (mouse PMΦ) [12]; 170, 95 (mouse PMΦ) [14,15]	+	+	+	−	12–15
OKM1 [IgG2b]	155, 94 (monocytes, PMN) [17]; 185, 105 (monocytes) [18]; 165, 95 (PMN) [19]	+	+	+	−	16–19
Anti-Mo1 (94; 44) [IgM; IgG2a]	155, 94 (monocytes, PMN) [17,20]; 165, 95 (PMN) [19]	+	+	+	−	17,20–22
OKM10 [IgG]	185, 105 (monocytes) [18]	+	+	+	−	18,19,23
OKM9 [IgG]	165, 95 (PMN) [19]	+	+	+	−	18,19,23
Anti-LFA-1 α (TS1/22; TS1/12) [IgG1]	177, 95 (monocytes, T cells, PMN) [19]; 150, 95 (PBL) [25]	+	+	+	+	19,24–26
Anti-LFA-1 (L1, L5, L11) [IgG1]	177, 94 (PMN, T cells) [28]	+	+	+	+	27,28
Anti-TA-1 [IgG2a]	170, 95 (T cells) [30]	+	−	NR[d]	+	29,30
Anti-LFA-1 β (TS1/18) [IgG1]	177, 150, 95 (T cells) [19]; 177, 165, 150, 95 (monocytes, PMN) [19]	+	+	+	+	19,24
MHM23	180, 94 (HSB-2 cells) [31]; 150, 130, 95 (PBL) [25]	NR	NR	+	+	25,31
60.3 [IgG2a]	150, 130, 95 (PBL) [25]	+	+	NR	+	25
IB4 [IgG2a]	185, 185, 153, 105 (monocytes) [18]	+	+	NR	+	18

[a] Monoclonal antibody designation; (clone number(s)); [antibody subclass].
[b] Kd, reducing conditions; underlined numbers, polypeptide containing the epitope recognized by the monoclonal antibody; (cellular lysates used for immunoprecipitation); multiple M.W. determinations indicate differing results from separate laboratories or antigenic differences between cell types [specific reference].
[c] Human cells.
[d] NR: Not reported.

Table 8.2. Inhibitory activity of monoclonal antibodies that identify Mo1 and LFA-1 glycoproteins.

Functional activity	Monoclonal antibodies (references)											
	Mac-1 (13,32)	OKM1 (16,18,19)	Mo1 (20,40b)	OKM10 (18,19)	OKM9 (18,19)	TS1/12, TS1/22 (19,33)	L1, L5, L11 (27,28)	TA-1 (30)	TS1/18 (19,26,33)	MHM23 (31)	60.3 (25,34–36)	IB4 (18)
Blocks CR3 ligand binding site	+	–/+[a]	+	+	–	–	–	NR	–	NR	+	–
Binds to CR3-related gp; does *not* block CR3 activity	–	+/–[a]	–	–	+	–	–	NR	+	NR	–	+
Blocks phagocytosis of C3/IgG opsonized particles	NR[b]	NR	+	NR	NR	NR	–	NR	NR	NR	+	NR
Blocks PMN, monocyte substrate adhesion/spreading	NR	NR	+[c]	NR	NR	NR	NR	NR	NR	NR	+	NR
Blocks PMN aggregation	NR	NR	+	NR	NR	NR	–[d]	NR	NR	NR	+	NR
Blocks PMN migration	NR	NR	+[c]	NR	NR	NR	NR	NR	NR	NR	+	NR
Blocks lymphocyte proliferation	NR	–	–	NR	NR	+	+	NR	+	NR	+	NR
Blocks CTL/NK cytotoxicity	–	–	–	NR	NR	+	+	–	+	+	+	NR

[a] Conflicting results between investigators.
[b] NR: Not reported.
[c] See footnote, p. 101.
[d] M.A. Arnaout and R.F. Todd III, unpublished results.

phils (12). In subsequent investigation carried out by Ault and Springer (13), it was found that anti-Mac-1 antibody also bound to human myeloid cells (monocytes, neutrophils, and NK cells) but would not immunoprecipitate the human Mac-1 homologue. Structural and biosynthetic analysis revealed that the two subunits of Mac-1 were synthesized independently from lower M.W. precursors (15), and that the lower-M.W. β structure (gp95, formerly 105) was structurally identical (by tryptic peptide mapping and immunological cross-reactivity) to the β subunit of another heterodimer, LFA-1 [gp180,95 (37)], expressed on the surface of most murine leukocytes (14,38,39).

Coincident with these observations relating to Mac-1, Breard et al., working in the human system, characterized a murine monoclonal reagent, OKM1, which, like anti-Mac-1, bound to a determinant found on human neutrophils, monocytes, and null cells, as well as on leukemia cells from several patients with acute myeloid leukemia (16). Meanwhile, Todd et al. reported the generation of a monoclonal antibody, anti-Mo1, whose distribution of reactivities for human myeloid cells was identical to that of OKM1 (21,22). In structural studies, anti-Mo1 immunoprecipitated a noncovalently associated two-subunit glycoprotein of 155 (α) and 94 (β) Kd from lysates of radiolabeled monocytes and neutrophils (17). OKM1, which failed to competitively block the binding of anti-Mo1, immunoprecipitated a heterodimer of identical M.W., suggesting that anti-Mo1 and OKM1 identify distinct epitopes on similar or identical structures (17).

Based on a similar distribution of antigen expression among myeloid leukocytes and similar M.W. characteristics, Mac-1 and Mo1 (OKM1) appeared to be homologous structures. This presumption was strengthened by the independent findings of three laboratories which demonstrated close functional similarities among these glycoproteins (Table 8.2). Beller et al. found that anti-Mac-1 specifically inhibited the rosetting of C3bi-coated particles to murine and human monocytes (32). Arnaout and his colleagues reported identical findings in blocking studies using anti-Mo1 antibody, which, as a result of immunoblotting experiments, was found to bind to the 155-Kd α subunit (20). In parallel to its inhibition of neutrophil EC3bi rosetting, anti-Mo1 also blocked C3bi-dependent neutrophil enzyme release (lysozyme, β-glucuronidase, and histaminase) (20). Wright et al. demonstrated that down-modulation of neutrophil determinants bound by substrate-fixed OKM1, OKM9, OKM10, and IB4 monoclonal antibodies resulted in an inability of these cells to bind C3bi-coated particles (18). However, only antibody OKM10 (in solution) could directly block receptor–ligand binding. Antibody competition experiments and sequential immunoprecipitations indicated that OKM1, OKM9, OKM10, and IB4 bound to distinct epitopes on a single heterodimer of 185 and 105 Kd (18); IB4 was additionally expressed on surface proteins featuring a distinct 185-Kd subunit as well as a polypeptide of 153 Kd (four polypeptides were detectable in IB4 immunoprecipitates) (18). The relationship between gp 185,105 and the C3bi receptor (CR3) was

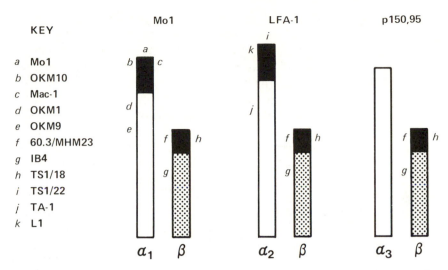

Fig. 8.1. Schematic representation of the Mo1, LFA-1, and p150,95 glycoproteins indicating the possible molecular location of epitopes identified by a panel of monoclonal reagents. Each two-subunit glycoprotein has a unique α polypeptide (α_1, α_2, α_3), which is noncovalently associated with a single β polypeptide. The black-shaded end of the α_1, α_2, and β polypeptides represents a functionally active region of the molecule as suggested by antibody blocking studies. Assignment of 60.3/MHM23 and IB4 epitopes to the β polypeptide remains speculative. The antibody Leu-M5 may identify an epitope on the α polypeptide of p150,95.

further strengthened by the demonstration that isolated gp185,105 (affinity purified by IB4 or OKM1 could bind to C3bi-coated erythrocytes in a cation-dependent (physiologic) fashion (18).

The results of these three groups, taken together, strongly suggested that anti-Mac-1, anti-Mo1, OKM1, OKM9, OKM10, and IB4 identify epitopes on a single glycoprotein, Mo1, that serves as the CR3 receptor on neutrophils and monocytes. The minor differences in M.W. estimation (Table 8.1) appear to relate to the behavior of the glycosylated α and β polypeptides under different electrophoretic conditions (acrylamide concentration). Mac-1, anti-Mo1, and OKM10 may bind to the actual ligand binding site (presumably on the higher-M.W. α subunit) since they each interfere directly with C3bi ligand binding, while OKM1, OKM9, and IB4 (failing to block ligand binding) probably bind to epitopes distant from the binding site (Fig. 8.1.)

Structural and Functional Similarity between Mo1 and LFA-1 Glycoproteins

Additional analyses conducted by several groups of investigators have focused on the relationship between CR3-associated heterodimer, Mo1, and the LFA-1 glycoprotein. LFA-1 in mouse (37) and man (24) is a two-subunit glycoprotein [gp170,95 in man (24)] expressed by both lymphoid and myeloid hematopoietic cells (26,40). Anti-LFA-1 antibodies block lymphocyte proliferation stimulated by mitogens and antigens, and target cell cytotoxicity produced by activated T lymphocytes and natural killer cells (26,27,33) (Table 8.2). Inhibition of cytotoxicity by these reagents appears to work at the effector cell level, probably by interfering with a nonspecific effector–target cell binding step during the cytolytic interaction (26). In structural studies featuring a series of seven murine anti–human LFA-1 reagents, six (including TS1/22 and TS1/12) were found to immunoprecipitate the 170/95-Kd heterodimer in lysates of lymphoid or myeloid cells, while one, TS1/18, precipitated four polypeptides of 170, 165, 150, and 95 Kd from lysates of monocytes or neutrophils (19). Under conditions that produce dissociation of the noncovalent bonds linking these polypeptides, the TS1/18 antibody was found to bind to an epitope on the 95-Kd β subunit shared by LFA-1 (gp177,95), Mo1 (OKM1) (gp165,95), and a third heterodimer of uncertain function, p150,95. Five of the other anti-LFA-1 reagents bound to epitopes on the 177-Kd α subunit of LFA-1, while OKM1, like anti-Mo1, was specific for the 165-Kd α structure of the CR3-associated heterodimer. No reagent has thus far been shown to identify the α subunit of p150,95.* These findings have suggested the existence of a family of three structurally related leukocyte surface glycoproteins (Mo1, LFA-1, and p150,95) each with a distinct α polypeptide (165, 177, and 150 Kd, respectively) but a common 95-Kd β subunit (19) (Fig. 8.1).

Other antibodies that are quite similar to the anti-LFA-1 reagents are TA-1 (29,30) and MHM23 (31). These reagents immunoprecipitate broadly expressed heterodimers of 170–180 (α) and 94 (β) Kd (30,31). While TA-1 does not demonstrate known inhibitory activity (30), MHM23, like anti-LFA-1, blocks both CTL and NK effector function (31). TA-1, analogous to the OKM1 and OKM9 epitopes of Mo1, may bind to a functionally inactive site on the LFA-1 molecule (Fig. 8.1).

Meanwhile, Beatty et al. reported the characterization of a murine monoclonal antibody, 60.3, which has a number of features in common with the TS1/18 reagent (25). 60.3 identifies an antigen which, like LFA-1, is broadly expressed among all human leukocytes (myeloid and lymphoid), and which also blocks lymphocyte proliferation and cytotoxicity (25). Moreover, 60.3 also inhibits several neutrophil functional activities including phagocytosis (34), substrate adhesion (endothelial cell monolayers) (34), spreading (34), directed motility (25), PMA-induced neutrophil aggregation (35), and the binding of C3bi-opsonized particles (36). On structural analysis, 60.3 immunoprecipitates a series of three polypeptides (150, 130, and 95 Kd) from lysates of PB leukocytes (reducing conditions), which exhibit similar electrophoretic mobilities to those exhibited

---

* Recent observations suggest that the monoclonal antibody Leu-M5 may identify an epitope on the α subunit of p150,95 (40a).

by the TS1/18 polypeptides isolated from human T lymphocytes (177, 150, and 95 Kd) (25). Direct comparison between immunoprecipitates isolated by 60.3 and the LFA-1 α reagent TS1/22 in fact suggests that the LFA-1 α polypeptide is identical or similar to the 60.3 150-Kd structure, and that the LFA-1 β subunit is the same as the 60.3 95-Kd polypeptide [binding competition experiments indicate that TS1/22 and 60.3 recognize distinct epitopes (25)]. Furthermore, 60.3 appears to recognize the MHM23 epitope, since MHM23 competitively blocks the binding of 60.3 and immunoprecipitates the same three bands from lysates of PBLs (25). In related structural studies, Arnaout has found that 60.3 and IB4 identify epitopes on the same molecule (by sequential immunoprecipitations from neutrophil lysates; M.A. Arnaout, P.G. Beatty, and S.D. Wright, unpublished). Thus, while no direct evidence is currently available, it is speculated that 60.3 (MHM23) and IB4 may in fact recognize epitopes on the TS1/18-associated β subunit common to the family of three leukocyte glycoproteins (Fig. 8.1). In the case of IB4, it has not been possible to check this theory directly since IB4, unlike TS1/18, fails to bind to its immunoprecipitated polypeptides in their dissociated state (S.D. Wright, personal communication).

In terms of its inhibitory properties on myeloid cell function, 60.3 is also similar to anti-Mo1, which, as noted above, identifies a CR3-associated epitope on the 155-Kd α subunit of Mo1 (gp155,94). Arnaout et al. had previously shown that anti-Mo1, like 60.3 (34), partially blocks the phagocytosis of C3 or IgG-opsonized particles by neutrophils or monocytes (20). More recent observations by Arnaout et al. indicate that anti-Mo1, like 60.3 (34,35), also inhibits neutrophil adhesion/spreading as well as neutrophil aggregation stimulated by fMLP or C5a (40b and *). Unlike 60.3, anti-Mo1 has no inhibitory effect on lymphocyte function.

Mo1/LFA-1 Deficiency Syndrome

The close relationship among the epitopes identified by these 12 monoclonal reagents was further strengthened by the discovery of a series of pediatric patients whose leukocytes were found to be deficient in the expression of these markers (reviewed in Ref. 41) (Table 8.3). Dana and her colleagues (42) first demonstrated the deficient expression of the epitopes identified by anti-Mo1 and TS1/18 by the neutrophils and monocytes of a 12 year old boy with an immunodeficiency syndrome characterized by recurrent severe pyogenic bacterial infections with *in vitro* defects in neutrophil phagocytosis and opsonized-zymosan-induced respiratory

---

* Arnaout, M.A., B. Styrt, J.D. Griffin, R.F. Todd III, M.S. Klempner and N. Dana. Mapping of a phagocyte surface glycoprotein (Mo1) using monoclonal antibodies: Identification of two functional domains, one involved in C3bi binding and another in granulocyte adhesion. (submitted for publication).

102   Robert F. Todd III and M. Amin Arnaout

Table 8.3. Patients with Mol/LFA-1 deficiency syndrome.

	Reported patients							
	1	2	3	4	5	6	7	8
Leukocyte antigen deficiencies								
Mac-1	+[a]							
OKM1	+		+		+	+	+	+
Mo1	+	+				+		
OKM10								
OKM9								
LFA-1 α (TS1/22, TS1/12)			+			+	+	+
LFA-1 (L1, L5, L11)	+	+						
TA-1	+							
LFA-1 β (TS1/18)	+	+				+	+	+
MHM23			+					
60.3	+		+	+				
IB4	+							
Leukocyte in vitro abnormalities								
PMN/Mo adherence/spreading		+	+	+	+	+	+	+
PMN aggregation		+			+	+	+	+
PMN/Mo chemotaxis	−	+	+	+	+	+	+	+
PMN/Mo phagocytosis	+	+	+	+		+	+	+
PMN C3-induced resp. burst	+	+	+	+		+		
PMN C3-induced degranulation	+					+		
PMN/Mo CR3 activity	−		+	+		+		
Lymphocyte proliferation	+	+	+					
CTL cytotoxicity	−	−						
NK cytotoxicity	−	−	+	+			+	+
References	42	44	46	46	47	48	49	49
	43	28	34	36		49		
	41	45	35					
			36					

[a] +, Antigen deficiency or leukocyte functional abnormality reported; −, no deficiency or abnormality observed; blank space, results not reported.

burst and degranulation (43). Subsequently, this child and a second boy with similar leukocyte abnormalities [that additionally included defects in neutrophil substrate adhesion/spreading and directed motility (28,44)] were found to be deficient in the epitopes identified by LFA-1 (L1, L5, and L11), TA-1, 60.3, and IB4 (28,41,45). The *in vitro* leukocyte abnormalities seen in these children therefore closely paralleled the functional defects observed in normal leukocytes exposed to anti-Mo1, anti-LFA-1, or 60.3 monoclonal antibodies. These two Mo1/LFA-1-deficient patients, far from being isolated quirks of nature, have since been joined by a growing number of individuals who share similar, if not identical, clinical traits and whose leukocytes are defective in a spectrum of adhesion-related phenomena (phagocytosis, substrate adhesion and spreading, neutrophil aggregation, lymphocyte proliferation) coincident with either deficiency or complete absence of epitopes identified by OKM1, Mo1, 60.3, MHM23, TS1/22, and TS1/18 (Table 8.3). Since this family of surface glycoproteins shares a common β subunit and because of the inherited nature of Mo1/LFA-1 deficiency, it is likely that a primary disorder involving the β subunit underlies this syndrome (41). Work is in progress to further clarify the molecular pathophysiology of the Mo1/LFA-1 deficiency syndrome.

Mo1 and LFA-1—Adhesion-promoting Molecules

Recent observations by Arnaout, Todd, and their colleagues tend to substantiate the role of Mo1 in neutrophil adhesion phenomena (40b,41,45, and *). Exposure of neutrophils to degranulating stimuli (e.g., calcium ionophore, C5a, fMLP, or PMA) under conditions that induce neutrophil aggregation or increased substrate adhesion (50,51) results in a 5–10-fold increase in the number of Mo1 binding sites per cell (28). A similar expansion in the surface expression of Mac-1 (52) and 60.3 (35) has also been reported. This phenomenon appears to be due to a rapid translocation of intracellular Mo1 antigen (both α and β subunits) from neutrophil secondary or tertiary granules to the plasma membrane (28,45), where it may augment neutrophil adhesion in the form of aggregation or adherence and spreading to substrates. The increase in neutrophil Mo1 expression occurs not only as a result of *in vitro* stimulation but also *in vivo* under conditions in which neutrophils are exposed to increased levels of endogenous C5a (in the setting of hemodialysis on unused cuprophane membranes) (40b). Surface LFA-1 expression by neutrophils is not significantly changed as a result of exposure to degranulating stimuli (28). These findings suggest that the Mo1 and LFA-1 glycoproteins, despite sharing a common β subunit, are packaged independently in the neutrophil cytoplasm with correspondingly distinct control mechanisms regulating their plasma membrane expression. The finding that anti-Mo1 antibodies block neutrophil spreading and aggregation provides direct evidence for the significance of Mo1 in neutrophil adhesion phenomena in a fashion that is analogous to the effect of anti-LFA-1 antibodies on lymphocyte adhesive interactions (19,41,45,53). Interestingly, activation of human T lymphocytes by mitogenic or alloantigenic stimulation results in a substantial increase in the surface expression of LFA-1 (28), which suggests that the expression of LFA-1 by lymphoid cells, like Mo1 in neutrophils, may relate to their state of activation.

The functional role of the common β subunit is less clear since TS1/18 interferes with lymphocyte function but does not inhibit CR3 activity (19), while 60.3 (presumed to identify another β subunit epitope) prevents the

---

* Arnaout, M.A., B. Styrt, J.D. Griffin, R.F. Todd III, M.S. Klempner and N. Dana. Mapping of a phagocyte surface glycoprotein (Mo1) using monoclonal antibodies: Identification of two functional domains, one involved in C3bi binding and another in granulocyte adhesion. (submitted for publication).

entire spectrum of myeloid and lymphoid adhesion-related activities (25,34). Meanwhile, there is little concrete information regarding the functional significance of the third member of the glycoprotein family, p150,95 (19). Recent preliminary data suggest that it may be involved in the binding of C3d-opsonized erythrocytes to the monocyte membrane (CR2 activity) (54). Further investigation employing monoclonal and polyclonal reagents will be required to fully elucidate the structure–function relationships exhibited by this interesting family of leukocyte surface glycoproteins.

Summary

The use of monoclonal antibodies has facilitated the characterization of several plasma membrane receptor structures expressed by human leukocytes. Among these, a pair of structurally related surface glycoproteins, Mo1 and LFA-1, has been identified by a series of 12 monoclonal reagents [anti-Mac-1, OKM1, anti-Mo1, OKM10, OKM9, anti-LFA-1 α (TS1/22), anti-LFA-1 (L1), TA-1, anti-LFA-1 β (TS1/18), IB4, 60.3, and MHM23] which bind to distinct epitopes. Both glycoproteins are heterodimers (α, β) with distinct higher-M.W. α subunits, but identical β polypeptides. In terms of function, Mo1 has complement receptor type 3 (CR3) activity and may play a role in myeloid cell adhesion phenomena (substrate adherence/spreading; aggregation); LFA-1 promotes lymphoid cell adhesion interactions that include lymphocyte proliferation and cytotoxic effector activity. The functional significance of these surface determinants has been demonstrated by antibody blocking experiments and is further indicated by abnormalities in *in vitro* leukocyte function displayed by a series of patients whose cells are selectively deficient in the expression of both glycoproteins. p150,95, a cell surface protein with possible phagocyte C3d receptor activity, may represent a third member of the Mo1/LFA-1 family.

Acknowledgments. This work was supported in part by NIH Grant CA39064. M. Amin Arnaout is an Established Investigator of the American Heart Association.

References

1. Meuer, S.C., R.E. Hussey, J.C. Hodgdon, T. Hercend, S.F. Schlossman, and E.L. Reinherz. 1982. Surface structures involved in target recognition by human cytotoxic T lymphocytes. *Science* **218**:471.
2. Meuer, S.C., O. Acuto, R.E. Hussey, J.C. Hodgdon, K.A. Fitzgerald, S.F. Schlossman, and E.L. Reinherz. 1983. Evidence for the T3-associated 90K heterodimer as the T-cell antigen receptor. *Nature* **303**:808.

3. Leonard, W.J., J.M. Depper, R.J. Robb, T.A. Waldmann, and W.C. Greene. 1983. Characterization of the human receptor for T-cell growth factor. *Proc. Natl. Acad. Sci. U.S.A.* **80:**6957.
4. Unkeless, J.C. 1979. Characterization of a monoclonal antibody directed against mouse macrophage and lymphocyte Fc receptors. *J. Exp. Med.* **150:**580.
5. Fleit, H.B., S.D. Wright, and J.C. Unkeless. 1982. Human neutrophil Fc receptor distribution and structure. *Proc. Natl. Acad. Sci. U.S.A.* **79:**3275.
6. Perussia, B., S. Starr, S. Abraham, V. Fanning, and G. Trinchieri. 1983. Human natural killer cells analyzed by B73.1, a monoclonal antibody blocking Fc receptor functions. I. Characterization of the lymphocyte subset reactive with B73.1. *J. Immunol.* **130:**2133.
7. Iida, K., R. Mornaghi, and V. Nussenzweig. 1982. Complement receptor (CR1) deficiency in erythrocytes from patients with systemic lupus erythematosis. *J. Exp. Med.* **155:**1427.
8. Gerdes, J., M. Naiem, D.Y. Mason, and H. Stein. 1982. Human complement (C3b) receptors defined by a mouse monoclonal antibody. *Immunology* **45:**645.
9. Skubitz, K.M., D.J. Weisdorf, and P.K. Peterson. 1985. Monoclonal antibody AHN-1 selectively inhibits phagocytosis by human neutrophils. *Blood* **65:**333.
10. Todd, R.F., III, S.C. Meuer, P.L. Romain, and S.F. Schlossman. 1984. A monoclonal antibody that blocks class II histocompatibility-related immune interactions. *Hum. Immunol.* **10:**23.
11. Cotter, T.G., P.J. Keeling, and P.M. Henson. 1981. A monoclonal antibody inhibiting FMLP-induced chemotaxis of human neutrophils. *J. Immunol.* **127:**2241.
12. Springer, T., G. Galfre, D.S. Secher, and C. Milstein. 1979. Mac-1: A macrophage differentiation antigen identified by monoclonal antibody. *Eur. J. Immunol.* **9:**301.
13. Ault, K.A., and T.A. Springer. 1981. Cross-reaction of a rat antimouse phagocyte-specific monoclonal antibody (anti-Mac-1) with human monocytes and natural killer cells. *J. Immunol.* **126:**359.
14. Kurzinger, K., M-K. Ho, and T.A. Springer. 1982. Structural homology of a macrophage differentiation antigen and an antigen involved in T-cell mediated killing. *Nature* **296:**668.
15. Ho, M-K., and T.A. Springer. 1983. Biosynthesis and assembly of the alpha and beta subunits of Mac-1, a macrophage glycoprotein associated with complement receptor function. *J. Biol. Chem.* **258:**2766.
16. Breard, J., E.L. Reinherz, P.C. Kung, G. Goldstein, and S.F. Schlossman. 1980. A monoclonal antibody reactive with human peripheral blood monocytes. *J. Immunol.* **124:**1943.
17. Todd, R.F., A. van Agthoven, S.F. Schlossman, and C. Terhorst. 1982. Structural analysis of differentiation antigens Mo1 and Mo2 on human monocytes. *Hybridoma* **1:**329.
18. Wright, S.D., P.E. Rao, W.C. Van Voorhis, L.S. Craigmyle, K. Iida, M.A. Talle, E.F. Westberg, G. Goldstein, and S.C. Silverstein. 1983. Identification of the C3bi receptor of human monocytes and macrophages by using monoclonal antibodies. *Proc. Natl. Acad. Sci. U.S.A.* **80:**5699.

19. Sanchez-Madrid, F., J.A. Nagy, E. Robbins, P. Simon, and T.A. Springer. 1983. A human leukocyte differentiation antigen family with distinct α subunits and a common β subunit: The lymphocyte function-associated antigen (LFA-1), the C3bi complement receptor (OKM1/Mac-1), and the p150,95 molecule. *J. Exp. Med.* **158:**1785.

20. Arnaout, M.A., R.F. Todd III, N. Dana, J. Melamed, S.F. Schlossman, and H.R. Colten. 1983. Inhibition of phagocytosis of complement C3- or immunoglobulin G-coated particles and of C3bi binding by monoclonal antibodies to a monocyte–granulocyte membrane glycoprotein (Mol). *J. Clin. Invest.* **72:**171.

21. Todd, R.F., III, L.M. Nadler, and S.F. Schlossman. 1981. Antigens on human monocytes identified by monoclonal antibodies. *J. Immunol.* **126:**1435.

22. Todd, R.F., III, and S.F. Schlossman. 1982. Analysis of antigenic determinants on human monocytes and macrophages. *Blood* **59:**775.

23. Talle, M.A., P.E. Rao, E. Westberg, N. Allegar, M. Makowshi, R.S. Mittler, and G. Goldstein. 1983. Patterns of antigenic expression on human monocytes as defined by monoclonal antibodies. *Cell. Immunol.* **78:**83.

24. Sanchez-Madrid, F., A.M. Krensky, C.F. Ware, E. Robbins, J.L. Strominger, S.J. Burakoff and T.A. Springer, 1982. Three distinct antigens associated with human T lymphocyte-mediated cytolysis: LFA-1, LFA-2, and LFA-3. *Proc. Natl. Acad. Sci. U.S.A.* **79:**7489.

25. Beatty, P.G., J.A. Ledbetter, P.J. Martin, T.H. Price, and J.A. Hansen. 1983. Definition of a common leukocyte cell-surface antigen (Lp95-150) associated with diverse cell-mediated immune functions. *J. Immunol.* **131:**2913.

26. Krensky, A.M., F. Sanchez-Madrid, E. Robbins, J.A. Nagy, T.A. Springer, and S.J. Burakoff. 1983. The functional significance, distribution, and structure of LFA-1, LFA-2, and LFA-3: Cell surface antigens associated with CTL–target interactions. *J. Immunol.* **131:**611.

27. Spits, H., G. Keizer, J. Borst, C. Terhorst, A. Hekman, and J. E. de Vries. 1984. Characterization of monoclonal antibodies against cell surface molecules associated with cytotoxic activity of natural and activated killer cells and cloned CTL lines. *Hybridoma* **2:**423.

28. Arnaout, M.A., H. Spits, C. Terhorst, J. Pitt, and R.F. Todd III. 1984. Deficiency of a leukocyte surface glycoprotein (LFA-1) in two patients with Mol deficiency: Effect of cell activation on Mo1/LFA-1 surface expression in normal and deficient leukocytes. *J. Clin. Invest.* **74:**1291.

29. LeBien, T.W., and J.H. Kersey. 1980. A monoclonal antibody (TA-1) reactive with human T lymphocytes and monocytes. *J. Immunol.* **125:**2208.

30. LeBien, T.W., J.G. Bradley, and B. Koller. 1983. Preliminary structural characterization of the leukocyte cell surface molecule recognized by monoclonal antibody TA-1. *J. Immunol.* **130:**1833.

31. Hildreth, J.E.K., F.M. Gotch, P.D.K. Hildreth, and A.J. McMichael. 1983. A human lymphocyte-associated antigen involved in cell mediated lympholysis. *Eur. J. Immunol.* **13:**202.

32. Beller, D.I., T.A. Springer, and R.D. Schreiber. 1982. Anti-Mac-1 selectively inhibits the mouse and human type three complement receptor. *J. Exp. Med.* **156:**1000.

33. Ware, C.F., F. Sanchez-Madrid, A.M. Krensky, S.J. Burakoff, J.L. Strominger, and T.A. Springer. 1983. Human lymphocyte function associated

antigen-1 (LFA-1): Identification of multiple antigenic epitopes and their relationship to CTL-mediated cytotoxicity. *J. Immunol.* **131:**1182.

34. Beatty, P.G., J.M. Harlan, H. Rosen, J.A. Hansen, H.D. Ochs, T.H. Price, R.F. Taylor, and S.J. Klebanoff. 1984. Absence of monoclonal-antibody-defined protein complex in boy with abnormal leucocyte function. *Lancet* **1984**(i):535.

35. Harlan, J.M., F.M. Senecal, R.F. Taylor, B.R. Schwartz, P.G. Beatty, and H.D. Ochs. 1984. The neutrophil membrane glycoprotein recognized by the monoclonal antibody 60.3 is required for phorbol ester-induced neutrophil adherence and aggregation. *Clin. Res.* **32:**464A (abstract).

36. Beatty, P.G., H.D. Ochs, R.D. Schreiber, J.M. Harlan, T.H. Price, H. Rosen, J.A. Hansen, and S.J. Klebanoff. 1984. Absence of a monoclonal antibody-defined leukocyte protein in patients with abnormal leukocyte function. *Ped. Res.* **18:**253A (abstract).

37. Davignon, D., E. Martz, T. Reynolds, K. Kurzinger, and T.A. Springer. 1981. Lymphocyte function-associated antigen 1 (LFA-1): A surface antigen distinct from Lyt-2,3 that participates in T lymphocyte-mediated killing. *Proc. Natl. Acad. Sci. U.S.A.* **78:**4535.

38. Sanchez-Madrid, F., P. Simon, S. Thompson, and T.A. Springer. 1983. Mapping of antigenic and functional epitopes on the α- and β-subunits of two related mouse glycoproteins involved in cell interactions, LFA-1 and Mac-1. *J. Exp. Med.* **158:**586.

39. Trowbridge, I.S., and M.B. Omary. 1981. Molecular complexity of leucocyte surface glycoproteins related to the macrophage differentiation antigen Mac-1. *J. Exp. Med.* **154:**1517.

40. Kurzinger, K., T. Reynolds, R.N. Germain, D. Davignon, E. Martz, and T.A. Springer. 1981. A novel lymphocyte function-associated antigen (LFA-1): Cellular distribution, quantitative expression, and structure. *J. Immunol.* **127:**596.

40a. Lanier, L.L., M.A. Arnaout, R. Schwarting, N.L. Warner, and G.D. Ross. P150/95, Third member of the LFA-1/CRIII polypeptide family identified by anti-Leu-M5 monoclonal antibody. *Eur. J. Immunol.* (in press).

40b. Arnaout, M.A., R.M. Hakim, R.F. Todd III, N. Dana, and H. Colten. 1985. Increased expression of an adhesion-promoting surface glycoprotein in the granulocytopenia of hemodialysis. *New England J. Med.* **312:**457.

41. Arnaout, M.A., N. Dana, J. Pitt, and R.F. Todd III. 1985. Deficiency of two human leukocyte surface membrane glycoproteins (Mo1 and LFA-1). *Fed. Proc.* **44:**2664.

42. Dana, N., R.F. Todd III, J. Pitt, T.A. Springer, and M.A. Arnaout. 1983. Deficiency of a surface membrane glycoprotein (Mo1) in man. *J. Clin. Invest.* **73:**153.

43. Arnaout, M.A., J. Pitt, J.J. Cohen, J. Melamed, F.S. Rosen, and H.R. Colten. 1982. Deficiency of a granulocyte-membrane glycoprotein (gp150) in a boy with recurrent bacterial infections. *New England J. Med.* **306:**693.

44. Crowley, C.A., J.T. Curnutte, R.E. Rosin, J. Andre-Schwartz, J.I. Gallin, M. Klempner, R. Snyderman, F.S. Southwick, T.P. Stossel, and B.M. Babior. 1982. An inherited abnormality of neutrophil adhesion: Its genetic transmission and its associations with a missing protein. *New England J. Med.* **306:**693.

45. Todd, R. F., III, M.A. Arnaout, R.E. Rosin, C.A. Crowley, W.A. Peters, and B.M. Babior. 1984. The subcellular localization of Mo1 (Mo1 α; formerly gp 110) a surface glycoprotein associated with neutrophil adhesion functions. *J. Clin. Invest.* **74:**1280.

46. Bowen, T.J., H.D. Ochs, L.C. Altman, T.H. Price, D.E. Van Epps, D.L. Brautigan, R.E. Rosin, W.D. Perkins, B.M. Babior, S.J. Klabanoff, and R.J. Wedgwood. 1982. Severe recurrent bacterial infections associated with defective adherence and chemotaxis in two patients with neutrophils deficient in a cell-associated glycoprotein. *J. Ped.* **101:**932.

47. Buescher, E.S., J. Nath, R.L. Roberts, B.E. Seligmann, J.A. Metcalf, N.L. Mounessa, and J.I. Gallin. 1984. Deficient OKM-1 antigen and absent polymorphonuclear leukocyte (PMN) aggregation in a patient with severe gingivitis, recurrent infections, and delayed separation of the umbilical stump. *Clin. Res.* **32:**365A (abstract).

48. Anderson, D.C., F.C. Schmalstieg, M.A. Arnaout, S. Kohl, M.F. Tosi, N. Dana, G.J. Buffone, B.J. Hughes, B.R. Brinkley, W.D. Dickey, J.S. Abramson, T. Springer, L.A. Boxer, J.M. Hollers, and C.W. Smith. 1984. Abnormalities of polymorphonuclear leukocyte function associated with a heritable deficiency of high molecular weight surface glycoproteins (GP138): Common relationship to diminished cell adherence. *J. Clin. Invest.* **74:**536.

49. Anderson, D., F. Schmalstieg, W. Shearer, S. Kohl, and T. Springer. 1984. Abnormalities of PMN/monocyte function & recurrent infection associated with a heritable deficiency of adhesive surface glycoproteins. *Fed. Proc.* **43:**1487 (abstract).

50. Craddock, P.R., D. Hammerschmidt, J.G. White, A.P. Dalmasso, and H.S. Jacob. 1977. Complement (C5a)-induced granulocyte aggregation *in vitro:* A possible mechanism of complement-mediated leukostasis and leukopenia. *J. Clin. Invest.* **60:**260.

51. Hoover, R.L., R. Folger, W.A. Haering, B.R. Ware, and M.J. Karnovsky. 1980. Roles of divalent cations, surface charge, chemotactic agents and substrate. *J. Cell Sci.* **45:**73.

52. Berger, M., J. O'Shea, A.S. Cross, T.M. Chused, E.J. Brown, and M.M. Frank. 1984. Human neutrophils increase expression of C3bi receptors upon activation. *Clin. Res.* **32:**364A (abstract).

53. Springer, T.A., D. Davignon, M-K. Ho, K. Kurzinger, E. Martz, and F. Sanchez-Madrid. 1982. LFA-1 and Lyt-2,3, molecules associated with T lymphocyte-mediated killing; and Mac-1, an LFA-1 homologue associated with complement receptor function. *Immunol. Rev.* **68:**111.

54. Wright, S.D., M.R. Licht, and S.C. Silverstein. 1984. The receptor for C3d (CR2) is a homologue of CR3 and LFA-1. *Fed. Proc.* **43:**1487 (abstract).

CHAPTER 9

Localization of the Human NA1 Alloantigen on Neutrophil Fc-γ-Receptors*

G. Werner, A.E.G. Kr. von dem Borne, M.J.E. Bos, J.F. Tromp, C.M. van der Plas-van Dalen, F.J. Visser, C.P. Engelfriet, and P.A.T. Tetteroo

Introduction

The neutrophil-specific-NA antigen system is a biallelic system. It comprises the antigens NA1 and NA2, which in Caucasian populations show a phenotype frequency of 46% and 88%, respectively (1). It is clinically an important system because it may be involved in diseases such as neonatal alloimmune neutropenia and autoimmune neutropenia as well as in blood transfusion reactions such as rigors, fever, and respiratory distress (1,2).

Although serologically well-defined, it is not yet known on which neutrophil membrane glycoproteins the antigens of the NA system are located. In this report we describe two monoclonal antibodies against the neutrophil Fc-γ-receptor, which we prepared in a similar way as the mAb described by Fleit et al. (3).

One of these antibodies reacted with neutrophils as well as K lymphocytes. However the other antibody was specific for neutrophils, reacted with the cells of only about half of the normal donors tested, and appeared to be specific for NA1. Thus by this antibody the NA antigens were located on the neutrophil Fc-γ-receptor.

---

* This paper is dedicated to Dr. Guy Werner, who died after a sad accident on the night of September 23 to 24, 1983.

Materials and Methods

Monoclonal Antibodies

CLB FcR gran 1 (M57) and CLB gran 11 (M82) were obtained from a fusion of spleen cells, from a mouse immunized with Ficoll–Isopaque-isolated human neutrophils and Sp 2/0-Ag14 myeloma cells. In a micro-ELISA-test (4) these antibodies reacted only with neutrophils, and not with other peripheral blood cells, and they inhibited EA-rosette formation on neutrophils in the micromethod described by Fleit *et al.* (3). Both antibodies were IgG2a. Stable hybrids were used to prepare antibody containing ascites in pristane-primed BALB/c mice and this was used for all further experiments.

Other mAbs applied were CLB gran/2 [CD15w (G.u) B4.3], an IgM antibody against the neutrophil-associated fucosyl-*N*-acetyllactosamine structure (anti-FAL) and CLB gran 10 (M81), an IgG1 antibody against a 170 Kd glycoprotein-antigen specific for mature neutrophils. mAbs 3G8 (3) and B73.1 (5) against the Fc-γ-receptor of neutrophils and K lymphocytes, respectively, were used as well.

Neutrophil-Specific Alloantisera

Anti-NA1 and anti-NA2 sera were obtained either from women immunized by previous pregnancies or from patients immunized by blood transfusion (6).

Serological Tests on Neutrophils

Agglutination and immunofluorescence tests were performed as described previously (6). For immunofluorescence the cells were fixed with 1% paraformaldehyde (PFA). As antiglobulin reagents we used FITC sheep anti–human Ig (SH 17-01-F08, CLB) for the detection of human alloantibodies and FITC goat anti–mouse Ig (GM 17-01-F02, CLB) for the detection of mouse mAbs. Fluorescence was quantified by cytofluorography. For cytotoxicity studies a double-color fluorescence test was applied (7).

Immunological Studies

All tests on other peripheral blood cells, bone marrow cells, human leukemic cell line cells, and cells from patients with various malignant blood diseases were also done after PFA fixation (8). Immunofluorescence was performed as described above, immunoperoxidase staining as described elsewhere (9).

EA-Rosette Inhibition

EA-rosettation of human neutrophils was tested with OR2R2 red cells sensitized with IgG-anti-D (8). Inhibition was studied in this assay by preincubation of the neutrophils (8×10^6/ml) with the monoclonal antibodies (end dilution 1/1000) or the alloantibody containing sera (end dilution 1/2).

Immunochemical Characterization of the Antigens

Intact neutrophils (10^7) were iodinated utilizing Iodogen. This was followed by lysis and isolation of the solubilized antigens by way of either preformed complexes of mAb and goat anti–mouse Ig (GM-17-02-P, CLB) (8) or mAb bound via goat anti–mouse Ig coupled to Sepharose by cyanogen bromide. The antigens were eluted by boiling for 5 min in sample buffer. They were reduced (15% 2-mercaptoethanol), analyzed by SDS–PAGE electrophoresis (10% acrylamide gels), and visualized by autoradiography.

Enzyme Treatment

The effect of enzyme treatment on the antigens was also studied. The enzymes applied were neuraminidase (cholera filtrate, Behringwerke, standard dilution), papain (4 mg/ml and 5 mM cystein), and bromelin (Difco, standard dilution), all at 37°C, cell concentration $0.5–1 \times 10^7$/ml.

Results

CLB FcR gran 1 (M57) and CLB gran 11 (M82) both reacted strongly with neutrophils in the immunofluorescence test and the granulocyte cytotoxicity test, but hardly reacted in the agglutination test (Table 9.1). CLB gran 2, an IgM antibody, was strongly positive in all three tests, while CLB gran 10 (M81), a non-complement fixing IgG1 antibody, reacted strongly in the agglutination and immunofluorescence test, but not in the cytotoxicity test. The antibodies were then tested with the neutrophils of

Table 9.1. Serological behavior of the mAs with neutrophils.

Antibody	Ig class	Immunofluorescence	Cytotoxicity	Agglutination
FcR gran 1	IgG2a	+ + + +	+ + + +	− / +
Gran 11	IgG2a	+ + + +	+ + + +	− / +
Gran 10	IgG1	+ + + +	−	+ + + +
Gran 2	IgM	+ + + +	+ + + +	+ + + +

Table 9.2. Reaction of the mAbs with the neutrophils of donors with different NA genotypes.

Donor genotype	Number of donors	Immunofluorescence results	
		FcR gran 1	Gran 11
NA1NA1	4	++++	++++
NA1NA2	16	++++	++++
NA2NA2	12	++++	−

32 normal donors, which had been typed for the neutrophil-specific antigens NA1 and NA2 with the allo-antisera. The results are shown in Table 9.2. Included are the results of seven members of a three-generation family, with different NA1NA2 genotypes.

From this study it appeared that CLB gran 11 is in fact a murine monoclonal anti-NA1, because it reacted only with the neutrophils of NA1NA1 and NA1NA2 donors. This was also found in the family (Fig. 9.1). The other antibody, CLB FcR gran 1, did not show NA-related specificity, nor did it show specificity for any of the other known neutrophil-specific

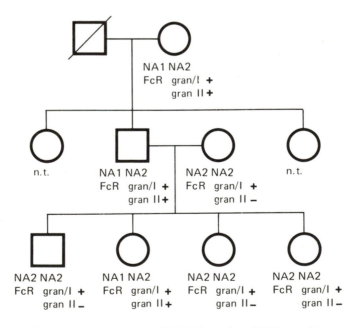

Family investigation with CLB FcR gran/I and CLB gran II

Fig. 9.1. A family study, in which the neutrophils of the different members were tested with anti-NA1 and anti-NA2 allo-antisera and the monoclonal antibodies gran 11 and FcR gran 1.

antigens (NB1, NC1, ND1 or 9[a]) when tested with the neutrophils of donors negative for these antigens (data not shown).

Inhibition Studies with Human Allo-Anti-NA1 and -Anti-NA2

When neutrophils of a NA1NA2 donor were first incubated with the allo-antisera and subsequently, after being washed, with the mAbs, it was found that only anti-NA1 inhibited the binding of gran 11 (anti-NA1 mAb) significantly, although not completely, but not that of FcR gran 1, which confirmed the specificity of gran 11 (Table 9.3).

In the reverse situation, i.e., incubation first with the mAb and subsequently with the allo-antisera, binding of both anti-NA1 and anti-NA2 was significantly inhibited (see also Fig. 9.2). The fact that the inhibition by gran 11 of the binding of anti-NA1 was complete, while the inhibition by anti-NA1 of the binding of gran 11 was not, could be explained by a greater affinity of gran 11.

The partial inhibition of the binding of anti-NA1 by FcR gran 1 and of the binding of anti-NA2 both by gran 11 and FcR gran 1 can be explained by steric hindrance.

EA Inhibition

Both gran 11 and FcR gran 1 strongly inhibited the rosettation of IgG-sensitized red cells to neutrophils, but gran 11 did so only when these neutrophils were NA1 positive (Table 9.4). Allo-anti-NA1 showed the same neutrophil group-related inhibitory activity as gran 11. However allo-anti-NA2 showed only weak and non-neutrophil-group-dependent inhibition, as did normal human (blood group AB) serum.

Presumably the antibodies in this serum are too weak and the weak inhibition which occurred (and as also occurred with normal serum) must

Table 9.3. Blocking studies with mAbs FcR gran 1 and gran 11, alloantibodies anti-NA1 and anti-NA2 utilizing neutrophils of a NA1NA2 donor.

First antibody	Second antibody	FITC anti-globulin[a]	Percentage fluorescence[b]
Anti-NA1	Gran 11	GAM	53
Anti-NA1	FcR gran 1	GAM	88
Anti-NA2	Gran 11	GAM	115
Anti-NA2	FcR gran 1	GAM	108
Gran 11	Anti-NA1	SAH	1
Gran 11	Anti-NA2	SAH	61
FcR gran 1	Anti-NA1	SAH	30
FcR gran 1	Ant-NA2	SAH	41

[a] GAM = goat anti–mouse Ig, SAH = sheep anti–human Ig.
[b] In a 2-step method, fluorescence was measured by cytofluorography and is expressed as percentage of the unblocked control.

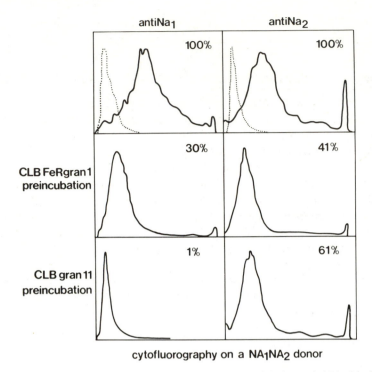

Fig. 9.2. Cytofluorography of the inhibition of allo-anti-NA$_1$ and -NA$_2$ binding to the neutrophils of a NA$_1$NA$_2$ donor by gran 11 and FcR gran 1.

be considered to be aspecific and possibly due to IgG aggregates present in the serum (induced by storage).

Cellular Expression of the Antigens

Various isolated peripheral blood cells were tested in the immunofluorescence test with gran 11 and FcR gran 1, as well as bone marrow cells and macrophages from human malignant ascites, pleural fluid, or lung lavage fluid. Also tested were eosinophils and basophils from patients with eosinophilia and basophilia, respectively, further purified by elutriation (in the department of cell-chemistry of our laboratory).

The bone marrow cells that reacted with the antibodies were further characterized by immunoperoxidase staining. The results are depicted in Table 9.5. Both antibodies reacted only with mature neutrophils, while the neutrophil precursors in the bone marrow were negative. Other granulocytes (eosinophils and basophils) as well as monocytes and macrophages were also unreactive.

Table 9.4. EA-Rosette inhibition studies with the mAbs and the allo-antisera.[a]

	Neutrophil genotype donor		
Inhibiting serum	NA1NA1	NA1NA2	NA2NA2
—	79%	76%	79%
Normal mouse	80%	63%	77%
Gran 11	0%	5%	88%
FcR gran 1	0%	1%	0%
Normal human	45%	34%	42%
Anti-NA1	0%	7%	78%
Anti-NA2	49%	21%	41%

[a] Percentage rosette formation, mean of 3 experiments shown.

FcR gran 1 reacted with a lymphocyte subpopulation in normal blood, varying from 3–36% in different donors (number of donors tested 19, mean 13.9%, standard deviation 8.3%). Presumably these were K lymphocytes, and similar percentages were found with the anti-Fc-γ-receptor mAbs 3G8 (3) and B73.1 (5). However gran 11 did not react with this lymphocyte subset, even when the donors were NA1 positive (in nine cases). This indicates that the NA1 antigen is not present on the K lymphocyte Fc-γ-receptor.

Other cells that were used for the investigations were red cells, platelets, endothelial cells (cultured from cord vein), fibroblasts (cultured from skin), and leukemic cell line cells (HSB, SB, HL-60, KG-1, ML-1, U-937, K562), and malignant cells from patients with leukemia and lymphoma (ALL, AML, CLL, CML, hairy cell leukemia, NHL). None of these cells were positive except the cells from CML patients, in which case only the mature neutrophils (stabs, segments) reacted with the two antibodies.

Table 9.5. Cellular expression of the antigens reactive with FcR gran 1 and gran 11.

Cell type	FcR gran 1 (anti-FcγR)	Gran 11 (anti-NA1)
Neutrophils	+++	+++
Segments	+++	+++
Stabs	+++	+++
Neutrophil precursors	−	−
Metamyelocytes	−	−
Myelocytes	−	−
Promyelocytes	−	−
Eosinophils	−	−
Basophils	−	−
Monocytes	−	−
Macrophages	−	−
Lymphocytes	+++ (3–36%)	−

Effect of Enzyme Treatment

Fc-γ-receptor structures on different cells are known to show a variable digestability by proteolytic enzymes. Therefore we studied the effect of papain and bromelin treatment, and as control that of the desialating enzyme neuraminidase, on the reactivity of neutrophils and K lymphocytes with gran 11 and FcR gran 1 (Table 9.6). The anti-Fc-γ-receptor mAbs 3G8 and B73.1 were also applied in these studies.

The reactivity of neutrophils with the four antibodies was not affected by neuraminidase treatment, but it was completely abolished by the proteolytic enzymes. The same result was obtained with FcR gran 1 and lymphocytes. mAb gran 11 remained negative with lymphocytes also after enzyme treatment. With allo-anti-NA1 and anti-NA2 similar results were obtained (data not shown), but these were less well interpretable, because of a marked background fluorescence after enzyme treatment.

Immunochemical Analysis

When neutrophils of NA1NA2 donors were used, both antibodies were found to bind proteins which formed a broad band in the 40–70 Kd region, known to represent the Fc-γ-receptor (3) (see Fig. 9.3). A similar result was obtained with 3G8 (not shown). However the glycoprotein bound by gran 11 (anti-NA1 mAb) appeared to be less heterogeneous, and formed a more narrow band in the lower part of the region, i.e., 40–60 Kd. This was also found with neutrophils of NA1NA1 donors (not shown). The analysis was repeated with FcR gran 1 and neutrophils of donors with different NA genotypes (Fig. 9.4).

It was then found that the broad band was characteristic for the cells

Table 9.6. The effect of enzyme treatment on the antigens detected by mAbs gran 11 and FcR gran 1.

mAb	Control	Neuraminidase	Papain	Bromelin
		Enzyme treatment		
	Reactivity of neutrophils with mAb[a]			
Gran 11	100% + +	100% + +	0%	0%
FcR gran 1	100% + + +	100% + + +	0%	0%
3G8	100% + + +	100% + + +	0%	0%
B73.1	100% (+)	100% (+)	0%	0%
	Reactivity of lymphocytes with mAb[a]			
Gran 11	0%	0%	0%	0%
FcR gran 1	8% + + +	10% + + +	0%	0%
3G8	10% + + +	14% + + +	0%	0%
B73.1	8% + + +	13% + + +	0%	0%

[a] Tested in immunofluorescence. The percentage of positively reacting cells are shown, as well as the reaction strength, evaluated by eye.

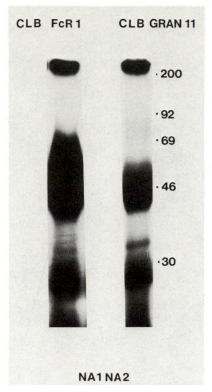

Fig. 9.3. SDS–PAGE of the neutrophil-antigens bound by gran 11 and FcR gran 1, under reducing conditions. Radioiodinated neutrophils from a NA1 NA2 donor.

from heterozygous NA1NA2 donors, while a narrower band was seen with the cells from both homozygous NA1NA1 and homozygous NA2NA2 donors. Moreover the NA1 protein had always a somewhat greater mobility than the NA2 protein. Thus there exists a NA-antigen-related electrophoretic heterogeneity of neutrophil Fc-γ-receptors, which results in a more heterogeneous protein on the cells of heterozygous NA1NA2 donors.

Quantitation of the Antigens on Neutrophils

The amount of antigen detected by FcR gran 1 and gran 11 on neutrophils of donors with different NA genotypes was measured by cytofluorography (Table 9.7). There was no difference in the amount of Fc-γ-receptors on neutrophils of the different types of donors. Moreover neither the amount of NA1 antigen nor the ratio of the amount of NA1 antigen to that of FC-γ-receptors was different between homozygous NA1NA1 donors and heterozygous NA1NA2 donors, i.e., a gene dose effect was not detectable.

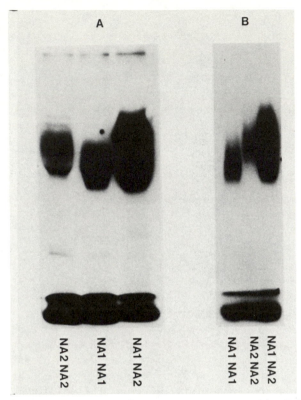

Fig. 9.4. SDS–PAGE of the neutrophil antigens bound by FcR gran 1, under reducing conditions. Radioiodinated neutrophils from NA1 NA1, NA1 NA2, and NA2 NA2 donors.

Discussion

We produced two murine IgG2a mAbs against the neutrophil Fc-γ-receptor. Both antibodies were strongly reactive with neutrophils in immunofluorescence and cytotoxicity tests, but did not agglutinate neutrophils. Neutrophil agglutination by IgG antibodies is an active process, which takes place optimally at 37°C and can be inhibited by the microfilament disrupter cytochalasin B (10). That agglutination is not induced by Fc-γ-receptor antibodies indicates that it is also receptor-mediated, i.e., that autorosettation takes place.

In fact, neutrophil agglutination by various neutrophil allo-antisera could be inhibited by these monoclonal antibodies (unpublished results). One of the antibodies, i.e., CLB FcR gran 1 (M57), had similar properties to monoclonal neutrophil Fc-γ-receptor antibodies prepared by other investigators (3,11,12): code names 3G8, VEP13, and NKP-15 or Leu 11a.

Table 9.7. The amount of Fc-γ-receptor and NA1 antigen on neutrophils of donors with different NA genotypes.[a]

		Donor genotype		
mAb	Specificity	NA1NA1	NA1NA2	NA2NA2
Gran 11	NA1 antigen	80.3	85.8	0
FcR gran 1	Fc-γ-receptor	80.8	91.5	96.2
Ratio	NA1/Fcγ	0.99	0.94	0.00

[a] Determined by cytofluorography with FITC goat anti–mouse Ig, monoclonal antibody diluted 1/3000. Values are mean fluorescence for two donors of each genotype tested. Fluorescence expressed as μG (1 μG = 10.000 FITC molecules).

It was reactive only with neutrophils and a subpopulation of lymphocytes, presumably K lymphocytes. It blocked EA-rosette formation with neutrophils. As for 3G8 (13) the expression of the antigen recognized by the antibody was restricted to mature neutrophils and the antibody bound a heterogeneous neutrophil membrane protein with a M_r of about 40–70 Kd. Basophils and eosinophils did not carry the Fc-γ-receptor recognized by this antibody. With regard to eosinophils this is in accordance with recent findings of Kulczycki (14). In contrast to Fleit *et al.* (3) we did not find neutrophil Fc-γ-receptor expression on the macrophage samples which we tested. The second antibody, CLB gran 11 (M82), appeared also to be directed against the same neutrophil Fc-γ-receptor. It had nearly the same characteristics as FcR gran 1, i.e., blocking of EA-rosette formation, reaction with mature neutrophils, and binding of a neutrophil membrane protein in the same M_r region. But it differed from FcR gran 1 in that it reacted only with the neutrophils of about half of the donors tested, and not with K lymphocytes. Moreover it appeared to show neutrophil-group specificity, and was in fact a murine monoclonal anti-NA1.

This implies that NA1 (and thus also NA2) is an alloantigen carried by the Fc-γ-receptor of neutrophils. However, it could also mean that neutrophils have two different receptors with comparable molecular weights, one shared with K lymphocytes not carrying the alloantigens, the other unique for this cell type and showing NA-system-determined polymorphism.

Our experimental results all pointed to the correctness of the first assumption and notably:

1. Not only gran 11, but also FcR gran 1 inhibited the binding of allo-anti-NA1 and allo-anti-NA2 to neutrophils. This indicated steric hindrance by our antibody bound to the same molecule, but not to the same epitope. Co-modulation could not have played a role, because paraformaldehyde-fixed neutrophils were used.
2. The antigens reactive with gran 11 and FcR gran 1 on neutrophils and with FcR gran 1 on K lymphocytes were both destroyed by proteases, not by neuraminidase.

3. The membrane protein bound by gran 11 from solubilisates of radioiodonated neutrophils of NA1NA1 and NA1NA2 donors had the same mobility as that bound by FcR gran 1 from NA1NA1 donors. The latter antibody bound a membrane protein with a somewhat lower mobility from neutrophils of the NA2NA2 genotype, while probably both proteins were bound in the case of the heterozygous NA1NA2 genotype, resulting in a much broader band.

Altogether it seems likely that there exists a genetically determined heterogeneity of the neutrophil Fc-γ-receptor, reflected in different NA antigen expression and in different electrophoretic mobility. There appear to be at least two different neutrophil Fc-γ-receptor proteins, i.e., Fcγ-F (fast) which is the NA1 protein and Fcγ-S (slow) which is the NA2 protein. These proteins are dependent on two different genes.

By measuring the binding of both mAbs to neutrophils of donors of different NA genotypes we tested whether this is accompanied by the existence of a gene dose effect. With the method used, cytofluorography, such an effect was not found, which could mean that the NA1 and NA2 genes are not responsible for the direct synthesis of the neutrophil Fc-γ-receptor, but that these genes induce a secondary, post-translational change in the receptor. The fact that the NA1 antigen is not expressed on K lymphocytes indicates that their Fc-γ-receptor is not identical to that of neutrophils, as has been postulated, but is a different, although closely related (antigenically and structurally) membrane protein.

Summary

Two new monoclonal antibodies against the Fc-γ-receptor of neutrophils were produced. One had similar characteristics to neutrophil Fc-γ-receptor antibodies produced by others (3G8, VEP13, Leu 11a); the other one was unique. It appeared to be an antibody with specificity for the neutrophil antigen NA1. Experiments performed with both antibodies indicated that the antigens of the neutrophil-specific NA system are located on the Fc-γ-receptor of neutrophils, but not of K lymphocytes. Moreover NA-antigen-related, genetically determined, heterogeneity of the receptor was found to exist. The absence of NA1 antigens on K lymphocytes indicated that their Fc-γ-receptor is different from that of neutrophils.

Acknowledgments. We thank Dr. H.B. Fleit and Dr. B. Perussia for their monoclonal antibodies.

References

1. Lalezari, P. 1984. Granulocyte antigen systems. In: *Immunohaematology*, C.P. Engelfriet, J.J. van Loghem, and A.E.G. Kr. von dem Borne, eds. Elsevier Science Publishers, Amsterdam, pp. 33–44.

2. Lalezari, P., and E. Radel. 1974. Neutrophil-specific antigens: Immunology and clinical significance. *Semin. Haemat.* **11**:281.
3. Fleit, H.B., S.D. Wright, and J.C. Unkeless. 1982. Human neutrophil Fcγ receptor distribution and structure. *Proc. Natl. Acad. Sci. U.S.A.* **79**:3275.
4. Lansdorp, P.M., G.C.B. Astaldi, F. Oosterhof, M. Janssen, and W.P. Zeijlemaker. 1980. Immunoperoxidase procedure to detect monoclonal antibodies against cell surface antigens. *J. Immunol. Methods* **39**:393.
5. Perussia, B., S. Starr, S. Abraham, V. Fannign, and G. Trichieri. 1983. Human natural killer cells analysed by B73.1, a monoclonal antibody blocking Fc receptor functions I. Characterization of the lymphocyte subset reactive with B73.1. *J. Immunol.* **130**:2133.
6. Verheugt, F.W., A.E.G. Kr. von dem Borne, F. Décary, and C.P. Engelfriet. 1977. The detection of granulocyte allo-antibodies with an indirect immunofluorescence test. *Brit. J. Haematol.* **36**:533.
7. Blaschke, J., C.D. Severson, N.E. Goeken, and J.S. Thompson. 1977. Microgranulocytotoxicity. *J. Lab. Clin. Med.* **90**:249.
8. van der Reijden, J.G., D.J. van Rhenen, P.M. Lansdorp, M.B. van 't Veer, M.M.A.C. Langenhuysen, C.P. Engelfriet, and A.E.G. Kr. von dem Borne. 1983. A comparison of surface markers and FAB classification in acute myeloid leukemia. *Blood* **61**:443.
9. Lansdorp, P.M., T.H. van der Kwast, M. de Boer, and W.P. Zeijlemaker. 1984. Stepwise amplified immunoperoxidase (PAP) staining. I Cellular morphology in relation to membrane markers. *J. Histochem. Cytochem.* **32**:172.
10. Verheugt, F.W.A., A.E.G. Kr. von dem Borne, J.C. van Noord-Bokhorst, E.H. van Elven, and C.P. Engelfriet. 1978. Serological, immunochemical and immunocytological properties of granulocyte antibodies. *Vox Sang.* **35**:294.
11. Rumpold, H., D. Kraft, G. Obexer, G. Bock, and W. Gebhart. 1982. A monoclonal antibody against a surface antigen shared by human large granular lymphocytes and granulocytes. *J. Immunol.* **129**:1458.
12. Philips, J.H., and G.F. Babcock. 1983. NKP-15: A monoclonal antibody reactive against purified human natural killer cells and granulocytes. *Immunol. Lett.* **6**:143.
13. Fleit, H.B., S.D. Wright, C.J. Durie, J.E. Valinsky, and J.C. Unkeless. 1984. Ontogeny of Fc receptors and complement receptor (CR3) during myeloid differentiation. *J. Clin. Invest.* **73**:516.
14. Kulczycki, A. 1984. Human neutrophils and eosinophils have structurally distinct Fcγ receptors. *J. Immunol.* **133**:849.

CHAPTER 10

Murine Monoclonal Anti-P

A.E.G. Kr. von dem Borne, M.J.E. Bos, N. Joustra-Maas,
M.B. van 't Veer, J.F. Tromp, and P.A.T. Tetteroo

Introduction

Monoclonal antibodies (mAbs) reactive with human red cells may show
specificity for the antigens of various human blood group systems, such as
the blood group ABH, Le, MN system (1). Because of their strength,
purity, and uniform quality they are excellent tools for serological studies
as well as for more basic hematological and biochemical studies on blood
group antigens.

A mAb with specificity for an antigen of the blood group P system, i.e.,
P^k, also known as globotriaosylceramide or Gb-3, was recently described
(2,3). Its usefulness for immunocytological investigations and for the
study of malignant cells was also amply demonstrated.

In this report we describe the preparation and characterization of a
mAb against another antigen of the blood group P system, i.e., P, also
known as globoside or Gb-4. Its applicability for immunocytological investigations and the typing of human malignant blood cells was studied as
well.

Materials and Methods

The Monoclonal Antibody

The mAb (CLB ery 2, Workshop code M23) was a by-product of a (suc-
cessful) experiment to prepare anti-CALLA mAbs. It was detected because it reacted only with erythrocytes during screening of the supernatants from the primary hybrid-cultures. It appeared to be an IgM
antibody. Ascites was prepared in pristane-primed BALB/c mice and
used for further experiments.

Erythrocyte Serology

The specificity of the antibody was tested on a panel of red cells from donors with the following blood groups: O, A, B, h, I, i, S-s̄-, P1, P2, P1^k, p̄, k̄, Jk(a-b-), Lu(b-), Lu(a-b-), Fy(a-b-), Rh null, En(a-), Co(a-), JHM-, Lan-, Gerb-, Yt(a-), Kn(a-).

The reaction of the antibody with red cells was tested by standard serological methods, i.e., saline agglutination and hemolysis of normal and bromelinized cells, with and without added human complement (normal AB serum), and the anti-globulin test with goat anti–mouse Ig serum (GM 17-01-F02 of our laboratory) absorbed with human blood group ABO red cells.

Immunocytology

The reaction of the mAb with different isolated human peripheral blood cells, nonseparated bone marrow and tonsil cells, leukemic cell line cells, and Ficoll–isopaque-enriched cells from patients with various hemato-oncological diseases (derived from blood, bone marrow, and/or lymph-node tissue) was tested as described (4).

The cells were always fixed with 1% paraformaldehyde (PFA). Indirect immunofluorescence was performed on cells in suspension or spun onto slides, with FITC goat anti–mouse Ig (GM 17-01-F02) as second antibody. Fluorescence was quantified, if necessary, by cytofluorography (FC-200, Ortho instruments). In some experiments the cellular expression of the antigen was visualized by a cytological immunoperoxidase method (5).

Bromelin and Neuraminidase Treatment of Cells

The cells were first treated for ½ hr at 37°C with 0.5% bromelin (Difco) or neuraminidase (cholerafiltrate, Behringwerke, 1U/ml) and then fixed with PFA.

Results

Red Cell Serology

mAb CLB ery 2 (M23) reacted with all red cells of the typing panel except those of donors with the blood group P1^k and p.

The tests with the red cells of 12 donors with different blood group P phenotypes are shown in Table 10.1. The results strongly indicated that the antibody is anti-P or anti-globoside (gb-4). Final evidence for this specificity was provided by Dr. D.M. Marcus (Baylor College of Medicine, Houston) (personal communication), who found in three different

Table 10.1. Antigenic specificity of CLB ery 2 (M23). I.

Blood group P phenotype	Reaction[a]	No. donors tested
P1	++++	3
P2	++++	2
P1[k]	−	2
p	−	5

[a] Bromelin agglutination test at 37°C.

Table 10.2. Antigenic specificity of CLB ery 2 (M23). II. Reaction with glycolipids.[a]

Trivial name	Abbreviation	Antigen	Titer
Globotriaosylceramide	Gb-3	P[k]	0
Globoside	Gb-4	P	20.000
Forsman	Gb-5	Forsman	0
Paragloboside	Nlc-4	—	0

[a] Results provided by D.M. Marcus, from an ELISA assay; antibody prediluted 1/100.

Table 10.3. Serological behavior of CLB ery 2 (M23).

Test	Reaction[a]
Saline agglutination	0
Bromelin agglutination	128.000
Indirect anti-globulin	1.000
Saline hemolysis	4.000
Bromelin hemolysis	64.000

[a] At 37°C, prediluted 1/100, expressed as titer.

assays on purified glycolipids, i.e., liposome lysis, immunostaining on a thin layer chromatogram, and an ELISA assay, that the antibody was globoside specific. The results of the ELISA assay are shown in Table 10.2. The serological properties of the antibody are depicted in Table 10.3. At 37°C (but also at lower temperature) it was mainly a nonagglutinating antibody, only detectable in the hemolysis and the anti-globulin test. However it became a strong agglutinin when the red cells were bromelinized or treated with neuraminidase (data not shown). Also its hemolytic activity was then markedly enhanced.

Reaction with Peripheral Blood, Bone Marrow, Lymph Node, and Tonsil Cells

Of other peripheral blood cells only platelets and a very small percentage (1–3%) of lymphocytes showed a positive reaction with CLB ery 2 (Table 10.4). Granulocytes and monocytes were nonreactive.

Table 10.4. Reaction of mAb CLB ery 2 (M23) with peripheral blood, bone marrow, lymph node, and tonsil cells in immunofluorescence.

	Percentage positive	Reaction strength[a]	Number of samples tested
Erythrocytes	100	++	10
Platelets	5–50	+	12
Granulocytes	0	−	10
Monocytes	0	−	10
Lymphocytes	1–3	++	5
T lymphocytes	1–3	++	3
Non-T lympho- cytes	1–3	++	3
Bone marrow cells	2–35	++	10
Lymph node cells	3–7	+	2
Tonsil cells	26–45	++	4

[a] Reaction strength evaluated by eye.

The strength of the reaction with platelets was much lower than that with red cells or with the few reactive lymphocytes. Moreover the proportion of platelets which reacted varied from donor to donor, but was relatively constant in the same donor.

Neither the reaction strength nor the percentage of positively reacting platelets was influenced by enzyme treatment. The small percentage of positively reacting lymphocytes was observed both in T cell as well as in non-T cell fractions (obtained by sheep erythrocyte rosette sedimentation). This was also not altered by enzyme treatment.

Bone marrow cells, from different donors, showed (apart from positively reacting erythrocytes) a variable percentage (2–35%) of positively reacting nucleated cells. In immune peroxidase stains these appeared to be erythroblasts, whereas proerythroblasts were negative. Megakaryocytes sometimes showed a weakly positive reaction.

Nonseparated lymph node cells contained a low percentage of positive cells (Table 10.4), but tonsil cells showed quite a high number (26–45%) of strongly positive cells in immunofluorescence. Immunoperoxidase staining demonstrated that these cells were mainly intermediate to large lymphocytic cells. A more extensive analysis including various other antibodies is shown in Table 10.5 More than half of the cells were definitely of B-lymphocytic nature, and only one third T-lymphocytoid. Moreover no contaminating red cells or platelets were present. However, the notion that the P-positivity of many tonsil cells is related to the high content of B cells, and thus that the P-antigen is a marker of a B lymphocyte subset, is not proven. Enrichment or depletion of the tonsil cells for either T or B cells by different methods (E- and EA-rosette sedimentation, panning with OKT3) did not significantly affect the percentage of P-positive cells. Treatment of the tonsil cells with anti-P mAb and rabbit complement did

Table 10.5. Results of the immunocytological analysis of three samples of tonsil suspension.

Antibody[a]	Percent positively reacting cells		
	1	2	3
Anti-K	37	31	28
Anti-λ	22	28	20
Anti-HLA-DR	81	67	53
Anti-T3	37	27	28
Anti–B lymph	67	63	61
Anti-P	45	43	26
Anti-glycophorin	0	0	0
Anti–platelet glycoprotein III[a]	0	0	0

[a] The anti-K and anti-λ were polyclonal antibodies and directly labeled with FITC; anti-HLA-DR (code CLB HLA II/1), anti-glycophorin (code CLB ery 1), and anti-glycoprotein III[a] (CLB gp III[a]/2 or C17) were monoclonal antibodies from our own laboratory; anti-T3 (OKT3) and anti–B lymph (Y29-55) were obtained from elsewhere.

not either significantly affect their composition. Double staining of the cells with a rhodamine-anti-K and anti-λ mixture and anti-P mAb, followed by FITC anti-Ig, showed that many P-positive cells did not express surface immunoglobulins.

When anti-P mAb was tested in the immunoperoxidase test on cryostate sections of lymphatic tissues (thymus, lymph node, and spleen) reactivity for P-antigen was found to be restricted to the follicular reaction centers (S. Poppema, personal communication).

Reaction with Various Human Leukemic and Lymphomatous Cell Line Cells

The mAb was tested with the T cell lines HSB, CEM, and PEER; the B cell lines SB, RAJI, and Thiel (a plasmoblastic cell line from our laboratory); the myeloid lines KG-1, ML-1, HL-60, and U 937; and the proerythroblastic line K562 and HEL. Only the plasmoblastic line and both proerythroblastic lines showed some positively reacting cells (1–3%). The results with the latter two cell lines are shown in Table 10.6, in which the results with some other red cell-reactive monoclonal antibodies are shown as well. Anti-P stained only few cells, while both cell types reacted at a much higher percentage with mAbs anti-glycophorin-A and anti-N (a glycophorin-A/B antigen).

Anti-P positivity was not enhanced by neuraminidase treatment, while reactivity with anti-glycophorin-A and anti-N was lost upon such treatment.

Table 10.6. Reaction of the proerythroblastic cell lines K562 and HEL with mAb anti-P and various other red cell-reactive mAbs.

mAbs[a]	K562[b]		HEL[b]	
	Untreated	N-ase treated	Untreated	N-ase treated
Anti-P	1% +	1% + +	1% +	3% +
Anti-P^K	–	–	–	–
Anti-H	2% + +	2% + +	100% +/+ +	100% + +/+ + +
Anti-A	–	–	–	–
Anti-B	–	–	–	–
Anti-glycophorin	18% +	–	34% +	–
Anti-N	24% + +	–	52% +/+ +	–

[a] Anti-P^K was from Wiels *et al.* (2); anti H (7C11 or M79), anti-glycophorin (CLB ery 1), and anti-N (D5 or M21 from 1982) were from our own laboratory; anti-A and anti-B were commercially available.
[b] Results with and without neuraminidase (N-ase) treatment of the cells.

Reaction with the Cells of Patients with Malignant Blood Diseases

These results are shown in Table 10.7. In patients with malignant lympho-proliferative diseases positive reactions were rarely encountered. An exception was non-Hodgkin's lymphoma: 5 out of the 17 patients (30%) had a significant number of positive cells. All were of the B or the null cell type. Only one case of B-ALL could be tested, which was positive. In acute myeloid leukemia a few cases (4 out of 37, i.e., 11%) were positive (unrelated to the FAB class), in chronic myeloid leukemia (CML) none. However, in CML blast crisis P-positivity of the leukemic cells was quite often encountered (6 out of 18, i.e., 33%).

Table 10.7. Reaction of anti-P mAb with the cells from patients with various malignant blood diseases.

Diagnosis[a]	Number tested	Number positive[b]	When positive, % positive cells
T-ALL	12	0	—
Common-ALL	17	1	24
B-ALL	1	1	56
AUL	4	0	—
CLL	22	0	—
Hairy cell L	14	0	—
B-prolymphocytic L	5	1	29
NHL	17	5	31, 42, 43, 70, 84
Multiple myeloma, macroglobulinemia	3	0	—
AML (M1 M2, M3, M4, M5)	37	4	28, 30, 35, 55
CML	15	0	—
CML blast crisis	18	6	31, 32, 41, 46, 62, 74

[a] Abbreviations: ALL = acute lymphatic leukemia, AUL = acute undefined leukemia, CLL = chronic lymphatic leukemia, NHL = non-Hodgkin's lymphoma, AML = acute myeloid leukemia, M1–M5 = FAB classification of AML, CML = chronic myeloid leukemia.
[b] 20% or more positive cells.

Table 10.8. Reaction of anti-P and other mAbs with the cells of patients with CML in metamorphosis (Blast Crisis).[a]

Patient no.	Anti-P	Anti-glycophorin-A	Anti-gp-III[a]	Anti-HLA-DR	Anti-FAL	Anti-CALLA
1	<u>31</u>	9	12	9	<u>26</u>	4
2	<u>32</u>	0	0	19	6	0
3	<u>41</u>	2	<u>33</u>	<u>78</u>	3	0
4	<u>62</u>	1	<u>65</u>	0	0	2
5	<u>74</u>	0	13	<u>53</u>	0	0
6	<u>46</u>	0	<u>26</u>	1	10	0
7-18	0-18	0-5	0-13	1-72	0-74	0-42

[a] Ficoll–Isopaque-enriched immature cells; percentage positive cells indicated, clearly positive results (>20%) underlined. Anti-glycophorin-A = CLB ery 1, anti-gp III[a] = CLB gp III[a]/2, anti-HLA-DR (CLB HLA II/1), anti-FAL (anti-granulocyte-fucosyl-N-acetyllactosyl-antigen) = CLB gran 2, anti-CALLA (common ALL antigen) = CLB-CALLA/1 were all from our laboratory.

A more detailed analysis, including other monoclonal antibodies, is shown in Table 10.8. None of the six P-positive cases had cells reacting with anti-glycophorin-A mAb, but three also had a significant proportion of cells expressing the thrombocytic marker glycoprotein III[a] (gp III[a]), and two had a significant proportion of HLA-DR-positive cells. In the immunoperoxidase method the P-positive cells appeared to be both erythroblasts and more immature blasts, while the gp III[a]-positive cells were all of blastic nature. This indicates that the P-antigen is a marker for erythroblastic or mixed erythroblastic–megakaryoblastic crises in this disease.

Discussion

We have produced a murine monoclonal IgM antibody which is specific for blood group P-antigen or globoside. Its reaction with red cells was markedly enhanced by treatment of the cells with bromelin or neuraminidase. This enhancing effect of enzyme treatment has also been shown with heterologous anti-globoside antisera (6). It has been attributed to a better availability of globoside on the red cell membranes by the induction of clustering of the glycophorin molecules, by cleavage of negatively charged groups from this membrane protein (7). Globoside was detected not only on mature erythrocytes but also on erythroblasts. However, proerythroblasts did not seem to express the P-antigen yet. The cells of the human proerythroblastic cell lines K562 and HEL were not either found to express significant amounts of globoside, a finding which is in accordance with previous results of Suzuki et al. (8) and Kannagi et al. (9). Thus globoside is a maturation marker of the red cell series. Of the other peripheral blood cells only platelets and some lymphocytes (1–3%) were found to react with the antibody, in contrast to granulocytes and monocytes. These results are in accordance with previous chemical studies (10–12).

The expression of globoside on platelets was much weaker than that on red cells, was not enhanced by enzyme treatment, and was restricted to a variable subpopulation (5–50%). The reason for this is not (yet) clear. Although peripheral blood lymphocytes were largely globoside negative, tonsil lymphocytes contained a large subset of strongly positive cells. In fact, a high globoside content of tonsil lymphocytes has been noticed before by chemical analysis (12).

Immunohistochemical studies on lymphatic tissues showed that the P-positive cells were present in the follicular centers, which mainly contain activated B lymphocytes. Indeed tonsils are rich in (activated) B lymphocytes, but in our cell separation studies it appeared that both tonsillar B and T cells expressed globoside. In this context it is therefore of interest that recently Gruner *et al.* (13) showed that alloantigen-activated mouse spleen T cells also express globoside. Thus globoside expression possibly occurs during the activation of both T and B cells in the follicular center reaction.

Anti-P appeared to be an interesting marker for the typing of some malignant blood diseases. Especially in non-Hodgkin's lymphoma (NHL) and in CML blast crisis many patients with a significant percentage of P-positive cells were encountered. The meaning of P-antigen positivity in NHL is not yet clear, and studies are under way in our laboratory to see whether it is linked to aggressive disease. In CML-BC it appeared to be associated with a erythroblastic or mixed erythroblastic/megakaryoblastic crisis, a type of crisis which seems to be quite frequent (14). Cellular P-positivity also occurred sometimes in acute myeloid leukemia (AML). Globoside expression in AML has been noticed previously (15,16) but its implication is not yet clear and it does not seem to be related to a certain FAB classification of the leukemia.

Summary

A murine monoclonal IgM antibody reactive with red cells appeared to show anti-P or anti-globoside specificity. Its reactivity with red cells was markedly enhanced by enzyme treatment of these cells. With this mAb it was shown that globoside is present on erythroblasts, but not, or only marginally, on proerythroblasts or human proerythroblastic cell line cells. Platelets were found to express globoside only weakly and partially. Peripheral blood lymphocytes were mostly P-negative, but tonsil lymphocytes contained many strongly P-positive cells. These seemed to be confined to the reaction centers of lymphocyte tissues in general and are probably of both T and B cell lineage. In human malignant blood diseases

cellular globoside expression was found to occur rarely in lymphoproliferative diseases except in non-Hodgkin's lymphoma. Globoside expression was detected in some cases of acute myeloid leukemia, but this was
not related to a particular classification of the leukemia. Quite often a
significant amount of P-positive cells was detected in CML blast crisis.
This then appeared to indicate an erythroblastic or mixed erythroblastic/
megakaryoblastic crisis.

Acknowledgment. This work was supported by grants from the Netherlands Cancer Society, the "Queen Wilhelmina Fund."

References

1. Voak, D., and D. Tills. 1983. Monoclonal antibodies in immunohaematology.
   Biotest Bulletin, Vol. 1, no. 4.
2. Wiels, J., M. Fellous, and T. Tursz, 1981. Monoclonal antibody against a
   Burkitt lymphoma-associated antigen. *Proc. Natl. Acad. Sci. U.S.A.* **78**:6485.
3. Nudelman, E., R. Kannagi, S. Hakomori, M. Parsons, M. Lipinski, J. Wiels,
   M. Fellous, and T. Tursz. 1983. A glycolipid antigen associated with Burkitt
   lymphoma defined by a monoclonal antibody. *Science* **220**:509.
4. van der Reijden, J.H., D.J. van Rhenen, P.M. Lansdorp, M.B. van 't Veer
   M.M.A.C. Langenhuysen, C.P. Engelfriet, and A.E.G. Kr. von dem Borne.
   1983. A comparison of surface marker analysis and FAB classification in
   acute myeloid leukaemia. *Blood* **61**:443.
5. Lansdorp, P.M., T.H. van der Kwast, M. de Boer, and W.P. Zeijlemaker,
   1984. Stepwise amplified immunoperoxidase (PAP) staining I. Cellular morphology in relation to membrane markers. *J. Histochem. Cytochem.* **32**:172.
6. Hakomori, S. 1969. Differential reactivity of fetal and adult human erythrocytes to antisera directed against glycolipids of human erythrocytes. *Vox
   Sang.* **16**:478.
7. Tillack, T.W., M. Allietta, R.E. Moran, and W.W. Young, 1983. Localization
   of globoside and Forsman glycolipids on erythrocyte membranes. *Biochem.
   Biophys. Acta* **733**:15.
8. Suzuki, A., R.A. Karol, S.K. Kundu, and D.M. Marcus. 1981. Glycosphingolipids of K562 cells: a chemical and immunological analysis. *Int. J. Cancer*
   28:271.
9. Kannagi, R., T. Papayannopoulou, B. Nakamoto, N.A. Cochran, T. Yokochi,
   G. Stamatoyannopoulos, and S. Hakomori. 1983. Carbohydrate antigen profiles of human erythroleukaemia cell lines HEL and K562. *Blood* **62**:1230.
10. Tao, R.V.P., C.C. Sweeley, and G.A. Jamieson. 1973. Sphingolipid composi-
    tion of human platelets. *J. Lipid Res.* **4**:16.
11. Macher, B.A., and J.C. Klock. 1980. Isolation and chemical characterization
    of neutral glycosphingolipids of human neutrophils. *J. Biol. Chem.* **255**:2092.
12. Stein, K.E., and D.M. Marcus. 1977. Glycosphingolipids of purified human
    lymphocytes. *Biochemistry* **16**:5285.
13. Gruner, K.R., R.V.W. van Eijk, and P.F. Mühlradt. 1981. Structure elucida-
    tion of marker glycolipids of alloantigen-activated murine T-lymphocytes.
    Biochemistry **20**:4518.

14. Ekblöm, M., G. Borgström, E. von Willebrand, C.G. Gahmberg, P. Vuopio, and L.C. Andersson. 1983. Erythroid blast crisis in chronic myelogenous leukaemia. *Blood* **62**:591.
15. Lee, W.M.F., M.A. Westrick, and B.A. Macher. 1982. Neutral glycosphingo- lipids of human acute leukaemias. *J. Biol. Chem.* **257**:10090.
16. Lee, W.M.F., M.A. Westrick, J.F. Klock, and B.A. Macher. 1982. Isolation and characterization of glycosphingolipids from human leukocytes. *Biochem. Biophys. Acta* **711**:166.

CHAPTER 11

NK-Associated and LFA-1 Antigens: Phenotypic and Functional Studies Utilizing Human NK Clones

Reinhold E. Schmidt, Gail Bartley, Thierry Hercend, Stuart F. Schlossman, and Jerome Ritz

Introduction

In recent years there have been numerous studies investigating the potential involvement of NK cells in a series of immunologic functions such as destruction of virally transformed cells and tumor cells (see Ref. 1 for review), control of hematopoietic differentiation (2), and regulation of immunoglobulin production by B lymphocytes (3). Although the precise role of NK active lymphocytes in the majority of these experiments remains to be clarified, there is now increasing evidence that NK activity represents a biologically important phenomenon. However, according to the operational definition of NK activity, i.e., the capacity to kill tumor cell lines *in vitro* without preimmunization, it has become apparent in both human and murine systems that NK cells are heterogeneous with respect to phenotypic characteristics (4–6). This heterogeneity may explain why monoclonal antibodies defining NK cell-specific antigen on one hand or antibodies delineating recognition structures involved in NK cytotoxic reactions on the other have not been readily identified. To help resolve these problems and perform studies with homogeneous populations of NK active lymphocytes, we have recently developed a series of human cloned NK cell lines (7,8). Analysis of human NK clones indicated that individual cloned cell lines had distinct phenotypes (8) and target specificity, and appeared to reflect the diversity of NK active cells in normal peripheral blood.

Since the heterogeneity and diversity of the small population of NK active cells in peripheral blood creates considerable difficulty in characterizing monoclonal antibodies defining NK cell antigens, we used these cloned human NK cell lines to test the NK-associated monoclonal antibodies of the Second International Workshop and Conference on Human Leukocyte Differentiation Antigens in order to identify antibodies with selective reactivity for NK effector cells. In addition, human NK clones have recently made possible the delineation of effector and target cell recognition structures involved in the NK cytotoxic process. For T cell-like NK clones, the T3 and Ti antigens are critical structures in the recognition function of these cytotoxic cells (9,10). TNK$_{TAR}$ represents an activation antigen involved at the target cell level of this process for the same cells (11). Therefore, using NK clones we were also able to test a large number of monoclonal antibodies for their ability to identify additional structures involved in the NK cell-mediated lytic mechanism. These studies identified anti-LFA-1 antibodies as being the only antibodies tested that were able to consistently block NK clone function. Because the antibodies defining the LFA-1 antigens revealed these inhibitory effects, we selected the LFA-1 antibodies for more detailed studies of NK function.

Materials and Methods

Isolation of Peripheral Blood Mononuclear Cells

Human peripheral blood mononuclear cells (PBMC) were isolated from healthy volunteer donors by Ficoll–Hypaque (F/H) density gradient centrifugation.

Monoclonal Antibodies

Monoclonal antibodies used in these studies have been submitted to the Second International Workshop and Conference on Human Leukocyte Differentiation Antigens. For designation of the various antibodies the Workshop code numbers were used. All antibodies from the myeloid section were tested for blocking activity with respect to cytotoxicity of various NK clones. For further phenotypic and functional studies only NK-associated and LFA-1 antibodies were selected.

Generation of Human Cloned Cell Lines

Methods for generation of human NK cloned cell lines have been described in detail elsewhere (7). Briefly, clones were obtained using a limiting dilution technique. Either PBMC or LGL were cloned at one cell per well on a feeder layer of autologous, irradiated (5000 rad) PBMC plus either PHA (2 μg/ml) or autologous, irradiated (5000 rad) EBV-transformed B cells. Selected colonies were expanded by the addition of culture medium containing lymphocyte conditioned medium (LCM) (10–15% final dilution) every 3 days. Culture medium was RPMI 1640 supplemented with 1% penicillin–streptomycin, 1% sodium pyruvate, and 20% human AB serum. All cell lines used in these studies have been subcloned at least four times at 100 cells per well on a feeder layer of autologous, irradiated PBMC plus irradiated EBV-transformed B cells. After subcloning procedures, both phenotypic and cytotoxic function have remained stable.

JT3, JT9 and JT10 clones have been previously described in detail. CNK1 and CNK2 were derived from an additional donor's peripheral blood lymphocytes stimulated by allogeneic cells. CNK3 was obtained from a bone marrow transplant patient's peripheral blood lymphocytes about 3 weeks after transplant and were initially stimulated with PHA. All three CNK clones were selected for their capacity to kill K562 cells. CNK1 cells have the following phenotype: $T3^+T4^-T8^-T11^+NKH1^+$ $NKH2^-$. CNK2 cells have the phenotype: $T3^-T4^-T8^-T11^+NKH1^+$ $NKH2^-$. CNK3 cells have the phenotype: $T3^+T4^-T8^+T11^+NKH1^+$ $NKH2^+$.

Cell Lines

Several continuously growing cell lines were used in these studies. Nalm 1 and Laz 221 are common acute lymphoblastic leukemia antigen (CALLA)-positive acute lymphocytic leukemia cell lines. MOLT-4, CEM, REX, and HSB are T cell leukemia cell lines. K562 was established from a patient with chronic myelogenous leukemia and HL60 from a patient with acute promyelocytic leukemia. KG-1 is a myeloid cell line. U937 is a histiocytic cell line.

Cytotoxicity Assays

All experiments were performed in triplicate using V-bottomed microtiter plates. Medium for cytotoxicity assays was RPMI 1640 plus 5% human AB pooled serum and 1% penicillin–streptomycin. Assays were performed at various E/T ratios using 5000 ^{51}Cr-labeled target cells per well. Cytotoxicity was measured following 4 hr of incubation at 37°C. Specific cytotoxicity was calculated according to a standard method previously described (7).

Phenotypic Analysis of Cell Surface Antigens

Phenotypic analysis was performed by indirect immunofluorescence with fluorescein-conjugated goat anti–mouse Fab IgG (Meloy) as previously described (7). Samples were analyzed on an Epics V or an Epics C flow cytometer (Coulter Electronics, Hialeah, FL). 10,000 cells were analyzed in each sample. Each histogram displays the number of cells (ordinate) versus the intensity of fluorescence (abscissa) expressed on a logarithmic scale. The negative control used to determine background fluorescence was an ascites derived from a nonreactive hybridoma. Monoclonal antibodies were always used at saturating concentrations (1 : 100–1 : 500).

Results

Distribution of NK-Associated Antigens

All seven monoclonal antibodies with designated specificity for NK-associated antibodies were tested for reactivity with four NK clones (Table 11.1). Immunofluorescence assays indicated that two of the NK-associated antigens, M2 and M86, were expressed on all NK clones regardless of whether they were T3⁻ or T3⁺. NK-associated monoclonal antibody M1 reacted with JT3 and CNK3 only. Antibody M54 was found to be positive on one NK clone only. M54 reacted with JT3. The antibodies M48, M78, and M105 did not show any reactivity with any of the NK clones. None of the NK-associated antibodies revealed any reactivity with any of the target cell lines tested (K562, REX, Nalm-1, MOLT-4, CEM, HSB, HL60, KG-1, U937, Laz 221) (data not shown).

Distribution of LFA-1 Antigens

All six monoclonal antibodies identified as being specific for LFA-1 were also tested for reactivity with four NK clones. As shown in Table 11.2, all monoclonal antibodies designated as LFA-1 except for M26 were strongly reactive with all NK clones. Antibody M26 did not show any reactivity with NK clones.

Table 11.1. Reactivity of NK-associated antibodies on NK clones.[a]

Antibody number	JT3	JT10	CNK1	CNK3
1 NKH2	+	−	−	+
2 NKH1A	+	+	+	+
42 D12	−	−	−	−
54 GO226	+	−	−	−
78 M3011	−	−	−	−
86 N901	+	+	+	+
105 Leu 7	−	−	−	−

[a] For analysis indirect immunofluorescence was used as described in Materials and Methods. An NK clone was considered to be positive with >30% reactivity with a specific NK-associated antibody.

Table 11.2. Reactivity of LFA-1 antibodies on
NK clones.[a]

Antibody number	JT3	JT10	CNK1	CNK3
26 VIM12	−	−	−	−
55 MHM23	+	+	+	+
56 MHM24	+	+	+	+
72 TS1/22	+	+	+	+
73 CLB-54	+	+	+	+
89 TS1/18.11	+	+	+	+

[a] For analysis indirect immunofluorescence was used as described
in Materials and Methods. An NK clone was considered to be
positive with >30% reactivity with a specific anti-LFA-1
antibody.

The expression of LFA-1 antigen on the cell lines varied widely (Fig.
11.1). Whereas some target cells like K562, Nalm-1, and Laz 221 were
completely negative, others reacted weakly (CEM, MOLT-4) and others
had a very strong expression of the LFA-1 antigen (REX, HSB, HL60,
KG-1, and U937).

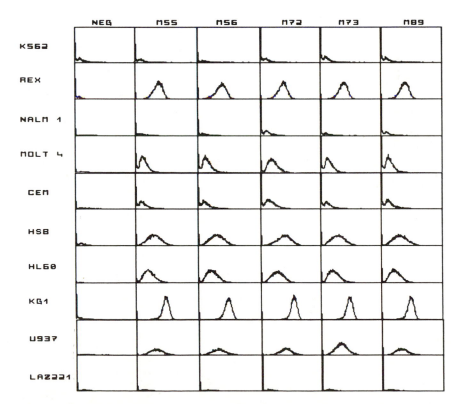

Fig. 11.1. Expression of LFA-1 antigen on different target cell lines. This analysis
was performed on the Epics V flow cytometer using indirect immunofluorescence
as described in Materials and Methods.

Inhibition of NK Cytotoxicity by LFA-1 Antibodies

All 120 myeloid antibodies were screened for their capacity to inhibit cytotoxicity against K562 and MOLT-4 by the NK clones JT3, JT10, and CNK2. All LFA-1 antibodies, again with the exception of M26, had a strong blocking effect on cytotoxicity, especially against MOLT-4 with all three NK clones. The cytotoxicity against K562 was less impaired (Table 11.3). None of the NK-associated antibodies or the other myeloid antibodies had a blocking effect on NK function of JT3, JT10, and CNK2. We also tested all myeloid antibodies for blocking of NK function when peripheral blood NK cells were used as effectors. Blocking effects by LFA-1 antibodies similar to those seen with NK clones could be demonstrated with unstimulated peripheral blood NK activity (Fig. 11.2). Again the inhibitory capacity is much more expressed when tested on MOLT-4 targets than on K562. No blocking activity was seen with any of the other myeloid monoclonal antibodies.

Inhibition of Cytotoxicity by LFA-1 Is Mediated at the Effector Cell Level

Immunofluorescence assays have shown that LFA-1 antigen is expressed on target cell lines as well as on NK clones. It was therefore important to assess whether the blocking effects of these antibodies were mediated at the effector or the target cell level. Experiments were conducted where either effector or target cells were preincubated with antibody excess which was subsequently washed out prior to cytolytic assays. Negative and positive controls were performed either in the absence of antibody or in the presence of excess antibody. As shown in Fig. 11.3, where cytotoxicity of JT10 effector cells was measured against MOLT-4 target cells, treatment of effector cells alone was as efficient in blocking cytotoxicity

Table 11.3. Inhibition of NK activity by LFA-1 antibodies on MOLT-4 and K562.

Antibody number	JT3		JT10		CNK2	
	MOLT-4	K562	MOLT-4	K562	MOLT-4	K562
26[a]	0[b]	0	0	3	0	0
55	51	54	61	15	74	39
56	100	58	72	17	94	55
72	80	56	74	12	78	38
73	100	54	72	8	80	38
89	100	54	76	13	74	52

[a] All experiments were performed using a saturating concentration of antibody (1:200 final dilution). See Table 11.3 for designation of LFA-1 antibodies.
[b] Cytotoxicity assays were performed as follows: effector cells were plated, and either medium or a saturating concentration of antibody was then added, incubated for 30 min at 20°C, and target cells then plated. The E:T ratio was 5:1. Percent inhibition expressed as means of triplicates; SD ≤5%.

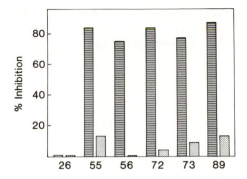

Fig. 11.2. Inhibition of cytotoxicity by different LFA-1 antibodies on peripheral blood NK cells. The effector cells were preincubated with the different LFA-1 antibodies (M26, 55, 56, 72, 73, 89) for 30 min at room temperature. Then MOLT-4 and K562 cells were added as targets. 5000 ^{51}Cr-labeled target cells were used per well. Specific cytotoxicity was measured 4 hr after addition of target cells as described in Materials and Methods. The percent inhibition compared to cells incubated in media alone was calculated for an E/T ratio of 60:1.

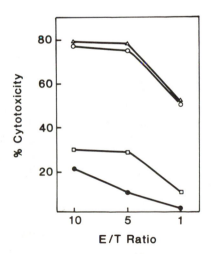

Fig. 11.3. Inhibition of cytotoxicity is secondary to the interaction between anti-LFA-1 antibody and the LFA-1 antigen on the effector cell. Cytotoxicity assays were performed using MOLT-4 target cells, JT10 effector cells, and M55 anti-LFA-1 antibody. Each point represents mean of triplicate cultures, SD ≤5%. ○, Cytotoxicity mediated by JT10 cells in the absence of antibody; ●, anti-LFA-1 antibody excess present during the entire cytotoxicity assay; □, JT10 cells preincubated with antibody for 30 min at room temperature, followed by removal of antibody excess prior to the assay; △, 51Cr-labeled MOLT-4 target cells (5000 cells/well), preincubated with antibody for 30 min at room temperature, followed by removal of antibody excess prior to the assay.

as was the presence of antibody excess during the entire NK assay. In contrast, antibody-treated target cells were killed as efficiently as control cells. Identical results were obtained when other LFA-1 antibodies were used.

Discussion

NK cells have been defined operationally as a population of cells killing allogeneic target cells without prior immunization (see Ref. 1 for review). According to this definition, it has become evident that the population of cells mediating NK activity is very heterogeneous. In human peripheral blood, the diversity of NK cells was suggested by phenotypic analysis of highly purified LGL fractions. Although LGL appear to be morphologically homogeneous and contain virtually the entire pool of NK active PBL, they were found to be very heterogeneous when tested with a series of monoclonal antibodies directed at antigens such as T8, T10, T11, and Mo1 (8,12).

Therefore, different approaches have been undertaken to develop monoclonal antibodies characterizing NK cells phenotypically. Monoclonal antibodies have been generated which supposedly define all or subsets of NK cells. HNK1 was considered to define most of the peripheral blood NK activity (13), but the antigen could not be demonstrated on activated NK cells. HNK1 also defines a population of non-NK active T cells (14). Moreover, this antibody also recognizes a myelin-associated antigen (15). Other monoclonal antibodies such as B73.1, VEP 13, NK-15, and 3G8 are directed against Fc receptors present on natural killer cells and polymorphonuclear cells (PMN)[16]. Whereas B73.1 seems to define an epitope mainly expressed on NK cells, the epitopes recognized by VEP 13, NK-15, and 3G8 are also present on PMN. Consistent with their specificity for Fc receptors, several of these antibodies have been shown to block antibody-dependent cellular cytotoxicity (ADCC) but are not able to block NK activity.

Another method for generating monoclonal antibodies against NK cells has been the use of human NK cell clones for immunization and initial screening. Using this approach, a pan-NK cell antigen, $NKH1_A$, which characterizes all peripheral blood NK cells and human NK clones has recently been described (17). $NKH1_A$ is directed against the same molecule as the previously described antibody termed anti-N901 (NKH1) (18). By using human NK cell clones, we also were able to generate a monoclonal antibody, termed anti-NKH2, delineating a subset of LGL that is functionally distinct from the subset defined by anti-$NKH1_A$ (17).

The present studies with NK-associated monoclonal antibodies submitted to the Second International Workshop and Conference on Human Leukocyte Differentiation Antigens were designed to characterize the structures defined by these antibodies on a series of human NK clones.

Two antibodies (M2 and M86) were found to be positive on all four NK clones regardless of whether they were T3$^+$ or T3$^-$. M42, M78, and M105 were negative on all NK clones, whereas M1 reacted with two clones and M54 with only one NK clone.

Considering the experience with NKH1$_A$ and NKH2 and other antibodies reactive with peripheral blood NK cells, it is possible to postulate that monoclonal antibodies M2 and M86, which reacted with all NK clones, may also define the majority of peripheral blood NK cells. The antigens defined by M42, M78, and M105 seem not to be present on activated NK cells. On the other hand, one would expect that antibodies such as M1, reacting only with clones JT3 and CNK3, and M54, positive on one NK clone only, may be useful for identifying subsets of peripheral blood NK cells.

In contrast to the NK-associated antibodies, all anti-LFA-1 antibodies with the exception of M26 exhibited a considerable blocking effect on NK cytotoxicity. None of the other 120 monoclonal antibodies submitted to the myeloid section tested had any effect on NK activity. LFA-1 is expressed on >97% thymocytes and >95% of blood lymphocytes and on 79% of bone marrow cells. Monoclonal antibodies to LFA-1 block the induction of T helper cell proliferative responses to antigen, but do not block several lymphocyte responses which are independent of cell interactions (19). LFA-1 antibodies have also been shown to inhibit CTL cytotoxicity and are able to block the NK activity of activated lymphocytes against K562. LFA-1 contains two noncovalently associated subunits, and α subunit of 180 Kd and a β subunit of 95 Kd molecular weight. There is a structural similarity to Mo1, as the β subunit of both antigens is identical or highly homologous (20,21).

In our studies, the inhibition of NK clone activity by anti-LFA-1 was shown to be target cell dependent and all NK clones were blocked to a greater extent in their activity against MOLT-4 than against K562. Peripheral blood NK cells were blocked in a similar pattern by the different LFA-1 antibodies. Although the LFA-1 antigen is expressed to a variable extent on the different target cells, the level of inhibition is at the effector cell.

Previous studies on the effector–target cell interaction of NK cells demonstrated that one minor population of NK active lymphocytes, represented by JT9 and JT10, interacts with target cells via 90-Kd clonotypic structures, intimately involved in determining target cell specificity (9,10). Although these NK active (T3,90 Kd)$^+$ lymphocytes expressed receptor structures which belong to the same family of molecules as those present on conventional MHC-restricted CTL, it was also evident that they did not interact with targets via recognition of either class I or class II MHC products. More recently, it has been possible to identify the target antigen (termed TNK$_{TAR}$) for this unique subset of T3$^+$ NK cells (11). A monoclonal antibody binding to this activation antigen on the target cell surface inhibited the cytotoxicity of JT9 and JT10 cells but had no blocking effect on other NK clones that had different phenotypes and specificities.

In the present experiments, we could demonstrate that the LFA-1 antigen is one additional molecule involved in the effector–target cell interaction of NK clones. With the LFA-1 antibodies we could show an inhibition of cytotoxic activity of $T3^-$ NK clones as well as of $T3^+$ NK clones. Moreover, the blocking of LFA-1 antibody affected the NK activity of peripheral blood NK cells and NK clones in a similar fashion, suggesting that this structure plays an important role in cytotoxic function of both activated and unstimulated NK cells. In further studies, the precise role of LFA-1 in the effector–target cell interaction with a variety of targets and in respect to other molecules described will be defined. Also, the importance of antibodies against either the α or β chain of LFA-1 and the relationship to the Mo1 antibodies will be of future interest.

Summary

A series of cloned human NK cell lines derived from normal individuals was used to characterize monoclonal antibodies specific for NK-associated antigens and LFA-1 antigens submitted to the myeloid section of the Second International Workshop and Conference on Human Leukocyte Differentiation Antigens. NK antibodies M2 and M86 were expressed on all NK clones; antibodies M1 and M54 were only expressed on some of the NK clones; and antibodies M42, M78, and M105 were not expressed on any clones tested. All LFA-1-specific antibodies except for M26 were strongly reactive with all NK clones. When tested on a series of NK target cells, none of the NK-associated antibodies demonstrated significantly reactivity, whereas LFA-1 antibodies reacted strongly with most myeloid and T cell targets. None of the NK monoclonal antibodies was able to block NK activity of the NK clones. In contrast, all of the LFA-1 antibodies except for antibody M26 blocked cytotoxicity of all NK clones. The extent of NK blocking was variable from one target cell to another but was found on NK clones with a mature T cell phenotype $(NKH1^+T11^+T3^+)$ as well as clones that do not have a mature T cell phenotype, i.e., those that expressed NKH1 and T11 antigens but were $T3^-$. Similar blocking of NK activity by anti-LFA-1 antibodies was found utilizing unstimulated peripheral blood NK cells as effectors. The blocking of anti-LFA-1 antibodies occurred at the effector cell level. Although anti-LFA-1 antibodies were not specifically reactive with NK active cells, these studies suggest an important role for this class of molecules in NK cell function. In contrast, two of the NK antibodies (M2, M86) were found to react specifically with all NK active clones and appeared to define pan-NK antigens, but did not inhibit NK function.

Acknowledgments. We would like to thank Mr. Herbert Levine, Mr. David Leslie, and Ms. Mary Kornacki for assistance in performing cell sorter analysis. This work was supported by NIH grants CA 19589 and CA 34183. Reinhold Schmidt is a recipient of a fellowship (Schm 596/1-1) from the Deutsche Forschungsgemeinschaft. Jerome Ritz is a Scholar of the Leukemia Society of America.

After the myeloid antibody code was broken, the antibodies used in these studies were identified as follows: M1, NKH2; M2, NKH1A; M26, VIM12; M42, D12; M54, GO22b; M55, MHM23; M56, MHM24; M72, TS1/22; M73, CLB-54; M78, M3D11; M86, N901; M89, TS1/18.11; M105, Leu 7.

References

1. Herberman, R.B., and J.R. Ortaldo. 1981. Natural killer cells: Their role in defenses against disease. *Science* **214:**25.
2. Abruzzo, L.V., and D.A. Rowley. 1983. Homeostatis of the antibody response: Immunoregulation by NK cells. *Science* **222:**581.
3. Hansson, M., M. Beran, B. Anderson, and R. Kiessling. 1983. Inhibition of *in vitro* granulopoiesis by autologous allogeneic human cells. *J. Immunol.* **129:**126.
4. Lust, J.A., V. Kumar, R.C. Burton, S.P. Barlett, and M. Bennett. 1981. Heterogeneity of natural killer cells in the mouse. *J. Exp. Med.* **154:**306.
5. Minato, N., L. Reo, and B.R. Bloom. 1981. On the heterogeneity of murine natural killers. *J. Exp. Med.* **154:**750.
6. Zarling, J.M., K.A. Crouse, W.E. Biddison, and P.C. Kung. 1981. Phenotypes of human natural killer cell populations detected with monoclonal antibodies. *J. Immunol.* **127:**2575.
7. Hercend, T., S.C. Meuer, E.L. Reinherz, S.F. Schlossman, and J. Ritz. 1982. Generation of a cloned NK cell line derived from the "null cell" fraction of human peripheral blood. *J. Immunol.* **129:**1299.
8. Hercend, T., E.L. Reinherz, S.C. Meuer, S.F. Schlossman, and J. Ritz. 1983. Phenotype and functional heterogeneity of human cloned natural killer cell lines. *Nature* **301:**158.
9. Hercend, T., S.C. Meuer, A. Brennan, M.A. Edson, O. Acuto, E.L. Reinherz, S.F. Schlossman, and J. Ritz. 1983. Identification of a clonally restricted 90KD heterodimer on two human cloned natural killer cell lines: Its role in cytotoxic effector function. *J. Exp. Med.* **158:**1547.
10. Hercend, T., S.C. Meuer, A. Brennan, M.A. Edson, O. Acuto, E.L. Reinherz, S.F. Schlossman, and J. Ritz. 1984. Natural killer-like function of activated T lymphocytes: Differential blocking effects of monoclonal antibodies specific for a 90KD clonotypic structure. *Cell. Immunol.* **86:**381.
11. Hercend, T., R.E. Schmidt, A. Brennan, M.A. Edson, E.L. Reinherz, S.F. Schlossman and J. Ritz. 1984. Identification of a 140KD activation antigen as a target structure for a series of human cloned NK cell lines. *Eur. J. Immunol.* **14:**844.

12. Ortaldo, J., S.L. Sharrow, T. Timonen, and R.B. Herberman. 1981. Determination of surface antigens on highly purified human NK cells by flow cytometry with monoclonal antibodies. *J. Immunol.* **127:**2401.

13. Abo, T., and C.M. Balch. 1981. A differentiation antigen of human NK and K cells identified by a monoclonal antibody (HNK-1). *J. Immunol.* **127:**1024.

14. van de Griend, R.J., B.A. van Krimpen, C.P.M. Routeltap, and R.L.H. Bolhuis. 1984. Rapidly expanded activated human killer cell clones have strong anti-tumor cell activity and have the surface phenotype of either Tγ, T-non-γ or null cells. *J. Immunol.* **132:**3185.

15. McGarry, R.C., S.L. Helfand, R.H. Quarter, and J.C. Roder. 1983. Recognition of myelin-associated glycoprotein by the monoclonal antibody HNK-1. *Nature* **306:**376.

16. Perussia, B., G. Trinchieri, A. Jackson, N.L. Warner, J. Faust, H. Rumpold, D. Kraft, and L.L. Lanier. 1984. The Fc receptor for IgG on human natural killer cells: phenotypic, functional, and comparative studies with monoclonal antibodies. *J. Immunol.* **133:**180.

17. Hercend, T., J.D. Griffin, A. Bensussan, R.E. Schmidt, A. Brennan, M.A. Edson, J.F. Daley, S.F. Schlossman, and J. Ritz. 1985. Generation of monoclonal antibodies to a human NK clone: Characterization of two NK associated antigens, NKH1. A and NKH2, expressed on subsets of large granular lymphocytes. *J. Clin. Invest.* **75:**932.

18. Griffin, J.D., T. Hercend, R.P. Beveridge, and S.F. Schlossman. 1983. Characterization of an antigen expressed by human natural killer cells. *J. Immunol.* **130:**2947.

19. Springer, T.A., D. Davignon, M. Ho, K. Kurzinger, E. Martz, and F. Sanchez-Madrid. 1982. LFA-1 and Lyt2,3 molecules associated with T lymphocyte mediated killing, and Mac-1, an LFA-1 homologue associated with complement receptor function. *Immunol. Rev.* **68:**171.

20. Krensky, A.M., F. Sanchez-Madrid, E. Robbins, J.A. Nagy, T.A. Springer, and S.J. Burakoff. 1983. The functional significance, distribution, and structure of LFA-1, LFA-2 and LFA-3: cell surface antigens associated with CTL target interactions. *J. Immunol.* **131:**611.

21. Hildreth, J.E.K., F.M. Gotch, P.D.K. Hildreth, and A.J. McMichael. 1983. A human lymphocyte-associated antigen involved in cell mediated lympholysis. *Eur. J. Immunol.* **13:**202.

CHAPTER 12

Reactivity Patterns of Monoclonal Antibodies Against Myeloid-Associated Antigens with Human Natural Killer Cells

Helmut Rumpold, Gabriele Stückler, Alois Fellinger, Renate Steiner, Elisabeth Faustmann, and Dietrich Kraft

Introduction

Natural killer (NK) cells (1) are defined by their ability to lyse certain tumor cell lines *in vitro* without previous exposure to them. These cells have been identified as large lymphocytes with an abundant cytoplasm, an indented nucleus, and azurophilic granules ("large granular lympho-cytes," LGL) (2). Although the origin and function of NK cells is still a matter of some controversy, it has been postulated that these cells play an important role in the host's defense against infections, the growth and dissemination of malignant tumors, and the regulation of normal hemato-poetic cells.

Various attempts were made to characterize human NK cells with the help of monoclonal antibodies (mAbs). It could be demonstrated that a considerable proportion of LGLs carry T cell-associated (T8, T10, T11A) as well as myelo-monocytic (Mac-1, M522, Mo1) antigens (3–6). The search for NK-specific mAbs led to the introduction of HNK-1 mAb (Leu 7), originally claimed to recognize all cells mediating NK activity (7). Experiments later performed showed that HNK-1$^+$ cells include some but not all NK activity (8), whereas the anti-NK/granulocytes FcγR mAbs 3G8 (9), VEP13 (10), B73.1 (11), anti–Leu 11a, and anti–Leu 11b (12) apparently detect all LGLs and therefore cover all NK activity (13,14). Another mAb, N901, has been reported to react with the majority of LGLs, but not granulocytes (15).

In the present Workshop investigation we tested the mAb panel against myeloid-associated antigens for reactivity with human NK cells from nor-mal human donors. Seven monoclonal antibodies have been stated in the

Workshop protocol to react with NK cells. In addition 12 mAbs were found in an initial screening to react with subpopulations of lymphocytes. Two of these 12 mAbs showed great variability in reacting with lymphocytes of different donors and because they are known to recognize antigens associated with certain lymphocyte functions they will be discussed by other Workshop participants. The remaining ten mAbs were tested, in addition to the NK-reactive antibodies, for reactivity with NK cells by employing FACS separation.

Materials and Methods

Reagents

Rabbit complement (C) and $Na^{51}CrO_4$ were obtained from Hoechst Austria, Ficoll–Paque and Dextran T500 from Pharmacia Fine Chemicals (Uppsala, Sweden), carbonyl iron S.F. from GAF (Manchester, UK), RPMI 1640 tissue culture medium from Flow Laboratories (Irvine, Scotland), fetal calf serum (FCS) from Seromed (Munich, BRD), anti–mouse IgG (H and L chain specific) $F(ab')_2$–FITC conjugated from Cappel Laboratories (Cochranville, PA, USA).

Preparation of Cells

Peripheral blood lymphocytes (PBLs) were obtained from human peripheral blood of healthy donors as described previously (16). Phagocytic cells were depleted by carbonyl iron and magnet treatment (16). Granulocytes were prepared by dextran sedimentation and Ficoll–Paque density gradient centrifugation with lysis of remaining blood cells by hypotonic shock.

Indirect Membrane Immunofluorescence (IMF) Test

3×10^6 cells were incubated with 300 μl mAb diluted in RPMI 1640 medium containing 10% FCS for 30 min at 4°C. Thereafter, the cells were washed with RPMI 1640 medium/10% FCS three times at 4°C and incubated for 30 min at 4°C with an FITC-conjugated anti–mouse Ig $F(ab')_2$ fragment previously shown to react with IgG and IgM. Thereafter the cells were washed three times at 4°C and resuspended in 300 μl RPMI 1640 medium/10% FCS and analyzed using a FACs III system. In each test normal mouse serum was used as a control. mAbs of the IgG class and the FITC conjugate were centrifuged for 30 min at 100,000 $\times$ g using a Beckmann Airfuge immediately before use.

For separation of fluorescent cells 6×10^7 cells were stained, increasing the amount of reagents used by the same factor. Cell separation was carried out at a flow rate of 2000 cells/sec.

Complement-mediated Lysis (CML)

2×10^7 PBLs were incubated with mAb diluted in RPMI 1640 medium/ 10% FCS for 30 min on ice and washed once with cold medium. Thereafter, rabbit C preabsorbed with human buffy coat cells diluted 1/5 was added, and cells were incubated for 1 hr at 37°C with gentle shaking every 10 min. Subsequently, an aliquot was taken and the percentage of killed cells was determined by the addition of 0.5% Trypan Blue solution in PBS. The remaining cells were layered onto Ficoll–Paque and centrifuged for 30 min at $400 \times g$ to remove dead cells. Interphase cells were washed three times in RPMI 1640 medium/10% FCS, counted, and adjusted to the cell concentration required for cytotoxicity assay.

Cytotoxicity Assay

Effector Cells

Lymphocytes were separated into mAb-reactive and -nonreactive cells by means of the FACS or depleted of reactive cells by CML as described above.

Target Cells

K562 cells were labeled with $100\text{-}\mu\text{Ci}$ $Na^{51}CrO_4$ for 1 hr at 37°C, washed twice, and adjusted to a final concentration of 5×10^4 cells/ml. NK activity was tested in a 3-hr ^{51}Cr release assay by mixing 100 μl effector cell and 100 μl target cell suspensions at various effector to target cell ratios into U-shaped microtiter plates. Supernatants were harvested using a Skatron supernatant-harvesting device. Maximum and spontaneous ^{51}Cr release was determined by the addition of 10% NP40 or medium, respectively, instead of effector cells.

%Specific lysis was calculated by the formula:

$$\frac{(\text{cpm experimental release} - \text{cpm spontaneous release}) \times 100}{\text{cpm maximum release} - \text{cpm spontaneous release}}$$

Results

Reactivity of mAbs with PBLs and Granulocytes

In an initial screening mAbs of the myeloid-associated antigen panel were tested by IMF for reactivity with PBLs. One mAb (86H1) was missing in the shipment. Fourteen mAbs were found to react with >95% of PBLs: JOAN-1, BW252/104, M101, GB3, CC1.7, MHM23, MHM24, TS1/22, CLB-54, CIPAN, 60.3, TS1/18.11, T10C6, M3C7.2A. A doubtful, only

Table 12.1. Reactivity of mAbs with PBLs.[a]

	Designated as anti–NK cell antibodies in the Workshop protocol						
	M1 NKH2	M2 NKH1A	M42 D12	M54 G022b	M78 M3D11	M86 N901	M105 Leu 7
Donor 1	0	23	18	13	0	14	16
Donor 2	0	42	40	39	2	32	30
Donor 3	0	28	21	20	2	14	39

	Not designated as anti–NK cell antibodies in the Workshop protocol									
	M22 BW243/41	M50 KD3	M51 BL-5	M53 UCHL1	M57 CLBFcRgran1	M58 VEP13	M59 BW209/2	M83 BL-1	M84 BL-2	M85 BL-3
Donor 1	13	44	23	44	18	12	10	28	29	25
Donor 2	33	57	35	34	45	27	32	38	32	38
Donor 3	18	48	29	52	18	20	20	28	31	28

[a] Values are given as percent reactive cells as revealed by FACS analysis.

weak positivity has been seen with the mAbs JD2, NHL30.5, SHCL3, and Ki-M5. Seven mAbs had been designated as anti-NK in the Workshop protocol. Of these, NKH1A, D12, G022b, N901, and Leu 7 mAbs were found to react with subpopulations of PBLs, whereas no reactivity was seen with NKH2 and M3D11 antibodies (Table 12.1). In the PBLs of three donors tested, similar percentages of positive cells were found in the case of NKH1A, D12, G022b, and N901 mAbs. However, up to 40% positive cells were found in the PBLs of donor two. Using Leu 7, the highest percentage of positive cells was seen in the PBLs of donor three.

In addition, 12 other mAbs were found to react also with subpopulations of PBLs. A great variability from donor to donor was seen with two of these antibodies (VIM12 and 60.1). Both of them had been designated in the Workshop protocol to recognize antigens associated with certain lymphocyte functions and were studied by other investigators.

The remaining ten mAbs were studied in more detail. BW243/41, BL-5, CLB FcR gran 1, VEP13, BW209/2, BL-1, BL-2, and BL-3 showed similar percentages of positive cells as the anti-NK antibodies NKH1A, D12, G022b, and N901. A higher percentage of positive cells was seen in all three donors in the case of KD3 antibody. In contrast to all other mAbs of this group, higher percentages of reactive cells were seen in donors one and three than in donor two with the use of UCHL1 mAb (Table 12.1).

Eight of these mAbs showed also reactivity with granulocytes (Table 12.2). FACS histograms of these mAbs are shown in Fig. 12.1. Clearly distinguishable subsets were observed in the case of G022b, BW243/41, KD3, BL-5, UCHL1, CLB FcR gran 1, VEP13, BW209/2, BL-1, BL-2, and BL-3, whereas the other Abs showed fluorescence intensities from dull to bright.

Table 12.2. Reactivity of mAbs with granulocytes.[a]

Reactive with granulocytes		Nonreactive with granulocytes	
M42	D12	M1	NKH2
M54	G022b	M2	NKH1A
M78	M3D11	M86	N901
M22	BW243/41	M105	Leu 7
M53	UCHL1	M50	KD3
M57	CLBFcRgran1	M51	BL-5
M58	VEP13	M83	BL-1
M59	BW209/2	M84	BL-2
		M85	BL-3

[a] Granulocytes were separated from peripheral blood and stained by means of indirect membrane immunofluorescence as indicated in Materials and Methods.

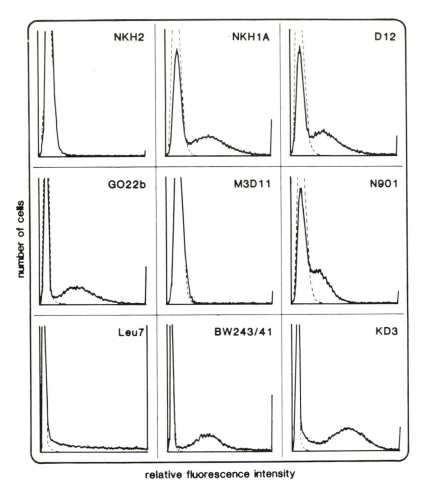

relative fluorescence intensity

Fig. 12.1. FACS histograms of PBLs stained with mAbs and anti–mouse Ig F(ab')$_2$–FITC. 10^5 cells were analyzed for each histogram. --- Autofluorescence of cells; —— mAb and anti–mouse Ig F(ab')$_2$–FITC-stained cells.

NK Activity of mAb and C-Treated PBLs

Nine mAbs (NKH2, NKH1A, D12, G022b, M3D11, Leu 7, BW243/41, CLB FcR gran 1, VEP13) were stated in the Workshop protocol to fix C. Therefore PBLs were treated with these mAbs and C and tested thereafter for their remaining NK activity against K562 target cells. No change of NK activity was seen in the case of NKH2, D12, and M3D11 mAbs. G022b, CLB FcR gran 1, VEP13 together with C treatment of PBLs caused a complete depletion of NK cells, whereas NKH1A and BW243/41 plus C treatment led to about 50% reduction. Only a slight decrease of NK activity was seen in the case of Leu 7 plus C-treated cells (Table 12.3).

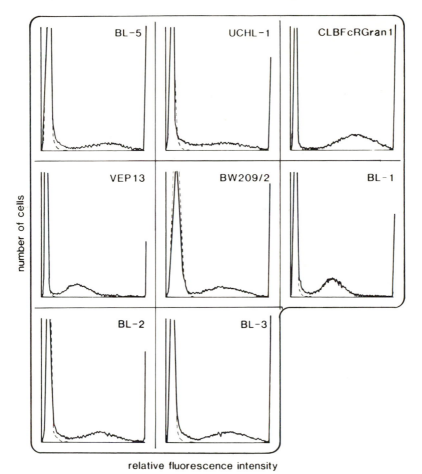

relative fluorescence intensity

Fig. 12.1. *Continued.*

Table 12.3. NK activity of PBLs treated with mAbs and C.[a]

E/T ratio	PBLs treated with C alone	M1 NKH2	M2 NKH1A	M42 D12	M54 G022b	M78 M3D11	M105 Leu 7
60	46.6	48.5	28.4	50.8	5.6	51.2	41.9
30	31.4	32.2	14.4	41.2	3.5	30.9	26.4
15	20.9	21.7	7.7	22.7	1.7	19.7	14.4
7.5	10.8	14.2	2.5	14.3	0.9	9.3	9.2

E/T ratio	PBLs treated with C alone	M22 BW243/41	M57 CLBFcRgran1	M58 VEP13
60	71.1	36.0	3.8	1.2
30	55.7	29.4	3.8	0.7
15	36.6	13.6	3.3	0.2
7.5	20.8	8.3	1.0	0.1

[a] Cytotoxicity was evaluated in a 3-hr ^{51}Cr release assay using K562 target cells. Values are given as percent specific lysis.

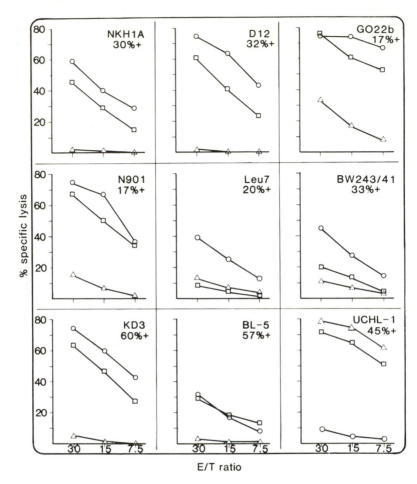

Fig. 12.2. NK activity of FACS-separated mAb-reactive and -nonreactive PBLs. NK activity against K562 target cells was measured in a 3-hr ^{51}Cr release assay. Values are given as % specific cytotoxicity. Purities of mAb-reactive cell preparations for fluorescence stain are given in Table 12.4. □ NK activity of unsorted PBLs stained with mAb and anti–mouse Ig F(ab')$_2$–FITC; ○ NK activity of FACS-sorted immunofluorescence-positive PBLs; △ NK activity of FACS-sorted immunofluorescence-negative PBLs.

NK Activity of FACS-Separated Lymphocyte Subpopulations

All mAbs listed in Table 12.1 which showed reactivity with subpopulations of PBLs were separated into fluorescence-positive and fluorescence-negative cells by means of the FACS and tested thereafter for their NK activity against K562 cells. An enrichment of NK activity in the positive cell population was seen with all fluorescence-positive antibodies of the anti–NK cell mAb group, whereas the negative cell population showed

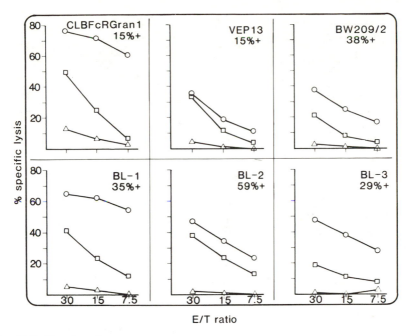

E/T ratio

Fig. 12.2. *Continued.*

depletion (NKH1A, D12) or a strong reduction (G022b, N901, Leu 7) of NK cell activity.

Testing the ten mAbs which have been found to react with subpopulations of PBL and were not designated as anti-NK in the Workshop protocol, in the case of nine mAbs (BW243/41, KD3, BL-5, CLB FcR gran 1, VEP13, BW209/2, BL-1, BL-2, BL-3) an enrichment of NK activity in the positive cell population was obtained, while the negative cells were devoid of it. Only with the use of UCHL1 mAb was the opposite phenomenon observed, namely NK activity was contained in the negative cell fraction, whereas UCHL1$^+$ cells were devoid of it.

Fluorescence-positive cells separated by means of the FACS and tested for NK activity (Fig. 12.2) were stained according to May-Grünwald-Giemsa and the percentage of LGLs evaluated (Table 12.4). In the case of eight mAbs only (NKH1A, D12, G022b, N901, BW243/41, CLB FcR gran 1, VEP13, BW209/2) LGL morphology was observed for 90% or more of the cells.

Discussion

The aim of the present Workshop investigation has been to test mAbs against myeloid-associated antigens for reactivity with NK cells. Seven of these mAbs had been classified in the Workshop protocol as anti-NK

Table 12.4. Percentages of LGL in FACS-sorted, immunofluorescence-positive cell preparations.[a]

mAB		%LGL	% Fluorescence-positive cells
M2	NKH1A	94	90
M42	D12	86	80
M54	G022b	90	>95
M86	901	92	80
M105	Leu 7	55	>95
M22	BW243/41	95	90
M50	KD3	65	>95
M51	BL-5	45	>95
M53	UCHL1	<5	>95
M57	CLBFcRgran1	91	90
M58	VEP13	90	90
M59	BW209/2	87	77
M83	BL-1	ND[b]	>95
M84	BL-2	37	>95
M85	BL-3	47	>95

[a] Cytocentrifuge preparations were stained according to May-Grünwald-Giemsa.
[b] ND: Not determined.

mAbs. The reactivity of five of these Abs with PBLs was confirmed by this study. However, two of the mentioned Abs (NKH2, M3D11) failed to react with PBLs in IMF and CML experiments. Since M3D11 reacted with granulocytes, damage of this mAb during shipment can be excluded; in the case of NKH2, this possibility has to be considered. Selection of positive or negative cell populations by using the FACS led to enrichment or reduction of NK activities in the case of NKH1A, D12, G022b, N901, and Leu 7, which confirms the Workshop designation. However, employing CML, the capability to deplete NK activity by Ab + C treatment was found to decrease in the following order: G022b > NKH1A ≫ Leu 7. No effects in CML were found with NKH2, D12, and M3D11 mAbs. Morphology of sorted cells represented nearly homogeneous populations of LGLs in the case of NKH1A, D12, G022b, and N901, whereas Leu 7[+] cells contained only 55% LGLs. This latter finding is in line with new reports that the Leu 7 antigen is not exclusively expressed on LGL (8).

Ten anti-myeloid mAbs, not designated as anti–NK cell mAbs, were detected to react with subsets of PBLs. Therefore, it was interesting to ask the question whether NK activity were attributable to the negative or positive subsets. FACS separation revealed that only in the case of UCHL1 was NK activity present in the negative subset, whereas in the case of BW243/41, KD3, BL-5, CLB FcR gran 1, VEP13, BW209/2, BL-1, BL-2, and BL-3 NK activities were found in the positive subpopulations. CLB FcR gran 1 and VEP13 were as effective as G022b in CML experiments, whereas BW243/41 showed a similar effect to NKH1A.

Looking at the cell morphology, homogeneous preparations of LGLs could be observed in the case of NKH1A, D12, G022bh, N901, BW243/

41, CLB FcR gran 1, VEP13, and BW209/2 but only lower percentages in the case of Leu 7, KD3, BL-5, BL-2, and BL-3. The latter mAbs apparently recognize LGLs as well as other lymphocyte subsets. However, since the data obtained relate to only a few experiments, these findings need further confirmation.

Previous reports on anti-NK cell antibodies have revealed a group of antibodies directed against the Fc-γ-receptor on LGLs and granulocytes (14). Within this group, VEP13, 3G8, anti–Leu 11a (NKP-15), and anti–Leu 11b (G022b) are directed against the same epitope, whereas B73.1 recognizes a different epitope which is expressed to a lesser degree on granulocytes than on LGLs. On the basis of the observed reactivities and the designation, the CLB FcR gran 1 mAb is likely to belong to the same group of anti-NK mAbs. In the case of D-12, BW243/41, and BW209/2, studies should be performed to clarify whether the mentioned antibodies can also be attributed to this group of mAbs or not.

Summary

The reactivity patterns of anti-myeloid mAbs have been evaluated with special reference to cells with NK activity and morphology of large granular lymphocytes (LGL). Fourteen out of 115 mAbs were found to react with the majority (80–100%) of PBLs, including LGLs, as tested by means of indirect membrane immunofluorescence (IMF). This group includes five antibodies which had been designated as anti-LFA$_1$. A second group of 15 mAbs, which showed reactivity with subpopulations of PBLs, has been studied in more detail using FACS sorting and C-mediated lysis (CML). In this group only five of seven mAbs designated as anti-NK antibodies showed reactivity with NK cells. Nine anti-myeloid mAbs, not designated as anti-NK mAbs in the protocol, were detected to react with NK cells, thus forming another group. Of these only the FACS-separated BW243/41, CLB FcR gran 1, VEP13, and BW209/2 positive subsets represented homogeneous populations with LGL morphology.

Acknowledgment. This work was supported by a grant from the Austrian Science Research Fund, Project No. 5399.

References

1. Herbermann, R.B. 1982. *NK cells and other natural effector cells.* Academic Press, New York.
2. Timonen, T., E. Saksela, A. Ranki, and P. Häyry. 1979. Fractionation, morphological and functional characterization of effector cells responsible for human natural killer activity against cell line targets. *Cell. Immunol.* **48**:133.
3. Zarling, J.M., and P.C. Kung. 1980. Monoclonal antibodies which distinguish between human NK cells and cytotoxic T lymphocytes. *Nature* **288**:394.

4. Ortaldo, J.R., S.O. Sharrow, T. Timonen, and R.B. Herberman. 1981. Determination of surface antigens on highly purified NK cells by flow cytometry with monoclonal antibodies. *J. Immunol.* **127**:2401.

5. Rumpold, H., D. Kraft, G. Obexer, T. Radaszkiewicz, O. Majdic, P. Bettelheim, W. Knapp, and G. Böck. 1983. Phenotypes of human large granular lymphocytes as defined by monoclonal antibodies. *Immunobiol.* **164**:51.

6. Kraft, D., H. Rumpold, and O. Eremin. 1984. Monoclonal antibodies as tools in phenotype characterisation of human NK cells. In: *Handbook of monoclonal antibodies,* M.P. Dierich and S. Ferrone, eds. Noyes Publications Publishers, Park Ridge, N.J., in press.

7. Abo, T., and C.M. Balch. 1981. A differentiation antigen of human NK and K cells identified by a monoclonal antibody (HNK-1). *J. Immunol.* **127**:1024.

8. Rumpold, H., D. Kraft, G. Gastl, and G. Huber. 1984. The relationship of HNK-1 (Leu 7) and VEP13 antigens on human cells mediating natural killing. *Clin. Exp. Immunol.* **57**:703.

9. Fleit, H.B., S.D. Wright, and J.D. Unceless. 1983. Human neutrophil Fc receptor distribution and structure. *Proc. Natl. Acad. Sci. U.S.A.* **79**:3275.

10. Rumpold, H., D. Kraft, G. Obexer, G. Böck, and W. Gebhart. 1982. A monoclonal antibody against a surface antigen shared by human large granular lymphocytes and granulocytes. *J. Immunol.* **129**:1458.

11. Perussia, B., St. Starr, S. Abraham, V. Fanning, and G. Trinchieri. 1983. Human natural killer cells analysed by B73.1, a monoclonal antibody blocking Fc receptor functions. I: Characterization of the lymphocyte subset reactive with B73.1. *J. Immunol.* **130**:2133.

12. Phillips, J.H., and G.F. Babcock. 1983. NKP15: A monoclonal antibody reactive against purified human natural killer cells and granulocytes. *Immunol. Lett.* **6**:143.

13. Perussia, B., and G. Trinchieri. 1984. Antibody 3G8, specific for the human neutrophil Fc receptor, reacts with natural killer cells. *J. Immunol.* **132**:1410.

14. Perussia, B., G. Trinchieri, A. Jackson, N.L. Warner, J. Faust, H. Rumpold, D. Kraft, and L.L. Lanier. 1984. The Fc receptor for IgG on human natural killer cells: phenotypic, functional and comparative studies with monoclonal antibodies. *J. Immunol.* **133**:180.

15. Griffin, J.D., Th. Hercend, R. Beveridge, and S.F. Schlossman. 1983. Characterisation of an antigen expressed by human natural killer cells. *J. Immunol.* **130**:2947.

16. Rumpold, H., D. Kraft, O. Scheiner, P. Meindl, and G. Bodo. 1980. Enhancement of NK but not K cell activity by different interferons. *Int. Arch. Allergy Appl. Immunol.* **62**:152.

CHAPTER 13

A Map of the Cell Surface Antigens Expressed on Resting and Activated Human Natural Killer Cells

Lewis L. Lanier and Joseph H. Phillips

Introduction

The major categories of human peripheral blood leukocytes were initially defined by morphology. Three classifications were identifiable: granulocytes, monocytes, and lymphocytes. Within the lymphocyte group, the cells were considered homogeneous and undistinguished, displaying a small roundish nucleus and scant cytoplasm. The vast heterogeneity within the lymphoid population was not appreciated until relatively recently, when it was shown that the presence or absence of certain cell surface antigens could be correlated with cellular lineage and function. The initial subdivision of lymphocytes was based largely on two properties. Expression of surface and/or cytoplasmic immunoglobulin became the standard criterion for the B lymphocytes. Since expression of immunoglobulin is exclusively a product of B lymphocytes and strictly relates to the function of antigen binding and triggering, it still is the most definitive marker for these cells. The presence of a cell surface receptor for binding sheep erythrocytes (E) was considered the benchmark for the human T cell population (1). However, as we discuss below, this receptor can be detected on cells not of thymic origin. With the recent discovery of the T cell-associated receptor for antigen, cytoplasmic or surface expression of the T cell antigen receptor will likely replace the E receptor as the ultimate indicator of the T lymphocyte lineage (2,3).

A third population of lymphoid cells, expressing neither surface immunoglobulin nor E receptors in high cell surface density, was called the "null cell" subset, simply for want of a better description. The function of these cells similarly was unknown, although it was observed that "natural cell-mediated cytotoxicity" and antibody-mediated cellular cytotoxicity (ADCC) activities were present in this cell population (reviewed in Refs. 4–7).

Natural killer cells are a functionally defined subset of lymphoid cells obtained from "unimmunized" hosts that are capable of lysing certain tumor cell lines and virus-infected cells. In contrast to T cell-mediated cytotoxicity, NK cytotoxicity is not restricted by the major histocompatibility complex (MHC). In the studies presented here, we demonstrate that the "null" cell population is far from devoid of surface markers. Moreover, we show that essentially all NK and ADCC activity is contained within this subset and that these cells are capable of activation in a mixed lymphocyte/tumor cell culture.

Antigens Associated with Human Natural Killer Cells: Analysis by Two-Color Immunofluorescence

Recently, a series of antibodies (including anti–Leu 11a, anti–Leu 11b, B73.1, 3G8, and VEP13) have been generated which apparently recognize the IgG-Fc receptor present on human natural killer cells (8–12). These antibodies react with the IgG-Fc receptor present on NK cells and neutrophils, but do not recognize the antigenically and structurally distinct Fc receptors present on monocytes, B lymphocytes, and eosinophils. All of these antibodies partially or completely inhibit the binding of EA-rosettes or aggregated IgG complexes to NK cells or neutrophils (11,12). Furthermore, some of these antibodies were extremely efficient inhibitors of ADCC function (11). The most significant aspect of these antibodies is that they are capable of specifically binding to essentially all peripheral blood cells which possess NK activity, as shown by antibody and complement depletion studies (anti–Leu 11b and VEP13) and cell sorting experiments using a fluorescence-activated cell sorter (FACS) (anti–Leu 11a, B73.1, and 3G8) (8–12). Morphological examination of the Leu 11 (B73.1, VEP13, 3G8) positive lymphocytes also was consistent with the observations of Timonen and Saksela suggesting that NK cells were predominately "large granular lymphocytes" (13).

Since essentially all NK activity was contained within the small percentage of lymphocytes reacting with these antibodies, it was of considerable interest to determine which other surface markers were present on this unique population. This could best be achieved by two-color immunofluorescence and detection using multi-parameter flow cytometry. In this study, peripheral blood from ten randomly selected donors was separated on Ficoll–Hypaque to isolate the mononuclear cell fraction. Monocytes were partially depleted by adherence on plastic tissue culture dishes (14). These cells were stained using an extensive panel of fluorescein isothiocyanate (FITC)- or phycoerythrin (PE)-conjugated monoclonal antibodies directed against human cell surface antigens. As shown in Table 13.1, a significant proportion of lymphoid cells expressing the Leu 11a antigen did not coexpress the Leu 4 (CD3), Leu 3a (CD4), DR, or

Leu M3 (monocyte-associated) antigens. Since the Leu 4 (CD3) antigen is present on essentially all peripheral blood T lymphocytes and is associated with the T cell antigen receptor (15,16), these findings indicate that freshly isolated NK cells significantly differ from peripheral T cells. The lack of constitutive DR antigen expression and the Leu M3 antigen similarly demonstrate the distinction between NK cells and B lymphocytes or monocytes, respectively.

By contrast, a varying proportion of the Leu 11$^+$ lymphocytes did coexpress the Leu 2a (CD8), Leu 7, and Leu 8 antigens (Table 13.1). Within the Leu 11$^+$ subset of lymphocytes, on average approximately 37% of the Leu 11$^+$ cells coexpressed Leu 2a, 50% Leu 7, and 27% Leu 8. Several examples of the immunofluorescent staining profiles are shown in Fig. 13.1. Of particular note is that the Leu 4$^-$,11$^+$ cells which coexpress Leu 2 have significantly lower cell surface density of the antigen that the Leu 2$^+$ (Leu 4$^+$) lymphocytes that lack the Leu 11 antigen (8). Perussia and coworkers also have reported T8 antigen in low density on NK cells (17).

Two other surface markers examined were present on the majority of Leu 11$^+$ lymphocytes. Anti–Leu 5b, a monoclonal antibody that recognizes the E-rosette receptor, reacted with the majority of Leu 11$^-$ lym-

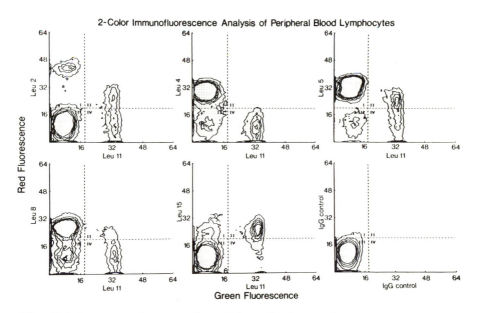

Fig. 13.1. Representative examples of NK cell subsets using two-color immunofluorescence. Nonadherent peripheral blood mononuclear cells were prepared and stained as described in Table 13.1, using the indicated combination of monoclonal antibodies. In the contour plots shown, green (FITC) fluorescence is displayed on the *x*-axis (log scale) and red (PE) fluorescence is displayed on the *y*-axis (log scale).

Table 13.1. Lymphocyte subsets identified by
two-color immunofluorescence and monoclonal
antibodies: Normal values determined from 10
random donors.[a]

Subset	Mean ± SD ($N = 10$) % of lymphocytes	% Range
Leu 11⁻,2⁺	16.6 ± 6.8	9.5–31.7
11⁺,2⁺	4.1 ± 2.2	1.6–7.1
11⁺,2⁻	6.9 ± 3.1	2.4–12.4
Leu 11⁻,3⁺	54.2 ± 9.6	31.7–65.4
11⁺,3⁺	<2	—
11⁺,3⁻	9.2 ± 4.4	2.7–17.3
Leu 11⁻,4⁺	78.6 ± 4.8	70.8–84.6
11⁺,4⁺	<2	—
11⁺,4⁻	9.0 ± 4.6	2.5–18.3
Leu 11⁻,5⁺	77.0 ± 8.2	70.5–85.9
11⁺,5⁺	5.8 ± 3.2	0.9–11.1
11⁺,5⁻	5.0 ± 2.9	1.7–11.1
Leu 11⁻,8⁺	52.3 ± 19.2	22.5–69.5
11⁺,8⁺	2.8 ± 1.6	0.5–4.8
11⁺,8⁻	7.7 ± 3.4	2.7–14.1
Leu 11⁻,15⁺	7.5 ± 1.7	6.1–10.5
11⁺,15⁺	7.9 ± 3.6	2.4–13.2
11⁺,15⁻	2.3 ± 1.3	0.7–4.5
Leu 11⁻,DR⁺	8.1 ± 2.4	5.4–13.1
11⁺,DR⁺	<2	—
11⁺,DR⁻	9.2 ± 4.4	2.8–16.8
Leu 11⁻,M3⁺	2.1 ± 0.4	1.7–2.9
11⁺,M3⁺	<2	—
11⁺,M3⁻	10.5 ± 5.1	3.6–20.4
Leu 11⁻,7⁺	7.6 ± 3.5	3.9–15.9
11⁺,7⁺	5.6 ± 4.0	1.1–14.4
11⁺,7⁻	4.6 ± 2.1	1.5–8.6
Leu 7⁻,2⁺	13.6 ± 5.0	7.7–25.0
7⁺,2⁺	5.1 ± 2.0	2.5–9.3
7⁺,2⁻	5.6 ± 3.3	2.0–10.9
Leu 7⁻,3⁺	54.5 ± 9.5	33.8–66.5
7⁺,3⁺	<2	—
7⁺,3⁻	10.2 ± 4.2	5.4–18.9
Leu 7⁻,4⁺	72.2 ± 6.7	62.0–83.3
7⁺,4⁺	6.2 ± 2.2	3.1–9.8
7⁺,4⁻	5.8 ± 4.4	2.1–13.6
Leu 7⁻,5⁺	7.20 ± 6.0	64.8–83.9
7⁺,5⁺	7.4 ± 1.7	4.0–9.2
7⁺,5⁻	5.2 ± 4.0	1.1–13.2
Leu 7⁻,8⁺	47.8 ± 19.3	20.2–69.7
7⁺,8⁺	2.2 ± 0.8	1.0–4.0
7⁺,8⁻	10.4 ± 4.7	4.8–17.5
Leu 7⁻,15⁺	9.5 ± 2.3	6.5–14.7
7⁺,15⁺	5.7 ± 1.9	3.6–8.8
7⁺,15⁻	6.3 ± 4.8	2.4–18.0

Table 13.1. *Continued.*

Subset	Mean ± SD (N = 10) % of lymphocytes	% Range
Leu 7⁻,DR⁺	8.3 ± 2.2	6.1–12.5
7⁺,DR⁺	<2	—
7⁺,DR⁻	11.0 ± 5.0	6.1–21.4
FITC Ig⁻,PE Ig⁺	0.5 ± 0.3	0.1–0.6
FITC Ig⁺,PE Ig⁺	0.1 ± 0.1	0.1–0.4
FITC Ig⁺,PE Ig⁻	0.2 ± 0.2	0.1–0.6

[a] Human peripheral blood was obtained from the American Red Cross, San Jose, CA. The blood was collected in standard CPDA-1 anticoagulant solution and the buffy coat was separated from the plasma and the majority of erythrocytes by centrifugation. The mononuclear cells were isolated using Ficoll–Paque (Pharmacia) according to standard methods. The cells were washed, resuspended in RPMI 1640 containing 10% fetal calf serum, and allowed to adhere to plastic tissue culture flasks (Falcon Plastics, Oxnard, CA) for 45 min at 37°C (14). The nonadherent cells were collected and washed in phosphate-buffered saline (0.1 M phosphate, pH 7.2) containing 0.1% sodium azide (PBS/azide). A predetermined optimal amount of the FITC- and PE-conjugated monoclonal antibodies were added to 1×10^6 cells in a total volume of 50 μl PBS/azide. Direct FITC conjugates of anti–Leu 11a and anti–Leu 7 were used in these studies. All other antibodies were direct PE conjugates, except for two-color detection of the Leu 7/Leu 11 subsets. For this analysis, cells were stained with FITC anti–Leu 11a and purified anti–Leu 7 antibody, and washed twice. Binding of anti–Leu 7 antibody was visualized by staining with PE rat anti-mouse IgM monoclonal antibody (anti–Leu 7 is an IgM antibody and anti–Leu 11a is an IgG1 antibody). FITC-Ig and PE-Ig controls were included for each individual to determine nonspecific binding. Cells were incubated for 15 min at 4°C, washed twice in cold PBS/azide, and fixed in 0.5 ml PBS containing 1.0% paraformaldehyde (pH 7.2). All cells were stained within a 5-day period, and analyzed on a FACS 440 system (BD FACS Division, Sunnyvale, CA) in the same experimental run (8,24). The fluorochromes were excited with 200-mW 488-nm light from an argon ion laser (Spectra Physics, Mountain View, CA). Fluorescence was measured using log amplifiers. The same PMT voltage settings and amplifier gains were used for all samples in the study. A compensation network was used to subtract the red emission of FITC from the PE detector channel. Data were collected into list mode files using a PDP11/23 Consort 40 computer system (BD FACS Division, Sunnyvale, CA). For data analysis, viable lymphocytes were identified on the basis of their characteristic low forward-angle and 90-degree light scatter profiles. A "gate" was set on the lymphocyte fraction, the data were reprocessed, and the red and green fluorescence profiles of lymphocytes were displayed as two-dimensional contour plots (8,24). Using the profile of the Ig control sample, markers were set on two-dimensional contour plots to define 4 regions: quadrant I, cells stained only with PE; quadrant II, cells stained with FITC and PE; quadrant III, unstained cells; and quadrant IV, cells stained only with FITC. In the control sample for each individual, the sum of the percentage of lymphocytes in quadrants I, II, and IV was not less than 1.0% or greater than 5.0% of total lymphocytes. The mean and range of "background" values are presented in the table. These values were not subtracted from the individual experimental samples. Identical "markers" were used for all individuals and samples in the study.

phocytes (i.e., T lymphocytes) and with most Leu 11$^+$ lymphocytes (Table 13.1, Fig. 13.1). It is apparent from the two-dimensional fluorescence display shown in Fig. 1 that the average amount of Leu 5b antigen on T cells was slightly higher than on the Leu 11$^+$ lymphocytes. This has been a consistent finding using this monoclonal antibody in all individuals examined. However, using another series of monoclonal antibodies against the E-rosette receptor, this has not been observed (11). Since the Leu 5 antigen was apparently in low density on Leu 11$^+$ cells, and there was no distinct boundary between Leu 5 antigen positive and negative cells within the Leu 11$^+$ subset, it was difficult to precisely enumerate the Leu 5$^-$,11$^+$ and Leu 5$^+$,11$^+$ populations. For these reasons, the percentages given for the Leu 5$^+$,11$^+$ population likely underestimate the actual values. This also was true for the CR$_3$ antigen. Anti–Leu 15, a monoclonal antibody which inhibits C3bi rosette formation and functionally defines the CR$_3$ receptor (G. Ross, A. Jackson, and L. Lanier, unpublished observation), was present in low density on lymphocytes. In Fig. 13.1, it is apparent that Leu 15 antigen is expressed on the majority of Leu 11$^+$ lymphocytes. Again, the percentages of Leu 11$^+$,15$^+$ cells given in Table 13.1 are underestimates. Using more sensitive techniques on Percoll gradient enriched NK cells we have evidence that CR$_3$ is present on essentially all Leu 11$^+$ cells (J. Phillips and L. Lanier, unpublished observations). These data confirm prior studies demonstrating the presence of Mac-1 on human NK cells (18).

Another marker associated with human natural killer cells is the antigen identified by the anti–Leu 7 monoclonal antibody (19). The Leu 7 antigen is a glycoprotein known as myelin-associated glycoprotein that is expressed on a subset of human NK cells (8), neuroectodermal tissue (20), prostate tissue, and small cell tumors of the lung (21). Recently, by two-color immunofluorescence and cell sorting using the FACS, we demonstrated that the Leu 11$^+$ population can be subdivided into two discrete populations expressing or lacking the Leu 7 antigen. The data in Table 13.1 indicate that on average approximately 50% of Leu 11$^+$ cells coexpressed Leu 7. The remaining half of Leu 7$^+$ cells lacked the Leu 11 antigen. A significant proportion (50%) of Leu 7$^+$ lymphocytes, presumably those which lacked the Leu 11 antigen, coexpressed the pan T cell antigen, Leu 4 (CD3) (Table 13.1, and Ref. 8). Leu 7$^+$ lymphocytes coexpressing the Leu 2a (CD8) antigen (48%) were observed in peripheral blood (Table 13.1 and Ref. 8). Unlike the low-density expression of Leu 2 on Leu 11$^+$ cells, the Leu 7$^+$,11$^-$ cells predominately expressed Leu 2 antigen in high cell surface density (8). Although a significant proportion of Leu 7$^+$ cells coexpressing Leu 3 were not seen in the peripheral blood from these normal donors (Table 13.1), Leu 3$^+$,7$^+$ cells have been observed in lymph node tissues and occasionally in the peripheral blood of certain individuals (L. Lanier and R. Fox, unpublished observation).

Cell Surface Antigens on Freshly Isolated Human NK Cells

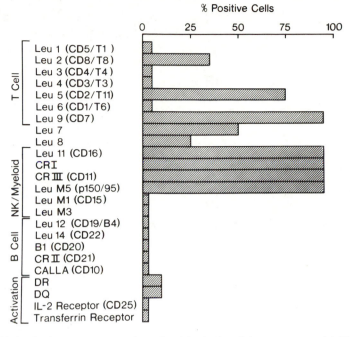

Fig. 13.2. Antigens expressed on freshly isolated human natural killer cells: a graphic summary.

A summary of the antigenic phenotype of freshly isolated human NK cells is graphically presented in Fig. 13.2. These data were complied from extensive two-color immunofluorescence studies using normal peripheral blood lymphocytes and Percoll gradient enriched large granular lymphocyte preparations. From these phenotypic studies the antigens can be divided into three general categories:

1. Antigens expressed on the majority of NK cells. Essentially all NK cells possessed the Ig-Fc receptor (Leu 11), CR_3, and CR_1 (defined using the 44D and 57F monoclonal antibodies). All three of these antigens are expressed in high density on neutrophils, but in low density or only on a very small proportion of Leu 4[+] T lymphocytes. The E-rosette receptor (Leu 5 epitope) was expressed on a majority of NK cells; however, the mean density of this antigen was lower than on T lymphocytes.
2. Antigens expressed on subpopulations of NK cells. NK cells were heterogeneous with respect to expression of the Leu 2, Leu 7, and Leu 8 antigens. In addition to NK cells, both Leu 2 and Leu 7 were expressed on a significant proportion of Leu 4[+] T lymphocytes (8).

Leu 8 antigen is present on granulocytes, monocytes, B lymphocytes, and some T lymphocytes (22). Studies by Gatenby *et al.* (23) have indicated that T cells with the phenotype Leu $3^+,8^-$ comprise the majority of "helper" cell activity in human peripheral blood. A small fraction of Leu 11^+ cells (27%) did coexpress this antigen in low density.

3. Antigens not expressed on resting NK cells. In general, we did not detect expression of several B cell-associated antigens [CR_2, DR, DQ (Leu 10) B1, Leu 12, and Leu 14] or T cell-associated antigens [Leu 1 (CD5), Leu 3 (CD4), and Leu 4 (CD3)] on a significant proportion of freshly isolated NK cells. Furthermore, these cells lacked the Leu M3 antigen associated with mature monocytes.

Correlation of Surface Phenotype with Natural Cell-Mediated Cytotoxicity

It is important to remember that NK cells are a *functionally* defined population, i.e., they must demonstrate the ability to kill certain sensitive targets to bear this designation. Although essentially all NK activity is encompassed within the Leu 11^+ subset, this does *not* imply that all Leu 11^+ lymphocytes are cytotoxic. Two-color immunofluorescence and FACS cell sorting were used to determine whether or not the antigenic heterogeneity observed within the Leu 11^+ population correlated with cytotoxic potential. Representative examples are shown in Fig. 13.3. Lymphocytes were stained with anti–Leu 11 and either anti–Leu 2a, anti–Leu 5, anti–Leu 7, or anti–Leu 8. The sorted subpopulations (>90% pure) were tested for cytotoxic activity against K562 target cells. All Leu 11^+ subsets (i.e., Leu $2^-,11^+$; Leu $2^+,11^+$; Leu $5^+,11^+$; Leu $5^-,11^+$; Leu 7^-11^+; Leu $7^+,11^+$; Leu $8^-,11^+$; and Leu $8^+,11^+$) mediated potent NK activity, demonstrating enrichment relative to the unsorted control cells. We cannot comment on whether the slight quantitative differences in cytotoxicity between the subsets identified by Leu 2, Leu 5, and Leu 8 are meaningful or simply due to individual variation or technical differences in the assays. Such interpretation would require numerous sorting experiments to establish a consistent pattern. However, with respect to the subpopulations identified on the basis of Leu 7 and Leu 11 expression, we have undertaken an extensive analysis of these cells. Based on more than 20 sorting experiments with different individuals, the cytotoxic potential of the subsets appears to be: Leu $7^-,11^+$ > Leu $7^+,11^+$ $\gg$ Leu $7^+,11^-$ $\gg$ Leu $7^-,11^-$. Whether or not the cells expressing the Leu 7 and Leu 11 antigens are developmentally related is currently under investigation.

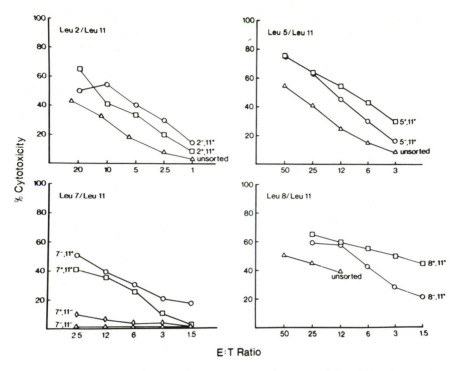

Fig. 13.3. NK activity of Leu 11⁺ lymphocyte subsets. Peripheral blood mononu-clear cells were isolated and stained with the indicated antibodies as described in Table 13.1. The indicated phenotypic subsets were separated by cell sorting using a FACS system. Reanalysis of the separated populations indicated >90% purity in all cases. The subsets or stained, unsorted cells (control) were assayed for NK activity against the NK-sensitive target K562 (8,24).

Activation of Leu 11⁺ Cells in a Mixed Lymphocyte/Tumor Cell Culture

We have reported that cells isolated from a mixed lymphocyte/tumor cell culture (MLTC) with the ability to lyse NK-sensitive targets can be identi-fied using the anti–Leu 11 antibody (24). These cells are large granular lymphocytes and most lack the pan T cell antigens Leu 1 (CD5) and Leu 4 (CD3). In a MLTC generated against a human EBV-transformed B cell line, CCRF-SB, approximately 10–15% of the lymphoblasts were Leu 11⁺ activated killer cells, whereas the rest were Leu 4⁺ lymphoblasts (24). In contrast to Leu 11⁺ cells isolated from control cultures (culture without the stimulator), a significant proportion of the activated Leu 11⁺ cells coexpressed class II MHC antigens, DR and DQ (Leu 10) (Table 13.2). Moreover, a proportion of the MLTC-activated Leu 11⁺ cells also ac-

Table 13.2. Activation of Leu 11$^+$ lymphocytes in a mixed lymphocyte/tumor cell culture: Appearance of activation antigens.

Culture[a]	Percentage of Leu 11$^+$ lymphoblasts coexpressing activation antigens[b]		
	HLA-DR	Transferrin receptor	IL-2 receptor
Control culture	2–3%	<2%	3–5%
SB-stimulated MLTC	60–70%	50–60%	40–50%
K562-stimulated MLTC	50–80%	60–70%	50–85%

[a] MLTC were obtained by co-culturing peripheral blood mononuclear cells with either mitomycin C-treated CCRF-SB or K562 cells at responder to stimulator ratios of 1:1 and 5:1, respectively. Control cultures consisted of peripheral blood mononuclear cells incubated without stimulator cells. After 5 days culture, viable cells were isolated and stained by two-color immunofluorescence as described in Table 13.1. For the MLTC-activated cells, a gate was set to include blast cells as identified by their characteristic low forward-angle and 90-degree light scatter patterns on the FACS. Since there was no significant blastogenesis in the control cultures, analysis gates were set to include all viable cells. Results are presented as the percentage of Leu 11$^+$ cells coexpressing either HLA-DR, transferrin receptor or IL-2 receptor. Anti-Tac monoclonal antibody was generously provided by Dr. T. Waldmann, NIH.
[b] Values reported represent the approximate range of percentages obtained from four independent experiments. In the control cultures and MLTC generated against CCRF-SB approximately 10–15% of the total lymphocytes isolated from the culture were Leu 11$^+$; whereas 70–80% of the total lymphocytes isolated from the K562-stimulated MLTC expressed the Leu 11 antigen.

quired expression of the IL-2 receptor and transferrin receptor. The cytotoxic capacity of the MLTC-activated Leu 11$^+$ cells was significantly enhanced relative to Leu 11$^+$ cells from control cultures (24).

Since it was possible to generate a small proportion of activated Leu 11$^+$ cells using an NK-insensitive stimulator, we questioned whether or not an NK-sensitive stimulator would preferentially favor expansion and activation of Leu 11$^+$ cells, rather than Leu 4$^+$ T lymphocytes. The majority of lymphoblasts (80%) isolated from a 6-day MLTC using K562 as the stimulator expressed the Leu 11 antigen. As with the SB-activated MLTC, a significant proportion of these cells now coexpressed class II MHC gene products, transferrin receptors, and IL-2 receptors, and possessed significantly higher cytotoxic activity than control cultured cells (25).

Depletion of Leu 11$^+$ lymphocytes prior to co-culture with either CCRF-SB or K562 prevented the generation of activated Leu 11$^+$ cells. Of particular interest was the observation that depletion of Leu 7$^+$ positive cells prior to culture neither qualitatively nor quantitatively diminished the generation of the Leu 11$^+$ lymphoblasts (24,25). Therefore, the peripheral blood precursor of the Leu 11$^+$ activated killer is Leu 7$^-$,11$^+$.

A Closer Look: Three-Color Immunofluorescence and Flow Cytometric Analysis of NK Subsets

The Leu 7$^-$,11$^+$ population, a small fraction of total lymphocytes (4.6 ± 2.1% of lymphocytes) appears to encompass the precursor population for

the MLTC-activated Leu 11⁺ cells. In order to investigate the antigenic
heterogeneity within this subset, Percoll enriched large granular lymphocytes (prepared as described in Ref. 14) were stained with: FITC anti–Leu
11a, allophycocyanin (APC) avidin/biotin anti–Leu 7, and either b-phyco-
erythrin (PE) anti–Leu 2a, r-phycoerythrin anti–Leu 5b, or r-phycoerythrin anti–Leu 8. In the upper panels of Fig. 13.4, the fluorescence profiles
of the large granular lymphocyte-enriched population are shown. Since
the major population of interest was the Leu 11⁺ subset, a gate was set on
FITC-stained Leu 11⁺ cells, then the APC and PE fluorescence distributions of only the Leu 11⁺ cells were displayed. It was evident from these
studies that within the Leu 7⁻,11⁺ population, there was still heterogeneity with respect to expression of the Leu 2, Leu 5, and Leu 8 antigens.
Whether or not the activated killer cell precursor will be preferentially
localized within these subgroups can be experimentally determined by
three-color cell sorting.

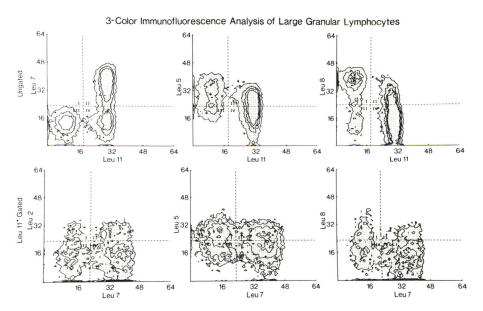

Fig. 13.4. Analysis of large granular lymphocytes using three-color immunofluorescence. Lymphocytes with large granular lymphocyte morphology and NK
activity were enriched by Percoll gradient separation as described previously (14).
1 × 10⁶ cells in 50 μl PBS/azide were stained with optimal amounts of FITC anti–
Leu 11a and r-PE anti–Leu 8, r-PE anti–Leu 5b, or b-PE anti–Leu 2a for 15 min.
The cells were washed once, stained with biotin anti–Leu 7 for 15 min, washed
twice, and then stained with allophycocyanin–avidin. After an additional 15 min,
the cells were washed twice, and fixed in PBS containing 1.0% paraformaldehyde
(pH 7.2). All procedures were carried out at 4°C. Analysis was performed using a
FACS 440 system equipped with an argon ion laser (488 nm excitation for FITC
and PE) and a dye laser (590 nm excitation for APC), as described previously (26).

Conclusions

These studies indicate that the non-T, non-B lymphoid population in human peripheral blood is hardly "null" with respect to expression of surface antigens. NK cells share certain antigens both with myeloid cells and T lymphocytes. In particular, NK cells share an antigenically similar Fc receptor with neutrophils, and share common complement receptors (CR_1 and CR_3) with both granulocytes and monocytes. However, like T lymphocytes, at least a proportion of NK cells bear the Leu 2 (CD8) and Leu 5 (CD2)/E-rosette receptors. Additionally, the Leu 7 antigen is apparently shared between a subset of NK cells and a subset of Leu 4^+ (CD3) T lymphocytes. Based on these studies and similar studies by other groups, it is evident that the array of the surface antigens expressed by these cells does not unequivocally resolve the question of whether or not NK cells arise from a T cell or myeloid lineage.

With respect to function, NK cells kill sensitive targets via a mechanism that is distinct from cytotoxic T lymphocytes. The cytotoxicity is not restricted by the MHC and does not demonstrate fine antigen specificity. In this respect, these cells are more similar to activated macrophages than T cells. We have shown that potent NK cytotoxicity can be mediated by several phenotypically distinct subsets of Leu 11^+ cells. Furthermore, it is possible to stimulate a particular phenotypic subset of NK cells, i.e., Leu $7^-,11^+$ lymphocytes, into proliferation and enhanced cytotoxic activity in a mixed lymphocyte/tumor cell culture. Like the T cells in these cultures, Leu 11^+ cells acquire the expression of IL-2 receptors and class II MHC antigens upon activation.

The most encouraging aspect of recent studies is that it is now possible to positively identify and isolate NK cells in high purity using monoclonal antibodies. These cells, once defined by what they were not, now are appreciated as a distinct and possibly important component of the immune system.

Acknowledgments. We thank Ms. An My Le for superb technical assistance with the functional studies and Mr. Keith Kelly for expert assistance with the flow cytometry.

References

1. Jondal, M., G. Holm, and H.J. Wigzell. 1972. Surface markers on human T and B lymphocytes. I. A large population of lymphocytes forming nonimmune rosettes with sheep red blood cells. *J. Exp. Med.* **136**:207.
2. Yanagi, Y., Y. Yoshikai, K. Leggett, S.P. Clark, I. Aleksander, and T.W. Mak. 1984. A human T cell-specific cDNA clone encodes a protein having extensive homology to immunoglobulin chains. *Nature* **308**:145.

3. Hedrick, S.M., D.I. Cohen, E.A. Nielsen, and M.M. Davis. 1984. Isolation of cDNA clones encoding T cell-specific membrane-associated proteins. *Nature* **308:**149.

4. Herberman, R.B., ed. 1982. *NK cells and other natural effector cells.* Academic Press, New York.

5. Herberman, R.B., J. Djeu, H.D. Kay, J.R. Ortaldo, C. Riccardi, G.D. Bonnard, H.T. Holden, R. Fagnani, A. Santoni, and P. Puccetti. 1979. Natural killer cells: characteristics and regulation of activity. *Immunol. Rev.* **44:**43.

6. Oldham, R.K. 1982. Natural killer cells: history and significance. *J. Biol. Response Modif.* **1:**217.

7. Roder, J.C., and H.F. Pross. 1982. The biology of the human natural killer cell. *J. Clin. Immunol.* **2:**249.

8. Lanier, L.L., A.M. Le, J.H. Phillips, N.L. Warner, and G.F. Babcock. 1983. Subpopulations of human natural killer cells defined by expression of the Leu 7 (HNK-1) and Leu 11 (NKP-15) antigens. *J. Immunol.* **131:**1789.

9. Rumpold, H., D. Kraft, G. Obexer, G. Bock, and W. Gebhart. 1982. A monoclonal antibody against a surface antigen shared by human large granular lymphocytes and granulocytes. *J. Immunol.* **129:**1458.

10. Perussia, B., S. Starr, S. Abraham, V. Fanning, and G. Trinchieri. 1983. Human natural killer cells analyzed by B73.1, a monoclonal antibody blocking Fc receptor functions. I. Characterization of the lymphocyte subset reactive with B73.1. *J. Immunol.* **130:**2133.

11. Perussia, B., G. Trinchieri, A. Jackson, N.L. Warner, J. Faust, H. Rumpold, K. Kraft, and L.L. Lanier. 1984. The Fc receptor for IgG on human natural killer cells: phenotypic, functional, and comparative studies with monoclonal antibodies. *J. Immunol.* **133:**180.

12. Perussia, B., and G. Trinchieri. 1984. Antibody 3G8, specific for the human neutrophil Fc receptor, reacts with natural killer cells. *J. Immunol.* **132:**1410.

13. Timonen, T., and E. Saksela. 1980. Isolation of human natural killer cells by density gradient centrifugation. *J. Immunol. Methods* **36:**285.

14. Phillips, J.H., N.L. Warner, and L.L. Lanier. 1984. Correlation of biophysical properties and cell surface antigenic profile of Percoll gradient-separated human natural killer cells. *Natl. Immun. Cell Growth Regul.* **3:**73.

15. Reinherz, E.L., S.C. Meuer, K.A. Fitzgerald, R.E. Hussey, J.C. Hodgdon, O. Acuto, and S.F. Schlossman. 1983. Comparison of T3-associated 49- and 43-kilodalton cell surface molecules on individual human T-cell clones: Evidence for peptide variability in T-cell receptor structures. *Proc. Natl. Acad. Sci. U.S.A.* **80:**4104.

16. Allison, J.P., L. Ridge, J. Lund, J. Gross-Pelose, L.L. Lanier, and B.W. McIntyre. 1984. The murine T cell antigen receptor and associated structures. *Immunol. Rev.* **81:**145.

17. Perussia, B., V. Fanning, and G. Trinchieri. 1983. A human NK and K cell subset shares with cytotoxic T cells expression of the antigen recognized by antibody OKT8. *J. Immunol.* **131:**223.

18. Ault, K.A., and T.A. Springer. 1981. Cross-reaction of a rat anti-mouse phagocyte-specific monoclonal antibody (anti-Mac-1) with human monocytes and natural killer cells. *J. Immunol.* **126:**359.

19. Abo, T., and C.M. Balch. 1981. A differentiation antigen of human NK and K cells identified by a monoclonal antibody (HNK-1). *J. Immunol.* **127:**1024.

20. Lipinski, M., K. Braham, J.-M. Caillaud, C. Carlu, and T. Tursz. 1983. HNK-1 antibody detects an antigen expressed on neuroectodermal cells. *J. Exp. Med.* **158:**1775.
21. Bunn, P.A., A.F. Gazdar, D.N. Carney, and J. Minna. 1984. Small cell lung carcinoma and natural killer cells share an antigenic determinant, Leu 7. *Clin. Res.* **32:**413A.
22. Lanier, L.L., E.G. Engleman, P. Gatenby, G.F. Babcock, N.L. Warner, and L.A. Herzenberg. 1983. Correlation of functional properties of human lymphoid cell subsets and surface marker phenotypes using multiparameter analysis and flow cytometry. *Immunol. Rev.* **74:**147.
23. Gatenby, P.A., G.S. Kansas, C.Y. Xian, R.L. Evans, and E.G. Engleman. 1982. Dissection of immunoregulatory subpopulations of T lymphocytes within the helper and suppressor sublineages in man. *J. Immunol.* **129:**1997.
24. Phillips, J.H., A.M. Le, and L.L. Lanier. 1984. Natural killer cells activated in a human mixed lymphocyte response culture identified by expression of Leu 11 and class II histocompatibility antigens. *J. Exp. Med.* **159:**993.
25. Phillips, J.H., and L.L. Lanier. 1984. In: *Mechanism for cytotoxicity by NK cells,* R.B. Herberman, ed. Academic Press, New York. In press.
26. Loken, M.R., and L.L. Lanier. 1984. Three-color immunofluorescence analysis of Leu antigens on human peripheral blood using two lasers on a fluorescence-activated cell sorter. *Cytometry* **5:**151.

CHAPTER 14

Inhibition of CFU-GM, BFU-E, and CFU-GEMM Colony Formation by Monoclonal Antibodies Selected from the Myeloid Panel

Anna Janowska-Wieczorek, P.J. Mannoni, M.J. Krantz,
A.R. Turner, and J.M. Turc

Introduction

The human hematopoietic stem cells, multipotential and committed, are being extensively studied with immunologic probes. Detailed characterization of antigenic phenotypes of stem cells is of major importance for further investigation of the events that separate self-renewing stem cells from their committed progenitors, as well as for the development of purification techniques that would be based on differential antigenic reactivity of the stem cells, their maturing descendants, and of neoplastic clones. No culture system has been proved to assay the true human pluripotent hematopoietic stem cell, although the mixed-colony assay described by Fauser and Messner and their colleagues (1,2) detects bi-, tri-, and multi-potential progenitor cells that may represent a developmental stage very close to the objective. The techniques most commonly used to investigate the reactivity of monoclonal antibodies with hematopoietic stem cells are: complement-dependent cytotoxicity, fluorescence-activated cell sorting, and immune adherence. All of these techniques have limitations. Moreover, alterations in membrane antigens do not necessarily indicate absolute gain or loss of an antigen: they may represent quantitative changes in the density of an antigen or, as proposed for glycolipid antigens, exposure at the cell surface (3). There is also evidence that the expression of antigens may relate to the proliferative state of the stem cell (4,5) or depend upon environmental inductive influences (6), or may be modulated by certain enzymes (7,8). These and other closely related factors, such as different techniques for culturing progenitors *in vitro,* may account at least partly for inter-laboratory differences in stem-cell phenotyping.

The aim of the present study was to test the reactivity of multipotential and committed progenitors by complement-dependent lysis and mixed-colony assay (1,2) with 23 monoclonal antibodies selected from the Myeloid Workshop panel, to further characterize stem-cell phenotype. In addition, because several recently generated "myeloid-specific" antibodies recognize very similar antigenic structures defined as the oligosaccharide 3-fucosyl-N-acetyllactosamine (7,9,10) or X-hapten, previously shown to be the embryonic stage-specific antigen 1 in the mouse (11), and because of reported differences in stem-cell reactivity with these antibodies (12–14), we made detailed studies of those Panel monoclonal antibodies that recognize X-hapten. Recently, Gooi *et al.* (7) reported that antigenic activity of X-hapten may be masked in the presence of the second fucose residue and sialic acid. Tabilio *et al.* (8) stated that common ALL blasts and several lymphoblastoid and Burkitt cell lines were positive with the above "myeloid-specific" antibodies and that sialidase significantly enhanced their antigenicity. These findings could indicate that sialic acid partly masks X-hapten during the early stage of hematopoietic differentiation. We tested this hypothesis by investigating X-hapten on stem cells from which the sialic acid had been removed.

Materials and Methods

Twenty-three monoclonal antibodies, selected from the Myeloid Workshop panel on the basis of their complement-binding properties, serological activity, and/or biochemical definition, were tested on multipotential and committed progenitors. They included monoclonal antibodies that recognize myeloid cells, natural killer cells, antigens of leukocyte-function-associated (LFA) type, or platelet-specific glycoprotein IIb/IIIa, or are known to react with X-hapten. In addition, we studied 80H5 and 82H6 (15,16), whose carbohydrate specificity as determined with a panel of synthetic antigens (synthesized by Chembiomed Ltd., University of Alberta) had shown their reactivity with X-hapten (unpublished). A monoclonal antibody that recognizes β_2 microglobulin (kindly provided by Dr M. Fellous, Paris) was used as control.

Bone marrow from the sternum of patients undergoing cardiac surgery was aspirated into preservative-free heparin and centrifuged. The buffy coat was suspended in Iscove's modified Dulbecco's medium (IMDM; Gibco), and light-density marrow cells were separated by centrifugation on 60% Percoll. Aliquots (100 μl) of 6×10^5 cells, supplemented with 5% fetal calf serum (FCS), were incubated with an equal volume of monoclonal antibodies (final concentration, 1:40) for 30 min at room temperature. Rabbit complement nontoxic for stem cells (Low-Tox-H-Rabbit

Table 14.1. Effects of neuraminidase (Nase) on colony formation.

	No. of colonies (mean ± SEM) per 2 × 10^5 cells plated		
	CFU-GM $n = 8$	BFU-E $n = 8$	CFU-GEMM $n = 2$
Cells	138 ± 23	116 ± 36	9 ± 1
Cells + C'	128 ± 23	115 ± 37	11 ± 1
Cells + Nase	138 ± 23	112 ± 36	9 ± 3
Cells + Nase + C'	121 ± 22	87 ± 41	9 ± 1

complement; Cederlane Lab.) was added to a final dilution of 1:4, and incubation was continued for a further 60 min at room temperature. Cells were washed twice before plating for mixed-colony assay (1,2), a technique that permits determination of growth of several types of colonies in one dish. Bone marrow cells were plated at a concentration of 2×10^5 cells/ml, or what was left after antibody and complement treatment, in 35-mm Lux standard tissue culture dishes containing a 1-ml mixture of IMDM, 0.9% methylcellulose with 5% medium conditioned by human leukocytes with 1% phytohemagglutinin added, 30% fresh human plasma, 5×10^{-5} M 2-mercaptoethanol, and 10% FCS. Human urinary erythropoietin (British Columbia Research Centre, Vancouver), 2.5 IU in 50 μl IMDM, was added to each dish just before the start of culture. Dishes were incubated at 37°C in a humidified atmosphere with 5% CO_2 in air. On days 13 to 16, each dish was viewed through an inverted microscope (Zeiss IM 35), 50–125× magnification, and CFU-GEMM, BFU-E, and CFU-GM colonies were counted.

In preliminary studies, prior treatment of the bone marrow cells with 0.5 U of neuraminidase (with or without complement) did not significantly alter the formation of CFU-GM, BFU-E, and CFU-GEMM colonies (Table 14.1); therefore, this method was used for the detailed studies of carbohydrate antigens on stem cells. Neuraminidase from the culture filtrate of *Vibrio cholerae* (Gibco, Grand Island, NY) was added in a concentration of 0.5 U/2 × 10^6 light-density bone marrow cells/ml. The cells were incubated for 60 min at 37°C, then washed four times, before incubation with monoclonal antibodies as above.

Results

In studies of the 23 Panel monoclonal antibodies tested by complement-dependent lysis of CFU-GM, BFU-E, and CFU-GEMM progenitor cells, the most significant killing was observed with M11 antibody (Tables 14.2 and 14.3). The M11 antigen was expressed on >95% CFU-GM cells, >90% BFU-E, and 100% CFU-GEMM cells; when tested without com-

Table 14.2. Reactivity of selected monoclonal antibodies with committed progenitors.

	Antigen definition	Cell specificity	Inhibition (%, mean + SD)	
			CFU-GM	BFU-E
	(80H5) X-hapten	Myeloid	8 ± 4	19 ± 8
	(82H6) X-hapten	Myeloid	5 ± 3	24 ± 8
M5	(82H5) X-hapten	Myeloid	29 ± 8	60 ± 8
M13	(1G10) X-hapten	Myeloid	12 ± 1	10 ± 3
M19	(HLC5) X-hapten	Myeloid	4 ± 4	11 ± 5
M25	(VIMC6) X-hapten	Myeloid	0 ± 0	14 ± 6
M1	(NKH2) 60 Kd	NK	0 ± 0	0 ± 0
M6	(86H1)	AML, T-ALL, myeloid	8 ± 1	1 ± 1
M10	(EDU3) IIb/IIIa	Platelets, U 937, HEL	0 ± 0	0 ± 0
M11	(JOAN-1)	Mo, T cells, AML, HL 60	96 ± 6	91 ± 12
M15	(T5A7)	GR, U 937, HEL	4 ± 5	4 ± 5
M16	(L1B2) 55 Kd	U 937, HL 60, HEL	16 ± 13	17 ± 23
M18	(TM.2.26) 120 Kd	Thymocytes, HEL, U 937	33 ± 2	29 ± 4
M24	(MHM 31) 190, 110 Kd	T and myeloid cell lines	2 ± 2	0 ± 0
M27	(VIM8) 175 Kd	GR, HL 60, U 937	0 ± 0	0 ± 0
M30	(VIM10) 165 Kd	GR	18 ± 20	0 ± 0
M33	(G7C5) 100 Kd	GR, Mo, HL 60, U 937	12 ± 17	0 ± 0
M42	(D12)	NK, GR	17 ± 23	0 ± 0
M44	(J.15) IIb/IIIa	Platelets	3 ± 4	0 ± 0
M46	(SHCL3) 150, 90 Kd	GR, thymocytes	10 ± 5	0 ± 0
M54	(GO22b) 38 Kd	NK, GR, U 937	17 ± 4	6 ± 8
M61	(29) 165 Kd	GR, thymocytes, U 937, K562	0 ± 0	0
M78	(M3D11)	NK, GR, U 937, thymocytes, K562	7 ± 10	10 ± 14
M75	(60.3) 95, 130, 150 Kd	AML, LFA-1-like	0 ± 0	5 ± 6
M76	(60.1) 95, 110 Kd	LFA-1-like	3 ± 4	5 ± 7

plement it did not affect the growth of colonies (Table 14.3). Some other monoclonal antibodies detected antigens on only a fraction of progenitor cells or were unreactive with stem cells. Of the six antibodies known to recognize X-hapten, M5 (82H5) was the strongest inhibitor, killing 29% of CFU-GM and 60% of BFU-E colonies. Other antibodies that recognize X-hapten inhibited the proliferation of less than 20% of progenitors or had no effect on colony formation.

M18 (TM.2.26), also, partly inhibited colony formation. The antigen recognized by this antibody was found on approximately 30% of CFU-GM, BFU-E, and CFU-GEMM (Tables 14.2 and 14.3). The reactivity of progenitor cells with M16 (L1B2), M30 (VIM10), M33 (G7C5), and M42 (D12) was variable in experiments on bone marrow cells from two donors.

Examination of CFU-GM, BFU-E, and CFU-GEMM colonies incubated with cytotoxic anti-NK antibodies (M1, M42, M54, and M78) revealed absence of M1 (NKH2) 60-Kd antigen on multipotential and committed progenitor cells and <20% inhibition (not significant) with M42,

Table 14.3. Effects of monoclonal antibodies (myeloid series), with and without complement, on colony formation.

	Mean no. of colonies per 2×10^5 cells plated[a]					
	CFU-GM		BFU-E		CFU-GEMM	
mAb	Alone	+ C	Alone	+ C	Alone	+ C
Controls						
Cells only	120	113	80	72	4	5
Anti-B$_2$m	115	1	79	1	5	0
M1	121	114	91	92	5	6
M6	116	102	74	72	3	7
M10	123	112	92	77	5	6
M11	120	0	88	1	5	0
M15	124	116	80	82	6	5
M16	121	85	86	49	6	2
M18	109	78	71	54	6	2
M24	114	110	83	84	5	8
M27	122	117	80	84	6	6
M30	119	77	90	91	6	7
M33	120	85	77	76	6	9
M42	116	75	78	77	6	7
M44	124	107	87	79	6	5
M46	121	98	83	74	6	6
M54	124	90	88	64	5	5
M61	115	113	76	ND	6	ND
M78	105	97	83	58	7	3

[a] Means of triplicate cultures from one representative bone marrow sample.

M54, and M78. Treatment with the cytotoxic antibodies M75 and M76, which recognize the LFA antigen family, did not inhibit colony formation. Also, antibodies M10 and M44, which are reactive with platelet-specific glycoprotein IIb/IIIa, did not inhibit the growth of CFU-GM, BFU-E, and CFU-GEMM. Antibody M6 (86H1), which reacts with AML and T-ALL blast cells (unpublished), did not inhibit multipotential and committed progenitor cells.

All six monoclonal antibodies that recognize X-hapten significantly inhibited the proliferation of CFU-GM, BFU-E, and CFU-GEMM colonies when tested on neuraminidase-treated progenitor cells (Figs. 14.1 and 14.2). In control cultures with antibody 80H3, which is IgG$_1$ glycoprotein (15), prior treatment of the bone marrow cells with neuraminidase did not influence expression of the antigen: CFU-GM, BFU-E, and CFU-GEMM growth was similar with and without neuraminidase. Antibody M5 (82H5) showed stronger reactivity than any other antibody of this group in cultures of untreated cells and of cells first incubated with neuraminidase. Of the six antibodies that recognize X-hapten, 82H6 produced the least inhibition.

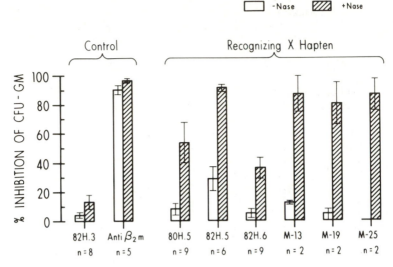

Fig. 14.1. Effects of neuraminidase and monoclonal antibodies that recognize X-hapten on the formation of CFU-GM colonies.

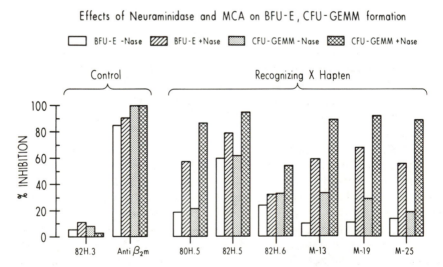

Fig. 14.2. Effects of neuraminidase and monoclonal antibodies that recognize X-hapten on the formation of BFU-E and CFU-GEMM colonies.

Discussion

In the studies of 23 monoclonal antibodies from the Myeloid Workshop by complement-dependent lysis of CFU-GM, BFU-E, and CFU-GEMM, only M11 (JOAN-1) completely inhibited proliferation of the human multipotential and committed precursors. Findings in concomitant studies (this volume, Chapter 29) indicate that this antibody is also a marker of monocytes, mature T and B lymphocytes, human myeloid cell lines, and AML blasts (Table 14.4), and in its serological pattern resembles M6 (86H1) and M75 (LFA-1-like). Both of these latter antibodies, however, although cytotoxic, were unreactive with progenitor cells. The M11 (JOAN-1) antibody has been raised against mononuclear cells from the peripheral blood of a patient who had infectious mononucleosis (Villela *et al.*, personal communication, 1984). Its pattern of expression is different from that of the other known stem-cell markers RFB1-1 and OKT10, both of which are unreactive with peripheral B-cells (17,18).

As expected, monoclonal antibodies against markers associated with the late stage of lineage differentiation (e.g., LFA-1 and platelet-specific glycoprotein IIb/IIIa) did not inhibit the growth of CFU-GEMM, BFU-E, and CFU-GM. The variable and/or part-inhibition induced by M16 (L1B2), M18 (TM.2.26), M30 (VIM10), M33 (G7C5), and M42 (D12) may reflect variable and/or low density of the antigen on stem cells obtained from different individuals or individual variation in lysis of accessory cells that are critical to colony proliferation. None of the monoclonal antibodies was clearly specific for progenitor cells committed to a certain lineage. Also, none of them inhibited the growth of CFU-GM, BFU-E, or CFU-

Table 14.4. Immunofluorescence studies (% positive cells) with monoclonal antibody M11 (JOAN-1).

Normal cells		Cell lines	
Granulocytes	0	U 937	99
Monocytes	76	HL-60	99
Erythrocytes	0	ML1	65
B cells	95	KG1	26
T cells	95	HEL	66
Thymocytes	0	K562	0
Leukemic cells		MOLT-4	40
AML : M1	94	1301	42
M2 (n = 3)	83 ± 9 (SD)	Ichigawa	0
M6	94	Namalwa	84
T-ALL	0	Daudi	65
c-ALL	97		

GEMM in the absence of complement, leading to the conclusion that they do not bind to membrane receptors. Inhibition of the colony formation by the six monoclonal antibodies reactive with the oligosaccharide structure 3-fucosyl-*N*-acetyllactosamine was highly variable—0–29% of CFU-GM and about 15–60% of BFU-E and CFU-GEMM. It is of note that colonies of the more immature BFU-E were more greatly inhibited than those of CFU-GM.

The reactivity of these antibodies with an oligosaccharide-containing X-hapten was significantly increased by removing sialic acid residues, a finding consistent with the observation that several factors may affect the reactivity of glycolipids at the cell surface (3). Factors that may contribute to crypticity include masking by other lipids, oligosaccharides, or proteins, differences in the fine structure of the carbohydrate chain beyond the region usually designated as hapten (such as the length, degree, or branching of this chain), and the structure of the lipid moiety. Differences in the degree of complement-dependent lysis of stem cells by monoclonal antibodies that recognize X-hapten could relate to the above factors. Determinants found on a fraction of erythroid and multipotential progenitors may reflect this variation in oligosaccharide structures shared by other cell types rather than uniqueness of the antigenic determinants. A further anomalous aspect of the specificity of these antibodies is their lack of reactivity with erythrocytes even though glycolipids containing the X-hapten structure have been purified from erythrocytes. X-hapten is not detectable on normal mature cells of the erythroid, lymphoid, or megakarocytic lineage; the only cells known to be reactive with X-hapten-specific antibodies during their maturation are those of myeloid lineage, an observation that underlines the significance of sialic acid in hematopoietic differentiation.

The results of this study show that the inhibition of colonies grown in mixed-colony assay can be used to identify monoclonal antibodies reactive with antigenic determinants expressed on progenitor cells from the myeloid and erythroid lineage as well as on multipotential stem cells. The use of cell-sorting techniques is needed to validate these results, to establish whether the findings relate specifically to the stem-cell population.

Moreover, as M6 (86H1) was strongly positive on all kinds of AML blasts and not on progenitor cells, such antibodies could be candidates for bone marrow purging or therapeutic trials. Also required is the identification of more monoclonal antibodies that define antigens which are stem cell specific, so that the entire antigenic structure of the progenitor cells can be mapped.

Acknowledgment.   A. Janowska-Wieczorek is in receipt of a Fellowship and research allowance from the Alberta Heritage Foundation for Medical Research.

References

1. Fauser, A.A., and H.A. Messner. 1978. Granuloerythropoietic colonies in human bone marrow, peripheral blood, and cord blood. *Blood* **53**:1243.
2. Messner, H.A., N. Jamal, and C. Izaguirre. 1982. The growth of large megakaryocyte colonies from human bone marrow. In *Cellular and molecular biology of hemopoietic stem cell differentiation* T.W. Mak and E.A. McCulloch, eds. Liss, New York, pp 45–51.
3. Hakomori, S.I., and R. Kannogi. 1983. Glycolipids as tumor-associated and differentiation markers. *J. Natl. Cancer Inst.* **71**:231.
4. Broxmeyer, H.E. 1982. Relationship of cell-cycle expression of Ia-like determinants on normal and leukemia human granulocyte-macrophage progenitor cells to regulation in vitro by acidic isoferritins. *J. Clin. Invest.* **69**:632.
5. Griffin, J.D., J. Ritz, R.P. Beveridge, J.M. Lipton, J.F. Daley, and S.F. Schlossman. 1983. Expression of My7 antigen on myeloid precursor cells. *Int. J. Cell Cloning* **1**:33.
6. Ball, E., P.M. Guyre, J.M. Glynn, W. Rigby, and M. Fanger. 1984. Modulation of class 1 HLA antigens on HL-60 promyelocytic leukemic cells by serum-free medium: re-induction by γ-IFN and 1,25-dihydroxyvitamin D_3. *J. Immunol.* **132**:2424.
7. Gooi, H.C., S.J. Thorpe, E.F. Hounsell, H. Rumpold, D. Kraft, O. Forster, and T. Feizi. 1983. Marker of peripheral blood granulocytes and monocytes of man recognized by two monoclonal antibodies VEP8 and VEP9 involves the trisaccharide 3-fucosyl-*N*-acetyllactosamine. *Eur. J. Immunol.* **13**:306.
8. Tabilio, A., C. Canizo, P. Mannoni, C.I. Civin, U. Testa, H. Rochant, J. Breton-Gorius, and W. Vainchenker. 1983. Modulation of the expression of the carbohydrate antigen (3-fucosyl-*N*-acetyl lactosamine) on normal and leukemic human hemopoietic cells. *Blood* **62**(suppl.1):155a (abstract).
9. Huang, L.C., C.I. Civin, J.L. Magnani, J.H. Shaper, and V. Ginsberg. 1983. My-1, the human myeloid-specific antigen detected by mouse monoclonal antibodies, is a sugar sequence found in lacto-*N*-fucopentaose. III. *Blood* **61**:1020.
10. Urdal, D., T.A. Brentnall, I.D. Bernstein, and S.I. Hakomori. 1983. A granulocyte reactive monoclonal antibody, 1G10, identifies the Gal-β1-4-(Fucα1-3)-GlcNAc (X determinant) expressed in HL60 cells on both glycolipid and glycoprotein molecules. *Blood* **62**:1022.
11. Gooi, H.C., T.Feizi, A. Kapadia, B.B. Knowles, D. Solter, and M.J. Evans. 1981. State-specific embryonic antigen involves α1-3 fucosylated type-2 blood-group chains. *Nature* **292**:156.
12. Janowska-Wieczorek, A., P. Mannoni, A.R. Turner, L.E. McGann, A.R.E. Shaw, and J.M. Turc. 1984. Monoclonal antibody specific for granulocytic-lineage cells and reactive with human pluripotent and committed haematopoietic progenitor cells. *Brit. J. Haematol.* **59**:159.
13. Andrews, R.G., B. Torok-Storb, and I.D. Bernstein. 1983. Myeloid-associated differentiation antigens on stem cells and their progeny identified by monoclonal antibodies. *Blood* **62**:124.
14. Strauss, L.C., R.K. Stuart, and C.I. Civin. 1983. Antigenic analysis of hematopoiesis. I. Expression of the My-1 granulocyte surface antigen on human marrow cells and leukemic cell lines. *Blood* **61**:1222.

15. Mannoni, P., A. Janowska-Wieczorek, A.R. Turner, L. McGann, and J.M. Turc. 1982. Monoclonal antibodies against human granulocytes and myeloid differentiation antigens. *Hum. Immunol.* **5**:309.
16. Mannoni, P., A. Janowska, P. Fromont, B. Weiblen, A.R. Turner, and J.M. Turc. 1984. Human myeloid differentiation studied by monclonal antibodies. In: Leucocyte typing, by A. Bernard, L. Boumsell, J. Dausset, C. Milstein, and S.F. Schlossman, eds. Springer-Verlag, Berlin, Heidelberg, pp. 410–419.
17. Bodger, M.P., G.E. Francis, D. Delia, S.M. Granger, and G. Janossy. 1981. A monoclonal antibody specific for immature human hemopoietic cells and T lineage cells. *J. Immunol.* **127**:2269.
18. Bodger, M.P., C.A. Izaguirre, H.A. Blacklock, and A.V. Hoffbrand. 1983. Surface antigenic determinants on human pluripotent and unipotent hematopoietic progenitor cells. *Blood* **61**:1006.

CHAPTER 15

Reactivity of Anti-Myeloid Monoclonal Antibodies with Committed Hematopoietic Precursor Cells

Christian Peschel, Günther Konwalinka, Dietmar Geissler, Kurt Grünewald, Kristof Liszka, Hannes Stockinger, Otto Majdic, Herbert Braunsteiner, and Walter Knapp

Introduction

The development of monoclonal antibodies against surface structures of myeloid cells has contributed to the elucidation of differentiation processes at different levels of hematopoiesis. Most anti-myeloid antibodies have been defined for expression on cells at the stage of terminal myeloid maturation as well as on leukemic cell populations (for review, see Ref. 1). Only in relatively few studies, however, have antigenic structures of hematopoietic precursor cells been investigated with myeloid antibodies (2–8). In addition, the study of hematopoietic precursor cells has been so far mainly restricted to the granulo-monocytic cell lineage. Studies with clonal assays for erythropoietic and megakaryocytic stem cells are lacking in most publications. Furthermore, different *in vitro* culture techniques may complicate the interpretation of clonal assays even if the antibodies used are directed against the same antigenic structure. For example, a number of antibodies recognize the carbohydrate structure 3-fucosyl-*N*-acetyllactosamine [1G10 (4), R1B19 (5), VIM-D5 (9)] exposed on committed myeloid progenitor cells. Still the extent of lysis of CFU-GM subpopulations observed in different laboratories with one or the other of these antibodies varies considerably (4,5,8), the main reason probably being differences in the experimental design. For clonal assays of erythroid progenitor cells the situation is even worse since culture techniques using methylcellulose (10), plasma clot (11),or agar (12) are even more difficult to compare.

Workshops like this one offer the unique opportunity to directly compare, under largely identical technical conditions, large numbers of antibodies from different laboratories. We therefore have selected 48 cytotoxic antibodies from the myeloid antibody protocol and have evaluated them with respect to their reactivity with myeloid, erythroid, and megakaryocytic progenitor cells.

Materials and Methods

Selection of Antibodies

From the total 118 antibodies in the myeloid Workshop we have selected 48 antibodies for evaluation in our progenitor cell assay systems. We have excluded from evaluation all antibodies which do not fix complement or where no information on that aspect was available. We have also excluded antibodies which were labeled as reacting with non-myeloid cells including NK cells or which were found to be directed against the 3-fucosyl-*N*-acetyllactosamine structure. Antibodies of the latter specifity have already been thoroughly tested in previous studies (4,5,8).

Antibody-Dependent Complement-Mediated Cytolysis

Bone marrow cells from informed and consenting adult volunteers were aspirated from the posterior iliac crest into preservative free heparin. Buffy coat cells were overlayered over Lymphoprep (density 1.077 g/ml, Nyegaard, Oslo). After washing in Hank's balanced salt solution (HBSS), mononuclear cells (MNC) were resuspended in modified McCoy's 5A medium (Gibco) at a concentration of 1×10^7 MNC/ml. A 0.2 ml volume of cell suspension was added to 0.2 ml of diluted antibody (final dilution 1:100) and incubated at room temperature for 30 min. A pretested single batch of rabbit complement (Behring Institut, K.-No 029980 A) was diluted with culture medium to a concentration of 1:4. Thereafter 0.6 ml of diluted complement was added and incubation continued for a further 60 min at 37°C. Cells were then washed twice in HBSS and resuspended in culture medium containing 20% FCS at a cell concentration of 2×10^6/ml based on the number of cells present in the sample prior to antibody treatment. Negative controls were cells incubated with medium only and cells incubated with medium and then with complement. Positive controls were ADCC test with VIP2b, an IgM antibody directed against the p45 structure also seen by the OKT10 antibody (13) and the myeloid antibody VIM-D5 (9) which has been proven to react with myeloid progenitors (8).

Clonal Assays

Committed Myeloid Progenitor Cell Assasy

In addition to CFU-GM day 7 and day 14, the number of cells which form clusters after 3 days of incubation was also evaluated. Cultures were performed according to a method previously described (14). MNC were suspended in modified McCoy's 5A medium with 20% FCS and 0.3% agar (Difco). Aliquots of 0.25 ml containing 1×10^5 cells were pipetted into multi-well tissue culture plates (Nunclon, Nunc, Denmark). After equilibration at room temperature, 0.25 ml of culture medium containing conditioned medium of the GCT cell line (Gibco) and 20% FCS were placed over the agar layer. Final GCT concentration for cultures scored after 3 and 7 days of incubation was 10%; CFU-GM day 14 were stimulated with 2.5% GCT. After 3, 7, and 14 days of incubation the agar cultures were processed as previously described (12) by fixation with 2.5% glutaraldehyde, washing in distilled water, and mounting on a microscopic slide. After drying on a warm plate the whole agar layer was stained with May-Grünwald-Giemsa and scored. Aggregates from 6–49 cells were defined as clusters and aggregates of more than 50 cells were scored as colonies.

Committed Erythroid Progenitor Cell Assay

Late and early erythroid progenitor cells (CFU-e, BFU-e) were stimulated in a miniaturized culture system as previously described (15). Aliquots of 50 μl containing 4×10^4 MNC were pipetted into 96-well micotiter culture plates (flat bottom). After solidification 50 μl of a liquid overlayer were added containing erythropoietin (EP, Step III, Connaught Medical Res. Lab, Ontario), bovine serum albumin (BSA, Behring Institut), and transferrin saturated with $FeCl_3$ (TR, Behring Institut). CFU-e and BFU-e were stimulated with 0.3 U/ml and 1.2 U/ml EP, respectively, and 1% BSA and 0.04% TR. In some experiments cultures for BFU-e growth were in addition stimulated with leukocyte-conditioned medium (L-CM) as an exogenous source of burst-promoting activity (BPA). CFU-e were defined as hemoglobinized aggregates containing no more than 65 cells at day 7; BFU-e were scored as hemoglobin-containing single or multiple colonies of more than 65 cells at day 14 of incubation. For staining and scoring, agar cultures were processed as described above.

Committed Megakaryocytic Progenitor Cell Assay

1.5×10^5 MNC suspended in aliquots of 0.25 ml were seeded in multi-well culture plates as previously described (16). In CFU-M assays FCS was replaced by 20% serum obtained from a pretested patient with severe aplastic anemia. The liquid overlayer contained 1×10^{-4} M 2-ME and 20%

aplastic serum. After 12 days of incubation, cultures were prepared as described and scored for CFU-M. Megakaryocytic colonies were identified in the May-Grünwald-Giemsa stained preparations by their typical morphological appearance, i.e., their large size, their multilobulated nucleus, and the presence of cytoplasmatic blebs. Each colony consisted of more than two megakaryocytes.

Analysis of in vitro Culture Experiments

All cultures were performed in triplicate or quadruplicate. Twelve antibodies as well as positive and negative controls were tested with the same bone marrow sample. Results are expressed as percentage of colony growth in negative controls. Antibodies which showed less than 40% reduction of either committed stem cell subpopulation were excluded from further tests. Reactive antibodies were tested in 3–4 subsequent experiments. For positive reactivity we decided to take into consideration only antibodies with more than 50% inhibition of colony growth in negative countrols. If marked reduction of more than one cell lineage was observed, antibody-treated cells were mixed with equal numbers of mitomycin C-treated peripheral blood cells and cultures as described.

Immunofluorescence Studies

Antibodies with pronounced reactivity (more than 40% inhibition) in one of the three progenitor cell assays were also tested for reactivity with immature leukemic cells from acute leukemia patients and with hematopoietic cell lines. The reactivity of the monoclonal antibodies with these cells was tested in indirect immunoflourescence using a secondary reagent, FITC-labeled goat F(ab')² anti–mouse IgG + IgM antibodies. Fluorescence of cells was evaluated using a fluorescence-activated cell sorter (FACS 440, Becton Dickinson, Sunnyvale, CA). The diagnosis of acute leukemia was made using standard clinical morphological and cytochemical criteria. The acute myeloid leukemia (AML) patients were subclassified according to the FAB scheme. The cell lines HL-60, KG-1, U 937, K562, Jurkat, and Reh-6 were used in the present study. The characteristics and references of these lines have recently been summarized (17).

Results

Reactivity of Myeloid Antibodies with Committed Myeloid Progenitor Cells

The antibodies which, in our hands, showed reactivity with myeloid progenitor cells are listed in Table 15.1. Antibodies 11, 12, 16, 17, 21, 48, 98, and 100 completely inhibited proliferation of clusters and colonies on day

Table 15.1. Reactivity of myeloid antibodies with committed myeloid progenitor cells.[a]

Antibody	Cluster day 3	CFU-GM day 7	CFU-GM day 14
M11	0	0	0
M12	0	0	0
M16	5.5 ± 8.6	12.7 ± 13	33.7 ± 36.4
M17	2.2 ± 2.3	7.3 ± 12.7	31.2 ± 30.5
M18	30.4 ± 12.5	88 ± 14.7	91.2 ± 6.8
M21	1.03 ± 1.7	7.4 ± 9.3	10.2 ± 9.1
M28	4.6 ± 2.7	35.8 ± 15.2	87.8 ± 8.8
M30	30.2 ± 9.5	112.3 ± 5.2	98.9 ± 9.1
M34	30.8 ± 13.7	44.8 ± 9.7	75.3 ± 24.3
M48	1.75 ± 1.86	11 ± 14.3	9.2 ± 7.5
M77	9.3 ± 4.0	39.2 ± 21.6	98 ± 12.4
M80	13.3 ± 2.8	71.3 ± 20.9	116 ± 22.3
M98	2.3 ± 2.5	3.5 ± 4.1	22.6 ± 19.8
M100	5.73 ± 2.5	0	0
M103	31.4 ± 7.2	111.4 ± 12.2	97.9 ± 5.2
M113	12.9 ± 3.6	53.9 ± 19.6	117 ± 18.2

[a] Values represent mean ± SD of 3–4 different experiments; results are expressed as percentage of colony growth in negative control experiments. Only antibodies with more than 50% inhibition of colony growth are considered in this table.

7 and colony formation on day 14 to a high extent. It must be stated that reactivity of some antibodies with CFU-GM day 14 was heterogeneous in bone marrow samples of different individuals. Antibodies 16, 17, 21, 48, and 98 completely inhibited colony formation in two experiments, whereas in two other tests some colony formation (up to 50% of negative controls) was observed. Antibodies 28, 77, 80, and 113 reacted with most cluster-forming cells and a substantial proportion of CFU-GM day 7; colony formation after 14 days of incubation, however, was not affected. The absolute number of clusters corresponded to the number of colonies on day 7, suggesting that all proliferating precursor cells up to the stage of CFU-GM day 7 were reactive. Antibody 34 significantly reduced the number of clusters but seems to recognize only a part of CFU-GM day 7. Antibodies 30 and 103 reduced the number of clusters but not proliferation of colony-forming cells. In one experiment Mit C-treated MNC were added to target cells after incubation with highly reactive antibodies; nevertheless, no significant augmentation of colony formation could be obtained in cultures with additional filler cells (data not shown).

Reactivity of Myeloid Antibodies with Committed Erythroid Progenitor Cells

Reactivity of myeloid antibodies with CFU-e and BFU-e is shown in Table 15.2. Antibodies 14 and 79 selectively recognized CFU-e. All other

Table 15.2. Reactivity of myeloid antibodies with erythroid progenitor cells.[a]

Antibody	CFU-e	BFU-e
M11	23.6 ± 20.9	0
M12	64.9 ± 12.9	15.7 ± 7.4
M14	34.3 ± 15.2	113 ± 17.1
M21	52.5 ± 24.3	37.2 ± 21.5
M28	72.4 ± 12.6	45.2 ± 22.1
M48	60.5 ± 29.7	55.9 ± 24.2
M79	0.7 ± 1.2	96 ± 9.2
M98	85.1 ± 7.7	52 ± 9.3
M100	44.3 ± 29.7	26.3 ± 22.5

[a] Values represent mean ± SD of 3–4 different experiments; results are expressed as percentage of colony growth in negative control cultures. Only antibodies with more than 50% inhibition of colony growth are considered in this table.

antibodies which exhibited clear-cut influences on erythroid stem cells also belonged to the group with high reactivity on myeloid progenitors. Antibody 11 completely inhibited burst formation and most CFU-e. Treatment with antibodies 12, 21, 48, and 100 produced partial inhibition of CFU-e proliferation. Examination of BFU-e revealed that incubation with antibodies 12, 21, and 100 resulted in a marked reduction of colony formation, whereas antibodies 28, 48, and 98 only partially inhibited the growth of BFU-e. For every antibody that inhibited BFU-e growth by more than 50%, additional control experiments have been performed, in which bone marrow cells were in addition stimulated with L-CM 2.5% as exogenous source of BPA (Table 15.3). In these experiments the partial reduction of BFU-e by antibodies 28, 48, and 98 could be largely abrogated by L-CM.

Table 15.3. Effect of stimulation on BFU-e growth after antibody treatment.[a]

Antibody	Stimulation	
	EP, BSA, TR	EP, BSA, TR + L-CM[b]
M11	0	0
M12	19.4	18.5
M21	9.5	14.1
M28	35	79
M48	33.9	84.6
M98	37.5	78.2
M100	11.2	46.3

[a] Values represent the mean of triplicate cultures from one bone marrow sample; results are expressed as percentage of colony growth in negative control cultures.
[b] Cultures were stimulated in addition with 2.5% L-CM (see text).

These results suggest that the reduction of burst colonies may be due in part to lysis of auxiliary cells rather than to direct effects on BFU-e cells.

Reactivity of Myeloid Antibodies with Committed Megakaryocytic Progenitor Cells

As shown in Table 15.4, the antibodies which exhibited strong reactivity with CFU-GM also significantly reduced proliferation of megakaryocytic progenitors. All except antibody 28 were expressed on the more immature CFU-GM which gave rise to colonies after 14 days of incubation. Treatment with antibody 34 resulted in a partial inhibition of CFU-M as found with CFU-GM day 7.

Binding of Progenitor Cell-Reactive Antibodies to Immature Leukemic Blasts and Cell Lines

Knowing which of the selected antibodies strongly (more than 40% inhibition) react with normal hematopoietic precursor cells, it was of course of interest to examine also the reactivity of these antibodies with malignant precursor cells. We therefore also evaluated the binding of these antibodies to leukemic blast cells from patients with acute blast cell leukemias and to hematopoietic cell lines. The results of these experiments are summarized in Table 15.5.

Table 15.4. Reactivity of myeloid antibodies with megakaryocytic progenitor cells.[a]

Antibody	CFU-M
M11	0
M12	0
M16	24.4 ± 8.9
M17	26.8 ± 6.8
M21	12.2 ± 9.3
M28	12.2 ± 4.8
M34	34.8 ± 12.4
M48	9.33 ± 13.6
M98	15.2 ± 2.1
M100	0

[a] Values represent mean ± SD of 3–4 different experiments; results are expressed as percentage of colony growth in negative control experiments. Only antibodies with more than 50% inhibition of colony growth are considered in this table.

Table 15.5. Binding of progenitor cell-reactive monoclonal antibodies to leukemic cells.

| | | Hematopoietic cell lines | | | | | | Leukemia cells | | | | | | | |
| | | Myeloid | | | | Lymphoid | | Myeloid (AML) | | | | | Lymphoid | |
Monoclonal antibodies		HL-60	KG1	U 937	K562	Jurkat	Reh-6	M1	M2	M3	M4	M5	T-ALL	C-ALL
11[a]	JOAN-1[b]	>90	>90	>90	<5	30	>90	>90	>90	>90	>90	>90	60	>90
12	Mo5	33	<5	>90	<5	15	<5	53	20	15	>90	>90	<5	26
14	5F1	<5	<5	>90	<5	<5	<5	29	16	39	77	>90	<5	<5
16	L1B2	68	<5	>90	<5	<5	<5	25	20	24	61	>90	<5	<5
17	L4F3	>90	<5	>90	<5	<5	<5	74	52	67	>90	>90	<5	21
18	TM.2.26	70	<5	>90	<5	45	<5	56	81	<5	>90	73	<5	21
21	BW252/104	>90	>90	>90	>90	>90	>90	49	13	47	>90	>90	18	47
28	VIM2	>90	20	>90	>90	<5	<5	45	62	65	75	>90	<5	<5
34	M5E2	<5	<5	51	<5	<5	<5	33	<5	<5	>90	>90	<5	<5
48	MY-9	67	<5	>90	<5	<5	<5	44	26	30	76	>90	<5	<5
77	G9F9	>90	48	>90	>90	<5	<5	<5	30	5	43	>90	<5	<5
79	7C11	<5	<5	22	<5	<5	28	20	37	54	14	80	<5	<5
80	CLB gran 7	>90	15	>90	61	<5	<5	16	<5	31	30	>90	<5	<5
98	M5D12	<5	<5	>90	<5	<5	<5	44	25	15	>90	>90	<5	<5
100	M3C7.2A	34	44	90	<5	<5	<5	71	>90	>90	>90	>90	<5	61
103	PMN-3	<5	<5	19	<5	<5	<5	<5	<5	<5	<5	<18	<5	<5
113	PMN-1	77	<5	39	27	<5	<5	<5	<5	<5	22	17	<5	<5

[a] Workshop no.
[b] Clone designation.

Discussion

In this study we have evaluated 48 antibodies of the myeloid panel for their reactivity with myeloid, erythroid, and megakaryocytic progenitor cells. Antibodies which exhibited a reduction of more than 40% of either tested stem cell subpopulation in the first round of experiments were examined in 3–4 repeated experiments. Due to this selection procedure, antibodies which produce only relatively low inhibitions of colony growth are possibly not considered in our results. Antibodies reactive with hematopoietic stem cells may be potentially applied for separation or enrichment of precursor cells or for *in vitro* deletion of leukemic cell populations. For these purposes strong specific reactivity is important and thus the limitation to highly reactive antibodies seems to be justified.

The reaction pattern of these selected antibodies is summarized in Table 15.6. As expected the more mature cluster-forming cells were recognized by most antibodies. Many of these antibodies also strongly reacted with CFU-GM day 7. As we have previously seen with VIM-D5 (8), the antibodies 28, 34, and 77 inhibited CFU-GM day 7 whereas CFU-GM day 14 remained unaffected. For some antibodies reactivity with CFU-GM day 14 was heterogeneous, while lysis of the more mature precursors in the same cell sample was constantly seen. The failure of complement-mediated killing may reflect low and variable density of antigen expression on early myeloid progenitors in some individuals.

Table 15.6. Reaction pattern of myeloid antibodies with hematopoietic precursor cells.

Antibody	Cl	CFU-GM 7	CFU-GM 14	CFU-e	BFU-e	CFU-M
M11	+++	+++	+++	++	+++	+++
M12	+++	+++	+++	+	+++	+++
M14	−	−	−	++	−	−
M16	+++	+++	++	−	−	++
M17	+++	+++	++	(+)	(+)	++
M18	++	−	−	−	−	(+)
M21	+++	+++	+++	++	+++	+++
M28	+++	++	−	−	(+)	+++
M34	++	+	(+)	+	+	++
M48	+++	+++	+++	+	(+)	+++
M77	+++	++	−	−	−	+
M79	−	−	−	+++	−	−
M80	+++	(+)	−	−	−	−
M98	+++	+++	++	−	+	+++
M100	+++	+++	+++	++	++	+++
M103	++	−	−	−	−	−
M113	+++	+	−	−	−	(+)

[a] Reduction of colony growth: +++, 80–100%; ++, 60–80%; +, 40–60%; (+) 20–30%; −, less than 20%.

Most antibodies with strong reactivity with immature myeloid precursor cells also significantly reduced the growth of BFU-e. The interpretation of these results may be crucial since this effect may be related to lysis of auxiliary cells which interact with BFU-e. This problem arises in particular in experiments with effective lysis of myeloid cells and relatively small reduction of BFU-e growth, as we have found with antibodies 16, 17, and 48. In fact, the antibody with code no. 17 (L4F3) has been described to react in part with BFU-e (4) whereas in our experiments this antibody has not been considered for positive reactivity. With certain antibodies the reduction of burst formation could be abrogated at least in part by the addition of L-CM to the culture medium as an exogenous source of burst-promoting activity. This suggests that lysis of auxiliary cells possibly plays a role in this assay system. To prove or exclude partial expression of an antigen at the level of early erythropoietic stem cells positive selection experiments with fluorescence-activated cell sorting would therefore be necessary.

The more mature erythroid progenitor cells (CFU-e) seem to display a completely different antigenic pattern than BFU-e cells. Such differences have also been reported in experiments with HLA-related antibodies (18). We observed marked reductions of CFU-e only with the antibodies 11 and 100, which kill all proliferating cells. Antibodies 14 and 79, however, reduced selectively CFU-e growth, whereas all other progenitor cells remained insensitive to these antibodies.

As found for BFU-e, we observed in most experiments a parallel reduction of megakaryocytic progenitors and of CFU-GM cells. There exists evidence of cooperation of activated T cells in early megakaryopoiesis (19). Relatively little information exists, however, about interactions between CFU-M and myelo-monocytic cells. Therefore an indirect reducion of CFU-M growth due to effects of antibodes on "growth factor"-producing myelo-monocytic cells cannot be excluded in cases where an inhibition of CFU-M and marked lysis of myelo-monocytic cells go in parallel.

Our complement-dependent negative selection experiments would suggest that the antibodies 11, 12, 21, 48, 98, and 100 are not restricted to the myeloid cell lineage but recognize structures which are also expressed by very early hematopoietic progenitor cells which belong to a differentiation stage shortly after the pluripotent stem cell level. It should be stressed again, however, that failure in colony growth can also result from rigorous reduction of viable cell number or lysis of critical auxiliary cells. These results would therefore have to be confirmed in positive selection experiments.

This Workshop enabled us to study a large number of myeloid antibodies, developed in several laboratories, for their reactivity with hematopoietic precursor cells. Our results may be of value for the classification of myeloid antibodies and for their possible application in diagnosis and therapy of hematological diseases.

Summary

Forty-eight monoclonal antibodies (mAbs) of the myeloid panel of Workshop antibodies were tested for reactivity with human hematopoietic precursor cells in complement-dependent cytotoxicity assays. This group included all mAbs with indicated lytic capacities except the ones directed against the 3-fucosyl-*N*-acetyllactosamine structure or with specificity for NK cells. In all instances we performed, subsequent to incubation with mAb and rabbit complement, *in vitro* culture assays for myeloid (cluster-forming cells, CFU-GM day 7 and 14), late and early erythroid (CFU-e and BFU-e), and megakaryocytic (CFU-M) progenitor cells. In parallel we also evaluated the reactivity of these antibodies with different well-defined myeloid and non-myeloid cell types by cell sorter analysis.

Different patterns of antigen expression on hematopoietic precursor cells could be observed. mAbs 30 and 103 reacted only with a proportion of cluster-forming cells. mAbs 18, 28, 77, 80, and 113 killed most cluster-forming cells and part of CFU-GM day 7. After incubation with mAbs 16, 17, 48, and 98 growth of also the more immature CFU-GM day 14 and of CUF-M was inhibited. mAbs 12 and 100 reacted with all progenitor cells except CFU-e, whereas mAb 11 completely inhibited growth of all cells tested. mAbs 14 and 79 selectively recognized CFU-e.

References

1. Knapp, W., O. Majdic, P. Bettelheim. 1984. The myeloid leukaemias. In: C.P. Engelfriet, J.J. van Loghem, and A.E.G.K. von dem Borne, eds. *Immunohaematology*. Elsevier Science Publishers, Amsterdam, p. 322.
2. Linker-Israeli, M., R.J. Billing, K.A. Foon, and P.I. Terasaki. 1981. Monoclonal antibodies reactive with acute myelogeneous leukemia cells. *J. Immunol.* **127**:2473.
3. Hanjan, S.N.S., J.F. Kearney, and M.D. Cooper. 1982. A monoclonal antibody (MMA) that identifies a differentiation antigen on human myelomonocytic cells. *Clin. Immunol. Immunopathol.* **23**:172.
4. Andrews, R.G., B. Torok-Storb, and I.D. Bernstein. 1983. Myeloid-associated diffentiation antigens on stem cells and their progeny identified by monoclonal antibodies. *Blood*. **62**:124.
5. Ferrero, D., H.E. Broxmeyer, G.L. Pagliardi, S. Venuta, B. Lange, S. Pessano, and G. Rovera. 1983. Antigenically distinct subpopulations of myeloid progenitor cells (CFU-GM) in human peripheral blood and marrow. *Proc. Natl. Acad. Sci. U.S.A.* **80**:4114.
6. Strauss, L.C., K.M. Skubitz, J.T. August, and C.I. Civin. 1984. Antigenic analysis of hematopoiesis: II Expression of human neutrophil antigens on normal and leukemic marrow cells. *Blood* **63**:574.
7. Griffin, J.D., J. Ritz, R.P. Beveridge, J.M. Lipton, J.F. Daley, and S.F. Schlossman. 1983. Expression of MY7 antigen on myeloid precursor cells. *Int. J. Cell Clon.* **1**:33.
8. Peschel, C., G. Konwalinka, D. Geissler, O. Majdic, H. Stockinger, H. Braunsteiner, and W. Knapp. 1985. Studies on differentiation of committed

hemopoietic progenitor cells with monoclonal antibodies against myeloid differentiation antigens. *Exp. Hematol,* in press.

9. Majdic, O., K. Liszka, D. Lutz, and W. Knapp. 1981. Myeloid differentiation antigen defined by a monoclonal antibody. *Blood* **58**:1127.

10. Iscove, N.N., F. Sieber, and K.H. Winterhalter. 1974. Erythroid colony formation in cultures of mouse and human bone marrow: analysis of the requirement for erythropoietin by gel filtration and affinity chromatography on agarose-Concanavalin A. *J. Cell. Physiol.* **83**:309.

11. Stephenson, I.R., A.A. Axelrad, D.L. McLeod, and M.M. Shreeve. 1971. Induction of colonies of hemoglobin-synthesizing cells by erythropoietin *in vitro. Proc. Natl. Acad. Sci. U.S.A.* **68**:1542.

12. Konwalinka, G., D. Geissler, C. Peschel, B. Tomaschek, F. Schmalzl, H. Huber, R. Odavic, and H. Braunsteiner. 1982. A micro-agar culture system for cloning human erythropoietic progenitors *in vitro. Exp. Hematol.* **10**:71.

13. Terhorst, C., A. Van Agthoven, K. Le Clair, P. Snow, E. Reinherz, and S. Schlossman. 1981. Biochemical studies of the human thymocyte cell surface antigens T6, T9, T10. *Cell* **23**:771.

14. Konwalinka, G., C. Peschel, D. Geissler, B. Tomaschek, J. Boyd, R. Odavic, and H. Braunsteiner. 1983. Myelopoiesis of human bone marrow cells in a micro-agar culture system, comparison of two sources of colony stimulating activity (CSA). *Int. J. Cell Clon.* **1**:401.

15. Konwalinka, G., C. Peschel, J. Boyd, D. Geissler, M. Ogriseg, R. Odavic, and H. Braunsteiner. 1984. A miniaturized agar culture system for cloning human erythropoietic progenitor cells. *Exp. Hematol.* **12**:75.

16. Geissler, D., G. Konwalinka, C. Peschel, J. Boyd, R. Odavic, and H. Braunsteiner. 1983. Clonal growth of human megakaryocytic progenitor cells in a micro-agar culture system: simultaneous proliferation of megakaryocytic, granulocytic and erythroid progenitor cells (CFU-M, CFU-C, BFU-E) and T-lymphocytic colonies (CFU-TL). *Int. J. Cell Clon.* **1**:377.

17. Majdic, O., P. Bettelheim, H. Stockinger, W. Aberer, K. Liszka, D. Lutz, and W. Knapp. 1984. M2, a novel myelomonocytic cell surface antigen and its distribution on leukemic cells. *Int. J. Cancer* **33**:617.

18. Robinson, J., C. Sieff, D. Delia, P.A.W. Edwards, and M. Greaves. 1981. Expression of cell surface HLA-DR, HLA-ABC and glycophorin during erythroid differentiation. *Nature* **289**:68.

19. Geissler, D., G. Konwalinka, C. Peschel, K. Grünewald, R. Odavic, and H. Braunsteiner. 1985. A regulatory role of activated T lymphocytes on human megakaryocytopoiesis in vitro. *Brit. J. Haematol.* **60**:233.

Note Added in Proof

In control experiments with monoclonal antibody Mo 5 (workshop code M12) which was kindly provided by R.F. Todd we observed no reactivity with the committed stem cells tested. These results are in agreement with the findings of R.F. Todd and J.D. Griffin (unpublished data). Therefore the antibody which we obtained with code M12 seems not to be identical with Mo5.

CHAPTER 16

Study of the Antigenic Profile of Normal Myelo-Monocytic Progenitors and Leukemic Cell Lines Using Monoclonal Antibodies

M.C. Alonso, R. Solana, A. Torres, R. Ramirez, C. Navarrete, J. Pena, and H. Festenstein

Introduction

Studies of hematopoietic differentiation have classically relied on the ability to identify mature hematopoietic cells and their precursors by morphological, histochemical, and functional criteria. However, the study of early myeloid differentiation has been difficult due to the small number of progenitor cells and their lack of distinctive features (1). To further characterize these progenitors, proliferative assays in the presence of specific growth factors have been applied (2,3). With the advent of monoclonal antibody (mAb) technology, it has been possible to define surface membrane markers which discriminate between subpopulations of lymphoid and myeloid cells and which relate to both the ontogeny and functional heterogeneity of these subpopulations. With reference to the myelo-monocytic lineage, the efforts of several laboratories have resulted in the development of mAbs that distinguish markers shared by myeloid cells and cells of other lineage (4), other markers specific for the myelo-mono-cytic series (5), and finally markers selective for fully differentiated mono-cytes (6). These mAbs have made it possible to associate membrane structural features with particular stages of normal myeloid differentiation and with leukemic cells that represent a malignant proliferation of a given stage of maturation (7,8).

Normal differentiation along the myelo-monocytic pathway is characterized by the spontaneous modulation of cell membrane antigens, e.g., a decrease in the amount of HLA-DR antigens when the progenitors mature into granulocytes and the acquisition of other specific markers have been described (9,10). The study of the hematopoietic progenitor phenotype is of particular interest not only for clinical bone marrow transplantation but also for the identification of normal and leukemic cell populations. There is evidence that chronic myelocytic leukemia (CML) and acute myeloblastic leukemia (AML) originate from pluripotent stem cells (1). Therefore it is important to determine whether leukemic differentiation mimics the normal myeloid pathway or whether the abnormalities observed will lead to the definition of a leukemic developmental pathway (11).

In this report we describe the expression of the antigenic determinants recognized by 38 mAbs against myeloid cells on normal myelo-monocytic progenitors (CFU-GM) using a complement-dependent cytotoxic assay. The modification of the antigenic profile of myeloid cell lines K562 and U 937 after treatment with differentiation inducer phorbol ester (TPA) is also described. K562 is a cell line derived from the acute blast crisis of a chronic myelocytic leukemia (12) which displays phenotypic characteristics of cells of both the erythoid (13–15) and the myeloid (15,16) lineages. Phorbol ester produces changes in the cell surface antigenic profile by suppressing the expression of certain erythroid markers and enhancing those of the granulocyte–monocyte lineage (17). U 937 is a cell line derived from a histiocytic lymphoma whose antigenic profile has been studied (7) and classified as the equivalent of a normal monoblast (8).

Materials and Methods

Monoclonal Antibodies

The following cytotoxic mAbs were used: Workshop myeloid panel (numbers indicated in the tables); W6/32: anti–monomorphic HLA-A,B,C (18); L 243: anti–monomorphic HLA-DR (19); TU-22: anti–monomorphic HLA-DQ (20); EDU-1: anti–monomorphic HLA-class II (21).

Preparation of Cell Suspensions

Mononuclear cells from peripheral blood obtained from normal donors was separated, washed with medium, and used for the preparation of Feeder-Layer. Normal bone marrow was aspirated from adult volunteers. Approximately 10 ml of the aspirated marrow were collected into 10 ml of McCoy's 5A medium containing 200 U preservative-free heparin. McCoy's 5A medium (Microbiological Assoc.), supplemented with 20% fetal calf serum (Gibco) and antibiotics (200 units/ml or penicillin and 100 ng/ml of streptomycin), was used in the preparation and culture of bone marrow cells. The cell suspension was layered on Ficoll–Hypaque (density 1080) and centrifuged at $800 \times g$ for 20 min. The mononuclear cell layer was collected, washed 3 times with medium, and used for the CFU-GM assay.

Preparation of Feeder-Layer

One million peripheral blood mononuclear cells were plated in 1 ml of 0.5% agar-medium on 35-mm petri dishes. After 3 days of incubation at 37°C in 7.5% CO_2 they were used as the source of GM-CSF.

Treatment of Bone Marrow Cells with mAb and Complement

Bone marrow cells were incubated with the relevant mAb (1/100 final dilution) for 1 hr in 200 µl of medium. The cells were washed three times and divided into two aliquots of 100 µl. One control was incubated without complement and the other was incubated for 2 hr with 100 µl of low-toxicity rabbit complement. After washing 5 times with medium the cells were used for the CFU-GM assay.

CFU-GM Assay

2×10^5 bone marrow cells were plated in 1 ml of medium with 0.5% agar (Bacto Agar, Difco) on petri dishes containing Feeder-Layer as a source of GM-CSF. After incubation for 14 days the number of granulocytic–macrophagic colonies (aggregates of more than 40 cells) were scored by examination through an inverted microscope (Olympus CK, Japan).

Preparation and TPA Treatment of Cell Lines

K562 and U 937 myeloid cell lines were incubated for 72 hr with 100 ng/ml or 12-0-teradecanoylphorbol 13-acetate (TPA) (Sigma).

The cell lines which were treated or not treated with TPA were collected from cultures growing in exponential phase. They were washed twice in phosphate-buffered saline (PBS) containing 0.02% sodium azide and 0.2% bovine serum albumin (BSA) and their concentration adjusted to 10^7 cel/ml for immunofluorescence.

Indirect Immunofluorescence

Half a million cells in 50 µl of PBS/BSA/azide were incubated for 1 hr at 4°C with the relevant mAb at the appropriate dilution. The cells were then washed three times and resuspended in 50 µl of PBS/BSA/azide. The suspension was then incubated for 20 min at 20°C with goat anti–mouse Ig labeled with fluorescein isothiocyanate (Meloy) to a final dilution of 1 : 20. After incubation the cells were washed again three times in PBS/BSA/azide. The pellet was resuspended in 50µl of butter and examined on a slide using a standard Zeiss immunofluorescence microscope.

Table 16.1. Number of bone marrow CFU-GM after pretreatment with mAb and complement.

| mAb code number | Number of CFU-GM | | % Inhibition[a] |
	Without C'	+ C'	
Control	168	161	4
1	164	166	—
2	166	171	—
5	136	77	44
6	142	0	100
8	155	132	15
10	147	135	8
11	142	0	100
12	124	119	4
13	138	136	1
14	153	144	6
15	140	138	1
16	119	36	70
18	155	142	8
19	151	169	—
20	168	177	—
22	147	137	7
25	131	0	100
27	137	12	91
32	126	0	100
33	138	0	100
34	146	139	5
38	156	161	—
40	143	107	25
42	137	131	4
44	128	140	—
48	172	177	—
49	148	145	2
60	135	98	27
72	139	23	83
77	111	12	89
78	161	156	3
80	140	2	99
91	127	0	100
97	143	97	32
98	133	0	100
99	138	138	—
105	139	5	96
106	139	0	100

[a] % Inhibition $= \left(1 - \dfrac{\text{Colonies with mAb} + \text{C}'}{\text{Colonies with mAb}}\right) \times 100$

Results

Reactivity of Cytotoxic mAbs with CFU-GM

As shown in Table 16.1 mAb numbers M6, M11, M16, M25, M27, M32, M33, M72, M77, M80, M91, M98, M105, and M16 produced, in the presence of rabbit complement, more than 70% inhibition in the number of marrow colonies. mAb numbers M1, M2, M8, M10, M12, M13, M14, M15, M18, M19, M20, M22, M34, M38, M40, M42, M44, M48, M49, M60, M78, M97, and M99 did not produce any inhibition in CFU-GM growth. mAb numbers M5 consistently produced 40–50% inhibition of CFU-GM growth.

Neither complement nor any of the mAbs significantly affect the normal marrow colony formation by themselves.

Reactivity of Anti-HLA-Class II mAbs with CFU-GM

Table 16.2 shows no inhibition in the number of CFU-GM when cells were treated with anti-DQ mAb (TU-22) in the presence of complement. More than 90% inhibition in colony growth was found when cells were treated with anti-DR (L 243) or anti-DR/DP (EDU-1) mAb plus complement.

Reactivity of Cytotoxic mAbs with K562 and TPA-treated K562 Line

As shown in Table 16.3, mAb numbers M5, M13, M19, M27, M33, M77, M80, M91, and M25 were positive (more than 50% positive cells) with the K562 cell line while mAb numbers M22, M38, M40, M42, M78, M97, M1, M2, M8, M10, M12, M14, M15, M18, M20, M34, M44, M48, M49, M60, M99, M6, M11, M72, M98, M16, M32, M105, and M106 were negative.

When K562 was treated with TPA only one of the negative mAbs (M6) became positive. None of the mAbs positive with K562 cell line became negative after treatment of the cells with TPA.

Table 16.2. Differential expression of class II antigens on CFU-GM.

mAb	Specificity	Inhibition of CFU-GM growth in presence of C'
L 243	Anti-HLA-DR	+++[a]
TU-22	Anti-HLA-DQ	−
EDU-1	Anti-HLA-DR/DP	+++
W6/32	Anti-HLA-ABC	+++

[a] +++ More than 90% inhibition.

Table 16.3. Reaction pattern of cytotoxic mAbs with K562 leukemic cell line and TPA-treated K562.

mAb code number	K562	K562 + TPA
1	−	−
2	−	−
5	+++[a]	+++
6	−	++[b]
8	−	−
10	−	−
11	−	−
12	−	−
13	+++	++
14	−	−
15	−	−
16	−	−
18	−	−
19	++	++
20	−	−
22	−	−
25	++	++
27	+++	+++
32	−	−
33	−	+++
34	−	−
38	−	−
40	−	−
42	−	−
44	−	−
48	−	−
49	−	−
60	−	−
72	−	−
77	+++	++
78	−	−
80	++	++
91	++	+++
97	−	−
98	NT	−
99	−	−
105	−	−
106	−	−

[a] +++ 80–100% positive cells.
[b] ++ 60–80% positive cells.

Reactivity of Cytotoxic mAbs with U 937 and TPA-treated U 937 line

Table 16.4 shows that mAb numbers M5, M13, M19, M22, M38, M40, M42, M78, M97, M27, M33, M77, M80, M91, M25, M6, M11, M72, and M98 were positive with the U 937 cell line (more than 50% positive cells)

Table 16.4. Reaction pattern of cytotoxic mAbs with U 937 leukemic cell line and TPA-treated U 937.

mAb code number	U 937	U 937 + TPA
1	−	−
2	−	−
5	++[a]	+++[b]
6	++	++
8	−	−
10	−	−
11	++	++
12	−	−
13	++	+++
14	−	−
15	−	NT
16	−	−
18	−	−
19	+[c]	+
20	−	−
22	++	++
25	++	−
27	++	++
32	−	−
33	+++	+++
34	−	−
38	++	+++
40	+++	++
42	++	−
44	−	−
48	−	−
49	−	−
60	−	−
72	+++	+++
77	++	+++
78	++	−
80	NT	+++
91	+++	+++
97	++	−
98	+++	−
99	−	−
105	−	−
106	−	−

[a] ++ 60–80% positive cells.
[b] +++ 80–100% positive cells.
[c] + 40–60% positive cells.

while mAb numbers M1, M2, M8, M10, M12, M14, M15, M18, M20, M34, M44, M48, M49, M60, M99, M16, M32, M105, and M106 were negative.

mAb numbers M25, M42, M78, M97, and M98 that were positive with U 937 became negative when this cell line was treated with TPA. None of the mAbs negative with U 937 became positive after TPA treatment.

Discussion

The study of the antigenic profile of human hematopoietic progenitor cells has relevance not only for the analysis of the differentiation of normal and leukemic cells, but also for clinical bone marrow transplantation. In this case, the selection of certain populations of cells from the graft using specific monoclonal antibodies becomes more and more feasible. Because committed hematopoietic progenitor cells cannot be distinguished from other immature bone marrow cells by morphological criteria, proliferation assays have to be applied to obtain information about membrane characteristics (3). In this sense, mAbs are ideal probes for this purpose, considering their homogeneity, specificity, and high titer.

In this report we study the antigenic profile of CFU-GM using 38 cytotoxic mAbs raised against myeloid cells and a panel of known class II specific mAbs. The characterization of the mAbs was done by negative selection experiments in which bone marrow cells were treated with antibody plus complement and the resultant cells cultured for CFU-GM. The reactivity of the mAbs directed against myeloid cells with committed hematopoietic progenitor cells is shown in Table 16.1. Twenty-four of the mAbs studied were negative with CFU-GM, while 14 produced more than 70% inhibition in the number of marrow colonies when incubated in the presence of complement. This indicates that the epitopes recognized by these 14 mAbs are present on the suface of the committed myelo-monocytic progenitors. These mAbs could be useful to positively select normal myeloid progenitor cells to study and the myeloid pathway and the relationship between mononuclear and polymorphonuclear phagocytic cells.

None of the mAbs significantly inhibited CFU-GM formation in the absence of complement, suggesting that although the molecules recognized by the positive mAbs are expressed on these cells, they do not seem to interfere in the recognition of growth factors during myelopoiesis.

Of particular interest are the results (Table 16.2) of the differential expression of class II antigens on these myelo-monocytic progenitors. Treatment of the bone marrow cells with mAbs L 243 and EDU-1 in the presence of complement produces more than 90% inhibition in the number of CFU-GM while treatment with mAb TU-22 plus complement did not inhibit CFU-GM growth. These mAbs are recognizing different class II molecules: while EDU-1 reacts with an epitope broadly expressed on class II antigens, L 243 reacts only with HLA-DR molecules and TU-22 only with HAL-DQ molecules (22). This means that HLA-DR antigens are expressed on inmature cells, probably as molecules regulating the proliferation, whereas HLA-DQ antigens are only expressed on the terminal phases of maturation of some cells of the immune system and have a different function from HLA-DR (23). These results also indicate that the expression of genes of the HLA-D region encoding for different class II products can be independently regulated in normal cells.

In order to further characterize the myeloid mAb panel we have studied the reactivity of these antibodies with two myeloid cell lines growing *in vitro*. Cell line K562 is derived from a chronic meyloid leukemia (12) and has the phenotypic markers associated with the erythroid (13,14) and myeloid lineages (15,16). U 937 is a cell line derived from a histiocytic lymphoma (24) and has been characterized as the equivalent of a normal monoblast (8). The reactivity of the mAbs with these cells was also studied after treatment with the phorbol ester TPA. TPA is known to induce morphological, functional, and phenotypic changes on several lymphoid (25) and myeloid (26) cell lines compatible with the features observed during normal differentiation (27).

As shown on Table 16.5, 15 of the 24 mAbs negative with CFU-GM were also negative with the cell lines K562 and U 937 either pretreated or not pretreated with the phorbol ester TPA, probably indicating that they are recognizing determinants expressed either on myeloid cells during very specific stages of maturation or on cells of other lineages (platelets, erthrocytes, etc.).

Six of the mAbs negative with CFU-GM were positive with the myeloid cell line U 937, classified as the equivalent of a normal monoblast (8). These mAbs may be reacting with determinants which appear only on cells of the monocytic lineage. Three of them lose the reactivity with U 937 when this leukemic line is treated with TPA, indicating that the recognized determinant is specific for a particular stage of monocytic differentiation.

mAbs numbers M13 and M19 reacted positively with K562 and U 937 which had been either treated or not treated with TPA, but they were negative with CFU-GM. These antibodies are particularly interesting since they appear to recognize a determinant present on the two leukemic cell lines in very different stages of maturation and absent from normal myelo-monocytic progenitors. mAb number M5 has a similar reaction pattern but produces 44% inhibition of CFU-GM growth.

When we consider the reactivity of the 14 mAbs that are positive with CFU-GM (Table 16.6) we see that six of them are also positive with the myeloid cell lines tested, indicating that they are recognizing determi-

Table 16.5. Reactivity of mAbs negative with CFU-GM with leukemic cell lines.

mAb code number	K562	K562 + TPA	U 937	U 937 + TPA
5, 13, 19	+	+	+	+
22, 38, 40	−	−	+	+
42, 78, 97	−	−	+	−
1, 2, 8, 10, 12, 14, 15, 18, 20, 34, 44, 48, 49, 60, 99	−	−	−	−

Table 16.6. Reactivity of mAbs positive with CFU-GM with leukemic cell lines.

mAb code number	K562	K562 + TPA	U 937	U 937 + TPA
27, 33, 77, 80, 91	+	+	+	+
25	+	+	+	−
6	−	+	+	+
11, 72	−	−	+	+
98	−	−	+	−
16, 32, 105, 106	−	−	−	−

nants expressed on cells of the myelo-monocytic lineage during most stages of differentiation. One of them (M25), which reacts with a 3-α-fucosyl-*N*-acetyllactosamine, lost the reactivity with U 937 after this cell line was treated with TPA. This indicates that, in the process of monocytic differentiation, the fucosyllactosamine structure is either lost or masked by sialinization.

mAb M6 recognizes a determinant which appears on K562 when it is treated with TPA that is also present on CFU-GM and U 937. TPA induces the maturation of myelo-erythroid cell line K562 to the myeloid lineage, thus enhancing the expression of myeloid markers (17). This mAb therefore seems to recognize a determinant specific for meyloid cells not expressed on the earlier developmental stage represented by K562.

Three mAbs are negative with K562 but positive with CFU-GM and U 937, suggesting a reaction with markers of particular stages of differentiation. Two of them (M11 and M72) are expressed on U 937 and TPA-treated U 937 while the other (M98) is only expressed on U 937.

mAbs numbers M16, M32, M105, and M106 are positive with CFU-GM but negative with the leukemic cell lines studied, suggesting that they are reacting with some specific markers of this particular stage of the normal myelo-monocytic differentiation. The confirmation of these results may have a relevance in clinical bone marrow transplant if we consider the importance of mAbs specific for CFU-GM in cell fractionation of the graft prior to the transplant.

Summary

The development of hybridoma techniques for the production of monoclonal antibodies has enabled the identification of cell surface determinants selectively expressed by myeloid cells at different stages of differentiation. The reactivity of cytotoxic mAbs with myelo-monocytic progenitors has been studied by negative selection experiments in which normal bone marrow cells were treated with antibody plus complement and the resultant cells cultured for CFU-GM. Fourteen of the Workshop mAbs were positive with CFU-GM whereas 24 were negative. In considering the reactivity of anti-class II mAbs with CFU-GM, we found that HLA-DR antigens are present on these progenitors but HLA-DQ molecules are absent.

Fifteen of the mAbs negative with CFU-GM were also negative with leukemic cell lines K562 and U 937, three were positive with both cell lines, and six were positive only with U 937. Considering the group of mAbs positive with CFU-GM, six were positive with both cell lines while four were negative and four were positive with U 937. TPA treatment modulated the reactivity of some mAbs with these leukemic cells.

Acknowledgments. We are very grateful to Mrs. Paz Bayon for technical assistance. This work has been supported by CAICYT (Spain) and Anglo-Spanish Joint Research in Higher Education (British Council and Ministerio de Educacion y Ciencia).

References

1. Mannoni, P., A. Janowska, P. Fromont, B. Weiblen, A.R. Turner, and J.M. Turc. 1984. Human myeloid differentiation studied by monoclonal antibodies. In: *Leucocyte typing,* A. Bernard, L. Boumsell, J. Dausset, C. Milstein, and S.F. Schlossman, eds. Springer-Verlag, Berlin, Heidelberg, p. 410–419.
2. Pike, B.L., and W.A. Robinson. 1970. Human bone marrow colony growth in agar-gel. *J. Cell. Physiol.* **76**:77.
3. Quesenberry, P., and L. Levitt. 1979. Hematopoietic stem cells. *New England J. Med.* **301**:755.
4. Berger, A.E., J.E. Davis, and P. Cresswell. 1981. A human leukocyte antigen identified by a monoclonal antibody. *Hum. Immunol.* **3**:231.
5. Mannoni, P., A. Janowska-Wieczorek, A.R. Turner, L. McGann, and J.M. Turc. 1982. Monoclonal antibodies against human granulocytes and myeloid differentiation antigens. *Hum. Immunol.* **5**:309.
6. Ugolini, V., G. Nuñez, R. Graham-Smith, P. Stastny, and D. Capra. 1980. Initial characterization of monoclonal antibodies against human monocytes. *Proc. Natl. Acad. Sci. U.S.A.* **77**:6764.
7. Minowada, J., H. Koshiba, K. Sagawa, I. Kubonishi, M.S. Lock, E. Tatsumi, T. Han, B.I.S. Srivastava, and T. Ohnuma. 1981. Markers profiles of human leukemia and lymphoma cell lines. *J. Cancer Res. Clin. Oncol.* **101**:91.
8. Minowada, J., E. Tatsumi, K. Sagawa, M.S. Lok, T. Sugimoto, K. Minato, L. Zgoda, L. Prestine, L. Kover, and D. Gould. 1984. A scheme of human hematopoietic differentiation based on the marker profiles of the cultured and fresh leukemia–lymphomas: The result of workshop study. In: *Leucocyte typing,* A. Bernard, L. Boumsell, J. Dausset, C. Milstein, and S.F. Schlossman, eds. Springer-Verlag, Berlin, Heidelberg, p. 519–527.
9. Winchester, R.J., G.D. Ross, C.I. Jarowski, C.Y. Wang, J. Halfer, and H.E. Broxmeyer. 1977. Expression of Ia-like antigen molecules on human granulocytes during early phase of differentiation. *Proc. Natl. Acad. Sci. U.S.A.* **74**:4012.

10. Sieff, C., D. Bicknell, G. Caine, J. Robinson, G. Lam, and M.F. Greaves. 1982. Changes in cell surface antigen expression during hemopoietic differentiation. *Blood* **60:**703.

11. Marie, J.P., C.A. Izaguirre, C.I. Civin, J. Mirro, and E.A. McCulloch. 1981. Granulopoietic differentiation in AML blasts in culture. *Blood* **58:**670.

12. Lozzio, C.B., and B.B. Lozzio. 1975. Human chronic myelogenous leukaemia cell line with positive Philadelphia chromosome. *Blood* **45:**321.

13. Gahmberg, C.G., and L.C. Andersson. 1981. K562 a human leukaemia cell line with erythroid features. *Semin. Hematol.* **18:**72.

14. Horton, M.A., S.H. Cedar, and P.A.W. Edwards. 1981. Expression of red cell specific determinants during differentiation in the K562 erythroleukaemia cell line. *Scand. J. Haematol.* **27:**231.

15. Marie, J.P., C.A. Izaguirre, G. Civin, J. Mirro, and E.A. McCulloch. 1981. The presence within single K562 cells of erythropoietic and granulopoietic differentiation markers. *Blood* **58:**708.

16. Drew, S.I., P.I. Terasaki, R.J. Billing, O.J. Bergh, J. Minowada, and E. Klein. 1977. Group-specific human granulocyte antigens on a chronic myelogenous leukaemia cell line with a Philadelphia chromosome marker. *Blood* **49:**715.

17. Horton, M.A., and S.H. Cedar. 1984. Expression of granulopoietic differentiation markers in the K562 cell line. In: *Leucocyte typing,* A. Bernard, L. Boumsell, J. Dausset, C. Milstein, and S.F. Schlossman, eds. Springer-Verlag, Berlin, Heidelberg, p. 395–397.

18. Parham, P., C.J. Barnstable, and W.F. Bodmer. 1979. Use of a monoclonal antibody (W6/32) in structural studies of HLA-A,B,C antigens. *J. Immunol.* **123:**342.

19. Lampson, L.A., and R. Levy. 1980. Two populations of Ia-like molecules on a human B cell line. *J. Immunol.* **125:**293.

20. Ziegler, A., B. Uchanska-Ziegler, G. Rosenfelder, D.G. Braun, and P. Wernet. 1981. Heterogeneity of established human haematopoietic cell lines: Surface antigens detected by monoclonal antibodies and glycosphingolipid patterns. In: *Leukemia markers,* W. Knapp, ed. Academic Press, London, p. 317.

21. Vilella, R., J. Yague, M.T. Gallant, and J. Vives. 1983. The efficacy of hybridoma technology in studying surface antigens in human T and B lymphocytes. *Inmunologia* **2:**58.

22. Fernandez, N., R. Solana, J. Sachs, C. Navarrete, and H. Festenstein. 1984. Characterization of workshop class II MoAbs using EBV transformed and mutant cell lines. *Disease Markers* **2:**9.

23. Navarrete, C., N. Fernandez, M.C. Alonso, and H. Festenstein. 1984. Study of class II antigens on normal and leukaemic cells. *Hum. Immunol.* (in press).

24. Sundstrom, C., and K. Nilsson. 1976. Establishment and characterization of a human histiocytic lymphoma cell line (U937). *Int. J. Cancer* **17:**565.

25. Guy, K., A.W.S. Ritchie, V. van Heyningen, and A.E. Dewar. 1983. Further serological characterization of the heterogeneity of expression of human MHC class II antigens: chronic lymphocytic leukaemia cells and monocytes differentially express DR and DC antigens. *Disease Markers* **1:**249.

26. Namikawa, R., S. Ogata, R. Ueda, I. Tsuge, K. Nishida, S. Minami, K. Koike, T. Suchi, K. Ota, S. Iijima, and T. Takahashi. 1983. Serological

analysis of cell surface antigens of HL-60 cells before and after treatment with a phorbol ester tumor promoter. *Leuk. Res.* **7:**375.

27. Cossman, J., L. Neckers, R.M. Braziel, A. Bakhshi, A. Arnold, and S. Kors-meyer. 1984. Induction of differentiation in B cell leukemia. In: *Leucocyte typing,* A. Bernard, L. Boumsell, J. Dausset, C. Milstein, and S.F. Schloss-man, eds. Springer-Verlag, Berlin, Heidelberg, p. 599–603.

CHAPTER 17

Expression of Antigens Present on Hematopoietic Progenitor Cells by Cells of Certain Hematopoietic Lineages

Masatoshi Takaishi and Shu Man Fu

Utilizing non-lymphoid cell lines as immunogens, a series of monoclonal antibodies have been established and shown to have selective reactivities to non-lymphoid cells. Two of them (K 15.1.9 and H 8.4.5) have been studied more extensively. They have been found to exhibit differing reactivities to CFU-GM and BFU-E progenitor cells as well as to more mature hematopoietic cells. This report describes our investigation with these two antibodies.

Materials and Methods

Monoclonal Antibody Production

Ten week old BC_3F_1 female mice were injected i.p. with 2×10^7 HEL, a human erythroleukemia cell line and K562, a CML line. Their spleen cells were fused with SP2/0 tumor cells. Hybridoma supernatants were screened by indirect immunofluorescence. The desired hybridomas were cloned on soft agar. Details of these procedures have been described previously (1).

Cell Preparation

Defibrinated blood from normal volunteers was used as a source of peripheral blood mononuclear cells (PBMC). PBMC were separated as previously described (2). Monocytes were separated by Percoll continuous gradient centrifugation. Granulocytes were isolated from the peripheral blood as the pellet fraction after Ficoll–Hypaque density gradient centrifugation. Contaminating erythrocytes were removed by a 60%, 65%, and 70% Percoll discontinuous density gradient centrifugation at 2200 r.p.m. for 20 min. The purity of the granulocyte fractions was more than 94% as determined by Wright Giemsa staining. Platelets were isolated from platelet-rich plasma and erythrocytes from the pellets of Ficoll–Hypaque gradients of peripheral blood cells.

Immunofluorescence Studies

Cells ($0.5–1.0 \times 10^6$) were first incubated with hybridoma culture supernatants for 20 min. After three washings with 0.01 M phosphate-buffered saline with 1% bovine plasma albumin, fluorescein labeled F(ab')$_2$ anti–mouse Ig was added and a 20-min incubation at 4°C was carried out. After three washings, cells were analyzed with a Coulter Epics V flow cytometer (Coulter Electronics, Hialeah, FL). Integrated fluorescence of the population gated by forward-angle light scatter and right-angle light scatter was measured and 10,000 cells were analyzed.

Bone Marrow Cells and Bone Marrow Cultures

Bone marrow cells were obtained from normal volunteers. They were separated by Ficoll–Hypaque density centrifugation to remove erythrocytes and mature granulocytes. Separated bone marrow cells were subjected to an indirect rosetting procedure (3) to obtain cells reactive with the monoclonal antibody of interest or they were subjected to a complement-mediated cytolysis procedure to deplete the reactive cells. For the complement-mediated killing procedure, 0.5 ml of 2×10^7/ml of bone marrow cells were incubated with 0.5 ml of hybridoma culture supernatant for 30 min at 37°C. 1.0 ml of rabbit baby complement (Pel Freeze Biologicals, Rogers, AR) was added and mixture was incubated for 45 min at room temperature. Dead cells were removed by centrifugation on a Ficoll–Hypaque gradient. Bone marrow cells were cultured in quadruplicates in Iscove's modified Dulbecco's medium containing 0.9% methylcellulose, 10% conditioned medium from human peripheral blood leukocytes stimulated with phytohemagglutinin (PHA-LCM), 30% FBS. One unit of erythropoietin (Connaught, Toronto, Ontario) was added to each plate. This procedure was essentially that described by Messner *et al.* (4). BFU-E were scored as hemoglobin-containing single or multiple colonies of greater than 64 cells on the 14th day. For CFU-GM colony formation, bone marrow cells were cultured in triplicates in RPMI 1640 medium containing 0.36% agar, 12% GCT condition medium (GIBCO, Grand Island, NY), and 20% FBS. Colonies containing more than 40 cells were counted on day 7.

Iodination and Immunoprecipitation

Cell iodination, as well as immunoprecipitation, gel electrolysis, and auto-radiography were performed as described previously (5).

Results

Characteristics of Monoclonal Antibodies K 15.1.9 and H 8.4.5

K562 and HEL were the two cell lines used as immunogens to generate K 15.1.9 and H 8.4.5, respectively. K 15.1.9 was typed to be IgG$_3$ and H 8.4.5, IgG$_2$. They both fixed complement.

K 15.1.9 precipitated a 21-Kd polypeptide from ^{125}I-labeled K562 cells. This is shown in lane 1 of Fig. 17.1. A similar peptide was precipitated from iodinated peripheral mononuclear cells. H 8.4.5 precipitated a molecule with a broad band between 65 Kd and 74 Kd (Fig. 17.2, lane 1). A similar peptide was detected on peripheral blood monocytes.

Fig. 17.1. A 21-Kd polypeptide was precipitated by antibody K 15.1.9 from ^{125}I-labeled HEL cells (lane 1). No protein was precipitated by a control IgG$_3$ antibody (lane 2).

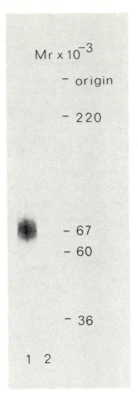

Fig. 17.2. A 65 Kd–74 Kd protein was precipitated by antibody H 8.4.5 from ^{125}I-labeled K562 cells (lane 1). No protein was precipitated by a control IgG$_2$ antibody (lane 2).

Cellular Distribution of Antigens K 15.1.9 and H 8.4.5

The cellular distributions of these antigens were analyzed by immuno-fluorescence with an Epics V instrument. Lymphocyte, monocyte, and granulocyte populations were isolated by a series of Ficoll–Hypaque and Percoll gradients. These preparations contained greater than 90% of the desirable cells. The cells were first gated by forward-angle and 90° light scattering. Integrated green fluorescence was measured. These results are shown in Fig. 17.3.

Panel A shows the staining patterns of antibody K. 15.1.9 and panel B shows those of antibody H 8.4.5. In the case of K 15.1.9 the majority of the lymphocytes and monocytes were positive. The staining intensity of the monocytes was less bright in comparison with lymphocytes. In the case of granulocytes, a vast majority of the cells were negative. However, a small percentage were positive. These positive cells were identified to be eosinophils. The reactivities of H 8.4.5 are less complex. It stained

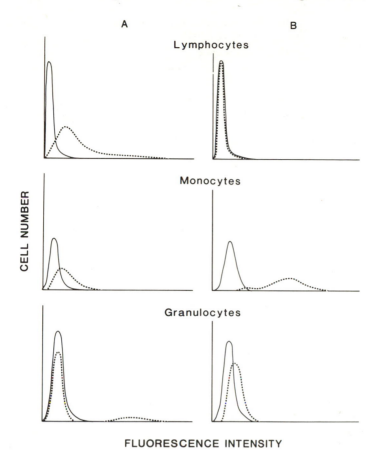

Fig. 17.3. Fluorocytometric analysis of staining patterns produced by antibody K 15.1.9 (panel A) and by antibody H 8.4.5 (panel B).

monocytes strongly and granulocytes weakly. The results of many separate experiments are summarized in Table 17.1.

Cell lines and leukemic cells of both lymphoid and myeloid origin were used to study these antigens further. These results are shown in Table 17.2. Although both antibodies stained all six non-lymphoid hematopoietic cell lines, studies on leukemic cells agreed with those on normal cells. K 15.1.9 stains leukemic cells of both lymphoid and myeloid origin while H 8.4.5 stains only leukemic cells of myeloid origin.

The reactivities of these two monoclonal antibodies with bone marrow cells were investigated. E^-mIg^- bone marrow cells were resolved into two populations by light-scattering analysis. The cell population with more light scattering was considered to be bigger in size. We term them large and small cell populations. K 15.1.9 reacted primarily with the small cell population while H 8.4.5 reacted with the large cell population

Table 17.1. Antigenic expression by normal hematopoietic cells detected by immunofluorescence.

	Staining monoclonal antibody	
Cell type	K 15.1.9	H 8.4.5
Lymphocytes (6)[a]	+++	−
Monocytes (6)	+	+++
Granulocytes (6)	+ (2–7%)[b]	+
Platelets (6)	−	−
Erythrocytes (6)	−	−
Thymocytes (3)	+	−

[a] The number of cell samples analyzed is given in parentheses.
[b] The stained granulocytes were predominantly eosinophils.

(Fig. 17.4). It is of considerable interest to note that K 15.1.9 stained a subpopulation of small cells very brightly. Analysis by log integrated green fluorescence indicated that this population was distinct.

Table 17.3 shows the characteristics of the two populations of bone marrow cells, resolved by light-scattering analysis. It is evident that the unfractionated bone marrow which was depleted of granulocytes and erythrocytes by Ficoll–Hypaque gradient was heavily contaminated by blood. In the small cell population, a large number of E receptor[+] and mIg[+] cells were detected. For this reason, bone marrow cells were depleted of E receptor[+] and mIg[+] cells by an immune rosette method. The resulting small cell population reacted primarily with monoclonal antibodies to HLA-DR and to K 15.1.9. In contrast, the large cell population stained predominantly with H 8.4.5.

Antigens K 15.1.9 and H 8.4.5 on Progenitor Cells

The question whether progenitor cells react to these two monoclonal antibodies in a different manner was investigated. A representative exper-

Table 17.2. Antigenic expression by various hematopoietic cell lines and leukemias.

	K 15.1.9	H 8.4.5
Non-lymphoid cell lines (6)[a]	++	++
T leukemia cell lines (5)	++	−
B lymphoblastoid lines (6)	++	−
Acute myelogenous leukemia (4)	++	++
Acute monomyelocytic leukemia (1)	++	++
Acute lymphocytic leukemia (4)	++	−
Chronic lymphocytic leukemia (3)	++	−

[a] The number of samples analyzed is given in parentheses.

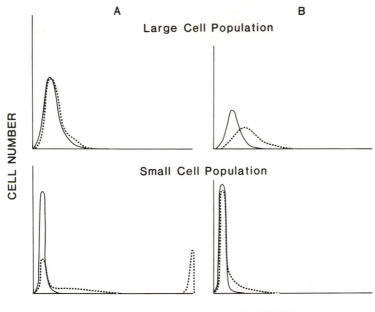

Fig. 17.4. Fluorocytometric analysis of staining patterns on bone marrow cells produced by antibody K 15.1.9 (panel A) and by antibody H 8.4.5 (panel B). The large cell population had more light-scattering characteristics in comparison to the small cell population.

iment is shown in Table 17.4. BFU-E and CFU-GM assays were used for the detection of the progenitor cells. Bone marrow cells were separated into two populations by an immune rosette method. As a control, the bone marrow cells were separated into HLA-DR$^+$ and HLA-DR$^-$ populations. Both BFU-E and CFU-GM progenitor cell populations were in the HLA-DR$^+$ population. In the case of H 8.4.5, the progenitor cells of CFU-GM were in the H 8.4.5$^+$ population and those of BFU-E were in the

Table 17.3. Antigenic expression by two populations of bone marrow cells resolved by light-scattering analysis.

	Unfractionated BM		E$^-$mIg$^-$BM	
Monoclonal antibody	Large	Small	Large	Small
Ab 23.1 (SRBC receptor)	<1.0%	53.0%	<1.0%	4.3%
Ab Josh 524 (HLA-DR)	6.3	25.5	6.1	48.3
Leu 10 (HLA-DC)	<1.0	18.6	<1.0	26.0
Anti-u	<1.0	11.1	<1.0	2.4
B1	<1.0	11.8	<1.0	10.4
K 15.1.9	3.6	70.4	4.0	59.5
H 8.4.5	>80.0	3.5	>80.0	10.1

Table 17.4. H 8.4.5 and K 15.1.9 positive bone marrow cells differ in the BFU-E assay.

Cell population	Colonies/10^5 cells plated	
	BFU-E	CFU-GM
Untreated	630	100
K 15.1.9$^+$	560	115
K 15.1.9$^-$	6	1
H 8.4.5$^+$	53	159
H 8.4.5$^-$	677	28
HLA-DR$^+$	539	127
HLA-DR$^-$	3	1

H 8.4.5$^-$ population. In contrast, K 15.1.9$^+$ contained the vast majority of progenitor cells of CFU-E and CFU-GM. Three additional experiments showed similar results. These results were confirmed with cell populations obtained by complement-mediated cytolysis.

Discussion

The antigen distribution as detected by antibodies K 15.1.9 and H 8.4.5 is summarized in Table 17.5. K 15.1.9 identifies a 21-Kd polypeptide. It reacts with the progenitor cells of BFU-E and BFU-GM. It also stains lymphocytes, monocytes, and eosinophils. This antigen appears to be similar to that described by Pessano *et al.* (6). In contrast, H 8.4.5 reacts only with CFU-GM progenitors. It stains monocytes and granulocytes as well as immature myeloid cells. It identifies a 65–74-Kd polypeptide.

The finding that the 21-Kd antigen is expressed by about 50% of small bone marrow cells (Table 17.3) and a subpopulation of these cells stained

Table 17.5. Summary of antigen distribution detected by K 15.1.9 and H 8.4.5.

	21 Kd (K 15.1.9)	67–74 Kd (H 8.4.5)
BFU-E	++	−
CFU-GM	++	++
Lymphocytes	Bright	−
Monocytes	Moderate bright	Bright
Granulocytes	Negative^a	Weak
Erythrocytes	−	−
Platelets	−	−

^{a} 2–7% of granulocytes were positive. This population was shown to be predominantly eosinophils.

brightly (Fig. 17.4) raised the possibility that the brightly stained cell population is markedly enriched for progenitor cells. It is also of considerable interest to determine the progenitor potential of the HLA-DR$^+$, K 15.1.9$^+$ population. This may offer a scheme for further enrichment for progenitor and stem cells.

The present results may suggest that the 21-Kd peptide is present on the multipotent stem cell. If this is the case, during the differentiation of the stem cells to cells of various lineages, K 21-Kd is preferentially retained by certain cells. This is of considerable interest in view of the finding that eosinophils express this antigen while neutrophils do not. The differential retention may have certain functional significance.

The lack of enrichment for BFU-E and CFU-GM progenitor cells in the K 15.1.9$^+$ cell population in the present experiment requires some comment. It is possible that the experimental procedure used to isolate these cells may be harmful to these cells or a suppressor cell population may be co-enriched in the present scheme. Further experimentation is needed to distinguish these possibilities.

Summary

A series of monoclonal antibodies were made against non-lymphoid hematopoietic cell lines to probe antigenic expression of hematopoietic cells. Two monoclonal antibodies, K 15.1.9 (IgG$_3$) and H 8.4.5 (IgG$_2$) were obtained by fusions between SP2/0 tumor cells and spleen cells from BC$_3$F$_1$ mice immunized with K562, a CML blast line, and HEL, a human erythroleukemia cell line, respectively. mAb K 15.1.9 precipitated a 21-Kd polypeptide. It stained 90–95% of lymphocytes and monocytes. It also stained eosinophils. About 50% of the small bone marrow cells were reactive. K 15.1.9 did not react with neutrophils, erythrocytes, or platelets. Large bone marrow cells showed little reactivity. H 8.4.5 identified a broad glycoprotein band in the 65Kd–74Kd region. Antibody H 8.4.5 reacted strongly with >95% monocytes and weakly with the majority of the granulocytes, greater than 80% of the large bone marrow cells. It did not stain lymphocytes, erythrocytes, platelets, or small bone marrow cells. Depletion of K 15.1.9$^+$ cells abolished both CFU-GM and BFU-E formation. However, depletion of H 8.4.5$^+$ cells abolished CFU-GM but not BFU-E formation by bone marrow cells *in vitro*. Thus these two antibodies identify antigens expressed on hematopoietic progenitor cells. The selective retention of antigen K 15.1.9 by certain cell lineages may suggest some functional role.

Acknowledgment. This work was supported in part by a Public Health Service Grant CA-34546 from the National Institutes of Health.

References

1. Yen, S.H., F. Gaskin, and S.M. Fu. 1983. Neurofibrillary tangles in senile dementia of the Alzheimer type share an antigenic determinant with intermediate filaments of the vimentin class. *Am. J. Pathol.* **113**:373.
2. Chiorazzi, N., S.M. Fu, and H.G. Kunkel. 1980. Stimulation of human B lymphocytes by antibodies to IgM and IgG: Functional evidence for the expression of IgG on B-lymphocyte surface membranes. *Clin. Immunol. Immunopathol.* **15**:301.
3. Gottlieb, A.B., S.M. Fu, D.T.Y. Yu, C.Y. Wang, J.P. Halper, and H.G. Kunkel. 1979. The nature of the stimulatory cell in human allogeneic and autologous MLC reactions: role of isolated IgM-bearing B cells. *J. Immunol.* **123**:1497.
4. Messner, H.A., H. Jamal, and C. Izaguirre. 1982. The growth of large megakaryocyte colonies from human bone marrow. *J. Cell. Physiol.* Suppl. 1:45.
5. Wang, C.Y., S.M. Fu, and H.G. Kunkel. 1979. Isolation and immunological characterization of a major surface glycoprotein (gp54) preferentially expressed on certain human B cells. *J. Exp. Med.* **149**:1424.
6. Pessano, S., L. Bottero, J. Faust, M. Trucco, A. Palumbo, L. Pegoraro, B. Lange, C. Brezim, J. Borst, C. Terhorst and G. Rovera. 1983. Differentiation antigens of human hemopoietic cell: Patterns of reactivity of two monoclonal antibodies. *Cancer Res.* **43**:4812.

CHAPTER 18

Heterogeneous Expression of Myelo-Monocytic Markers on Selected Non-Lymphoid Cells

Giorgio Cattoretti, Domenico Delia, Luciana Parola, Raffaella Schirò, Costante Valeggio, Giuseppe Simoni, Lorenza Romitti, Nicoletta Polli, Giorgio Lambertenghi DeLiliers, Maurizio Ferrari, and Angelo Cantù-Rajnoldi

Introduction

We have analyzed the early steps of the myelo-monocytic commitment and the dissection of the granulocytic from the monocytic differentiation pathway by using acute leukemias as a model. We have selected fresh acute leukemias and cell lines because of their peculiar phenotypic characteristics: the ML-3 cell line and two acute non-lymphoblastic leukemias (ANLL) showed a mixture of granulocytic and monocytic features, both morphologically and cytochemically; two cases of acute undifferentiated leukemias (AUL) had no reliable morphological or, under the light microscope, cytochemical myeloid character, but showed immunological and ultrastructural non-lymphoid markers. We compared these selected cases with fresh leukemia samples (representative of granulocytic, monocytic, and lymphoid lineage), cell lines, and normal tissues by using a morphological, ultrastructural, cytochemical, immunological, and cytogenetical characterization. This panel of cells was screened with the anti-myeloid monoclonal antibodies of the second International Workshop on Human Leukocyte Differentiation Antigens.

Materials and Methods

Cell lines ML-3, U 937, HL-60, and K562 were maintained in complete RPMI culture medium. Bone marrow (BM) samples from pediatric patients, containing more than 90% leukemic cells, were smeared for light

Table 18.1. Monoclonal antibodies used for the diagnosis of acute leukemias.

mAb	M.W.(Kd)	Specificity	Source	References
A3D8	80	Leukocyte common antigen	M.J. Telen	*J. Clin. Invest.* 1983. **71**:1878
F10.44.2	105	Leukocyte common antigen	R. Dalchau	*Eur. J. Immunol.* 1980. **10**:745
7.2	29–33	HLA-DR monomorphic	J. Hansen	*Immunogen.* 1980. **10**:247
AA3.84	29–33	HLA-DR monomorphic	F. Malavasi	*Diagnostic Immunology* 1984. **2**:53
Tü 22		HLA-DC monomorphic	A. Ziegler	*Leukemia markers*, W. Knapp, ed. 1981. Academic Press, New York, p. 317
Tü 39		HLA-DR+SB monomorphic	A. Ziegler	*Leukemia markers*, W. Knapp, ed. 1981. Academic Press, New York, p. 317
0.1.65		HLA-ABC monomorphic	F. Malavasi	*Diagnostic Immunology* 1984. **2**:53
BA-1	30	Pan B + PMNs	J.H. Kersey	*J. Immunol.* 1981. **126**:83
BA-2	24	CD9(nT,nB,p24)	J.H. Kersey	*J. Exp. Med.* 1981. **153**:726
FMC 27		Plts, some AML and ALL	H. Zola	*Pathology* 1984
VIL-A1	100	CD10(nTnB,p100)	W. Knapp	*Leuk. Res.* 1982. **6**:137
OKM1	94–155	C3bR	J. Breard	*J. Immunol.* 1980. **124**:1943
C3RTo5	205	C3b R	J. Gerdes	*Immunology* 1982. **45**:645
UCHM1	55	Monocytes	N. Hogg	*Blood* 1984. **63**:566
FMC17		CDw14(M,u)	H. Zola	*Pathology* 1983. **15**:45
5F1	85	Monocytes, plts, erythroid	I.D. Bernstein	*J. Immunol.* 1982. **128**:876
TG1	155	CDw15(G,u)	P.C.L. Beverley	*Nature* 1980. **278**:332
IG10	200–240	CDw15(G,u) anti LNFIII	I.D. Bernstein	*J. Immunol.* 1982. **128**:876
My-1		Anti-LNF III	C.I. Civin	*Blood* 1981. **57**:842
FMC 11		PMNs	H. Zola	*Brit. J. Haematol.* 1981. **48**:481
WT1	40	CD7(T,p41)	W.T. Tax	*Protides of the biological fluids.* E.H. Peters, ed. 1982. Pergamon Press, Oxford, p. 701
IHery-1		Glycophorin A	P. Lansdorp	unpublished
C17.27	110	Platelet gpIII°	P.A.T. Tetteroo	*Brit. J. Haematol.* 1983. **55**:509

Table 18.2. Light microscopy morphology, ultrastructural morphology, and cytochemistry of the cell panel.

Case	L. M. Morphology	E. M. Morphology	E. M. Cytochemistry (MPO)
ANLL-1	80% atypical hypergranular cells with a monocytoid chromatine meshwork. 20% more undifferentiated cells with high n/c ratio and a few azurophilic cytoplasmic granules.	Large cells with big nuclei and evident nucleoli; <5% irregular nuclear and cytoplasmic outlines.	MPO⁺ large granules and diffuse positive material in the cytoplasm. <5% low MPO reactivity, mainly in a few small granules.
ANLL-2	Mixed population with atypical granular myeloid cells (50%), monocytoid cells (35%), and more undifferentiated blasts 28% of the myeloid cells show basophilic Toluidine Blue-positive cytoplasmic granules.	Some of the blasts have typical monocytic characteristics, others have granulocytic features. There is a third population of immature "intermediate" cells which coexpress the features of both cell lineages.	Large granules and some of the strands of the endoplasmic reticulum are MPO⁺
ML-3	Hypergranular myeloid cells with hyperbasophilic cytoplasm.	Monocytic cell line.	
HL-60	Hypergranular myeloid cells with hyperbasophilic cytoplasm.	Blastic cells plus a small percentage of cells with monocytoid features.	High percentage of MPO⁺ cells.
M3	Atypical promyelocytes (FAB M3).	Hypergranular promyelocytes.	
U 937	Atypical cells resembling monoblasts.	Macrophage cell line.	MPO negative.
M5b	Atypical mature monocytes.		
AUL-1	Large undifferentiated cells with a variable amount of basophilic cytoplasm, sometimes with vacuolization. Few cells (1%) show fine azurophilic cytoplasmic granules.		
AUL-2	Lymphoblastoid cells (FAB L2).	Blastic cells without any characteristics of peculiar cell lineage.	<5% of blasts are MPO⁺.

microscopy and cytochemical evaluation. A mononuclear cell suspension was then obtained after Ficoll–Hypaque density gradient separation. Samples, frozen in 30% fetal calf serum plus 10% dimethyl sulfoxide and Hank's balanced salt solution, were kept at $-80°C$ unless utilized within a few hours. Thawed samples were more than 80% viable as assessed by Trypan Blue dye exclusion test.

Smears of fresh samples and cell lines were stained with May-Grünwald-Giemsa stain and with routine cytochemical methods (MPO, αNBE, CAE). The diagnosis was made according to the French/American/British classification (FAB).

Methanol-fixed cytocentrifuged cells were stained with rabbit anti-terminal transferase (BRL, Bethesda, MD, USA) and rabbit anti-lysozyme (Behring Institute, Italy) in indirect immunofluorescence.

Monoclonal antibodies (mAbs) from our local diagnostic panel (see Table 18.1) and from the Second International Workshop were used at saturating concentrations. Briefly, the cells were incubated for 30 min at 4°C with the mAb, washed twice with PBS-BSA, and counterstained with a fluorescein-conjugated sheep anti–mouse $F(ab')_2$ antiserum (New England Nuclear, Italy). After two final washes, 200 and 10,000 cells were scored with a fluorescence microscope (Leitz, Germany) and a FACS IV cell sorter (Becton Dickinson, USA), respectively.

For ultrastructural analysis, the cells were resuspended in gold buffer (PBS containing 1% BSA, 0.2 M NaN_3, and 1% decomplemented human AB serum). The immunogold method has been described in detail elsewhere (1). Then the samples were processed as previously described (1) for electron microscopy (EM) morphology and cytochemistry (MPO).* Ultrathin sections were examined with a Philips 410 EM.

For cytogenetic analysis, chromosome preparations were obtained from either cultured cell lines or from unstimulated fresh leukemic BM cultured in complete RPMI for 24 hr at 37°C. Karyotype studies were carried out by the QFQ banding technique (2).

Results

The combined morphological, cytochemical, and ultrastructural analysis of the cell panel is shown in Tables 18.2 and 18.3. Three groups of cell types can be identified:

1. ANLL-1, ANLL-2, and ML-3. These two fresh leukemic samples and the ML-3 cell line represent rather mature non-lymphoid cells with a mixture of granulocytic and monocytic features (Figs. 18.1 and 18.2).

---

* Abbreviations used in this paper: MPO, myeloperoxidase; CAE, chloroacetate esterase; αNBE, alpha naphthyl butyrate esterase.

Table 18.3. Cytochemical and immunological phenotype of the panel of leukemic cells.[a]

	ANLL-1	ANLL-2	ML-3	HL-60	M3	U 937	M5b	AUL-1	AUL-2	PMNs	Monos	K562
MPO	74	88	100	20	100	0	0	0	0	+	+	−
αNBE	90	90	100	0	0	100	87	0	0	−	+	−
CAE	70	30	100	100	100	0	0	0	0	+	−	−
Lysozyme	63	53	100	100	100	100	52	32	0	+	+	−
HLA-DR	6	8	0	0	0	0	86	67	30	−	+	−
HLA-ABC	ND	47	100	92	100	ND	96	94	ND	+	+	−/w+
OKM1	38	48	16	0	12	0	55	25	1	+	+	0–17
LFA-1	40–64	30–70	78–100	29–98	0–22	32–98	89–97	29–70	0–54	+	+	−
UCHM1	38	36	0	0	0	0	86	9	0	−	+	−
IG10	80	53	100	76	2	55	22	30	75	+	−/w+	+
A3D8			90	90	90	90				+	+	ND
F10.44.2	97	96					98	95	95	+	+	ND
BA-1	16	7	ND	0	1	ND	ND	46	6	+	−	+
BA-2	0	25	0	0	0	0	ND	3	ND	ND	ND	+

[a] Heterogeneous results for one marker on different cell populations are enclosed in the boxes.
[b] ND: Not determined.

The sum of CAE and αNBE exceeds 100% in each case, indicating that some, if not all, cells coexpress both markers. Lysozyme positivity confirms their late myelo-monocytic precursor nature (3).

2. AUL-1 and AUL-2 (Figs. 18.3 and 18.4). Their non-lymphoid lineage was assessed by the lysozyme positivity of AUL-1 in the absence of cALL and B4 antigens (not shown) and by the EM MPO positivity in AUL-2 (Figs. 18.5 and 18.6).

3. U 937, M5b, HL-60, and M3. A third group composed of leukemic cells phenotypically close to a normal progranulocytic or monocytic cell. Two monocytic populations (the histiocytic cell line U 937 and a fresh M5b leukemia) show intense αNBE, but no CAE or MPO. The reverse is found for the myeloid HL-60 cell line and for a fresh promyelocytic leukemia (M3) Table 18.3 shows for comparison the phenotype of normal granulocytes, monocytes, and of the erythro-megakaryocytic cell line K562. HL-60 cells show by EM a minor percentage of monocytic cells (Figs. 18.7 and 18.8).

The immunological phenotype of the cell panel shows that HLA-DR antigen is absent from the first group of cells and also from U 937, HL-60, and M3. OKM1 antigen is weakly or not expressed on ML-3, HL-60, M3, U 937, and AUL-2. A group of the Workshop mAbs directed against a leukocyte-function-associated antigen (LFA-1) strongly stain all the samples, except M3, AUL-2, and K562, which are weak or negative.

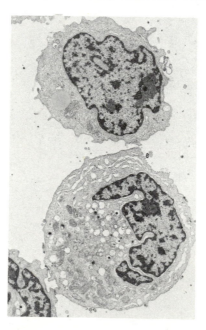

Fig. 18.1. ML-3: cells (especially the lower) showing monocytic features. 4800× (PbUa).

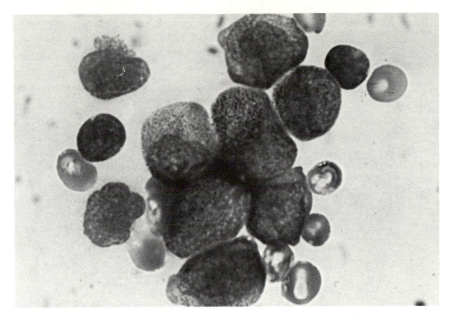

Fig. 18.2. ANLL-2: bone marrow smear. MGG stain. 1000×.

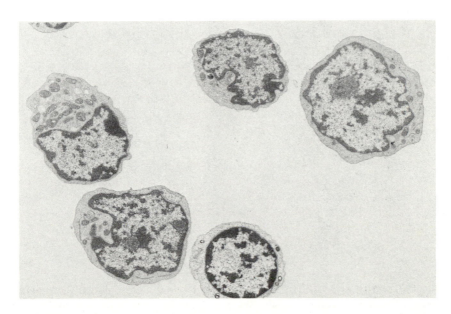

Fig. 18.3. AUL: overview showing undifferentiated leukemic blasts. 6700× (PbUa).

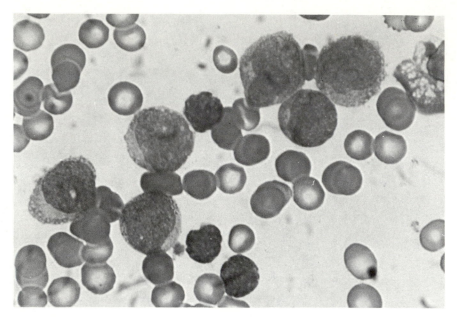

Fig. 18.4. AUL-1: bone marrow smear. MGG stain. 1000×.

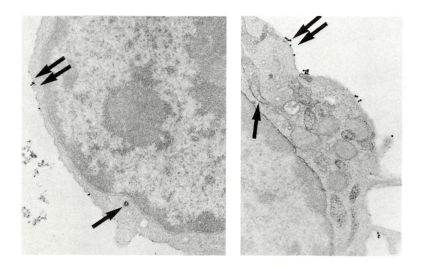

Fig. 18.5 and 18.6. AUL: particulars of cells showing very light MPO reactivity (arrow) localized in tiny granules or in the endoplasmic reticulum and in the perinuclear cisternae, and showing also a positive reaction for the TG1 mAb (double arrow) by the immunogold method. Left, 24,000×; right, 33,000×.

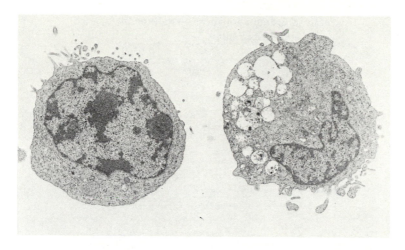

Fig. 18.7. HL-60: two cell types present in this sample: on the left a less differentiated cell with few granules in the cytoplasm and two nucleoli in the nucleus; on the right a cell undergoing maturation. 15,000× (PbUa).

The expression of the anti-monocyte mAb UCHM1 is even further restricted to the ANLLs and M5b cells and minimally to AUL-1.

A group of monocyte-restricted mAbs (Table 18.4) positively distinguish AUL-1 from AUL-2. The BA-1 and TG1 mAbs give differential results on the two AULs. FMC11, BA-2, and FMC27 mAbs stain ANLL-2 but not ANLL-1. The 116 mAbs from the myeloid panel of the Workshop are tentatively grouped according to the their reactivity pattern

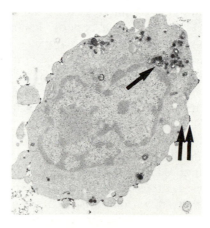

Fig. 18.8. HL-60: monocyte positive for myeloperoxidase reaction and reactive with the M8 mAb by the immunogold method (double arrow). 8100×.

Table 18.4. Immunological phenotype of selected
non-lymphoid leukemias.

	ANLL-1	ANLL-2	AUL-1	AUL-2
5F1	10	23	12	0
FMC17	23	31	13	0
C3RTo5	27	26	10	0
TG-1	90	63	8	70
My-1	74	46	6	ND
FMC 11	0	28	3	ND
FMC 27	0	43	3	ND

against the panel of leukemias and normal cells. Table 18.5 reports the
results for six clusters and a few isolated mAbs.

Cluster I (M8, M63, M67, M69, M99) reacts with monocytes and
weakly with granulocytes. All the samples showing a monocytic component (either by morphology, immunology, or cytochemistry) bind to this
group of mAbs, with the exception of ML-3 and U 937. However M63, the
most strongly reacting mAb of the cluster, stains ML-3 cells as well. At
the ultrastructural level, M8 stains mainly the monocytic component of
HL-60 (Fig. 18-6). M46, a monocyte–macrophage-specific mAb stains all
the Cluster I-positive cells, in addition to ML-3 and U 937.

Cluster II mAbs (M34, M64, M104) recognize a granulo-monocytic antigen expressed by both ANLL cases, the U 937 and M5b leukemias, and
AUL-1 but not the other samples. A somewhat complementary reactivity
with Cluster II is given by M30: its granulocytic-specific reactivity is
confirmed by the lack of reactivity with mono-hystiocytic leukemias and
by the positive recognition of granulocytic and granulo-monocytic cells. It
also parallels the CAE figures on our panel of leukemias. However two
fresh M3 leukemias are unreactive with this lineage-restricted mAb.

A large group of mAbs from the Workshop and mAbs IG10 and My-1
from our local panel show an anti-myelo-monocytic specificity compatible
with the characteristics of anti-lacto-*N*-fucopentaose III° (LNF III°) antibodies. They react with granulocytes, thymic Hassal corpuscles, myeloid
precursors, a few monocytes, and K562 cells but not erythroid cells,
platelets, and lymphocytes. Two subgroups result from their reactivity
with M5b and AUL-1 cases: Cluster III (M5, M13, M19, M33, M52, M77,
M91, M110, M111, and M117) and Cluster IV (M25, M80, and M116).
M61 has an intermediate pattern of reactivity between the two clusters. A
major difference arises from the comparison between the two clusters: the
former does not distinguish between granulocytic and monocytic leukemias; the latter is weakly or not reactive at all with monocytic cells, both
AULs, and the K562 cell line. The remaining leukemias of the panel do

Table 18.5. Reactivity of the anti-myeloid panel of mAbs from the second international workshop with leukemic cells and cell lines.

mAb[a]	ANLL-1	ANLL-2	ML-3	HL-60	M3	U 937	M5b	AUL-1	AUL-2	PMNs	Monos	K562
Cluster I	16–32	21–36	0[b]	12–28	0	0	48–87	14–19	0	w+	+	–
M46	38	37	33	57	0	61	87	32	0	–	+	–
Cluster II	30–39	13–39	0	0	0	0–18	20–95	0–32	0	+	+	–
M30	69	25	91	60	0	0	0	0	0	+	–	–
Cluster III	76–84	47–67	92–100	54–100	0–22	47–81	16–51	20–52	74–88	+	–/w+	86–95
M61	78	58	98	75	0	23	0	14	66	+	–	79
Cluster IV	67–81	34–52	98–100	59–100	0	15–24	0	0	32–60	+	–	73–80
Cluster V	21–68	0	0	0	0	12–28	64–76	0–17	0	w+	+	–
Cluster VI	0	10–16	0[c]	0	0[d]	70–98	0–28	13–89	57–95	–	–	–

[a] Cluster I (mAbs 8, 63, 67, 69, 99) reacts with germinal centers in the lymph node (weak). Cluster II (mAbs 34, 64, 104) reacts with germinal centers (M34 negative). Hassal corpuscles. Cluster III (mAbs 5, 13, 19, 33, 52, 77, 91, 110, 111, 117) reacts with Hassal corpuscles, sinus histiocytes in the lymph node, germinal centers (weak) (M77, M91). Cluster IV (mAbs 25, 80, 116) reacts as Cluster III. Cluster V (mAbs 93, 95, 114) reacts with histiocytes, vessels (M95, M114), lymphoid cells (M95), and platelets (M114). Cluster VI (mAbs 50, 51, 83, 84, 85) reacts with germinal center and lymph node histiocytes (M50, M51), B lymphocytes, and some T lymphocytes (M50).
[b] M63 = 37% weak.
[c] M51 = 30% weak.
[d] M50 = 51%.

Table 18.6. Cytogenetic findings in the leukemia cell panel.

	ANLL-1	ANLL-2	ML-3	HL-60	M3	U-937	M5b	AUL-1	AUL-2	K562
Chromosome 4	NI[a]	NI	polys(3)[b]	NI	NI	NI	NT[c]	NT	t(4;11)(q21;q23)	NI
Chromosome 6	NI	NI	polys(6)[d]	NI	NI	t(6;?)(p22;?)[e]	NT	NT	NI	polys(3)[f]
Chromosome 11	t(5;11)(q21;q24)	NI	polys(4)	NI	NI	t(10;11)(p21;q13)	NT	NT	t(4;11)(q21;q23)	polys(3)
Chromosome 12	NI	NI	polys(6)	NI	NI	t(6;12)(p12;p11)[e]	NT	NT	NI	polys(4)[g]
Chromosome no.	46	46	97	45	46	59	NT	NT	46	67

[a] NI: Not involved.
[b] polys = polysomy (no. of chromosomes).
[c] NT: Not tested.
[d] Four chr. 6 of six present in the karyotype show a deletion of the long arm del(6)(q22).
[e] Two clones are present in this cell line. The first shows a chr. 6 with a rearranged short arm (6pt), a normal chr. 6, and an additional 6p on the short arm of chr. 12. The second shows in addition a further 6p rearrangement.
[f] One chr. 6 shows a rearrangement of the short arm in the p22 region.
[g] One chr. 12 had a t(5;12)(q11;p11) rearrangement.

Table 18.7. Assignment of surface membrane markers and enzymes to human chromosomes.

Marker	Chromosome location	Ref.
ESAT (esterase activator)	4	*Proc. Natl. Acad. Sci. U.S.A.* 1972. **69**:3273
HLA-ABC	6p21 → p23	*Cytogen. Cell Genet.* 1970. **25**:32
HLA-DR	6p21 → p23	*Cytogen. Cell Genet.* 1970. **25**:32
F10.44.2 (MIC4)	11	*Eur. J. Immunol.* 1982. **12**:659
A3D8	11p	*Somat. Cell Genet.* 1983. **9**:333
anti-LNF III° mAbs (UJ308, VIM-D5, IG10, B4.3, My-1, FMC10)	11	*Proc. Natl. Acad. Sci. U.S.A.* 1983. **80**:3748
ESA 4 (esterase A-4)	11cen → q22	*Cytogen. Cell Genet.* 1979. **25**:47
BA-2	12	*Hematol. Blood Transf.* 1983. **28**:3

not show any variation of reactivity within the two clusters; however the overall positivity for ANLL-2 and for HL-60 is less than for ANLL-1 and for ML-3, respectively.

Two clusters distinguish ANLL-1 from ANLL-2. Cluster V (M93, M95 and M114) stains ANLL-1 and shows mainly an anti-monocyte specificity. It weakly stains granulocytes and heterogeneously recognizes lymphocytes and platelets. Cluster VI (M50, M51, M83, M84, M85) stains B cells, germinal center and corona in the lymph node, hystiocytes, and U 937 cells; it positively recognizes ANLL-2 but not ANLL-1.

The cytogenetic results of the leukemia panel are reported in Table 18.6. In this study our attention was drawn to chromosomes 4, 6, 11, and 12, because they carry the genes encoding for some of the membrane antigens and enzymes under investigation (see Table 18.7). The two M3 cases analyzed had a t(15q;17q)(q25;q22) rearrangement; no such marker was found in the HL-60 cell line. The latter shows, however, two differently rearranged chromosomes 16 (data not shown).

Discussion

The comparison of different characterizations of non-lymphoid cells (morphological, cytochemical, immunological, and cytogentical) reveals heterogeneous behavior of traditional cell markers in the selected leukemic samples.

HLA-DR antigen and related DC and SB determinants are absent from all three myelo-monocytic samples (both ANLLs and ML-3) and from U 937. DR antigens are usually expressed on the vast majority of monocytic leukemias (4). For the myelo-monocytic leukemias of this panel, we can rule out a loss or a rearrangement of the short arm of chromosome 6 (encoding for HLA antigens) as the cause for the failure to express HLA-DR. Moreover all the cases are HLA-ABC positive (also coded by chr. 6). The U 937 cell line does show a rearranged and translocated chromosome 6p, but it has recently been shown that under experimental conditions this cell line can express HLA-DR (5). The HL-60 cell line and DR-negative normal granulocytic precursors can also be induced to express HLA-DR under appropriate culture conditions (6,7). The same type of cells are able to express *de novo* monocytic markers (adherence, phagocytosis, esterases, and membrane markers) with and without the combined synthesis of HLA-DR (6). HLA-DR-negative, nonspecific esterase-positive granulocytic cells are found *in vivo* in preleukemic conditions (8) but have never been found in normal bone marrow. Finally, some of the M3 leukemias can show esterase positivity and are invariably DR-negative (0/14 M3 cases is DR$^+$; personal observation).

We studied this peculiar type of cell by screening the three granulomonocytic samples (ANLL-1, ANLL-2, and ML-3) with a large panel of mAbs. Our results show that anti-monocyte mAbs (Cluster I and II and M46) stain equally well fresh monocytic leukemias (M5b and AUL-1) and both ANLLs. The same group of mAbs is heterogeneous in its reactivity with the ML-3 and U 937 cell lines and stains in HL-60 a higher number of monocytic cells than expected (on the basis of EM morphology and cytochemistry). EM after immunogold staining of M8 confirms the monocytic significance of Cluster I antibodies on HL-60 cells (see Fig. 18.8). The heterogeneous results cited above may be due to the differential reactivity of the three groups of mAbs on different maturative stages of the monoblasts; in fact, we analyzed an immature M5a monoblastic leukemia, found positive for M46 but negative for Cluster I mAbs (not shown). However the results obtained on a rather mature type of cells such as ML-3 hardly fit with the ultrastructural and cytochemical phenotype.

The finding of positive mature monocytic markers on an esterase-negative case such as AUL-1 or HL-60 cells is also unexpected. Anti-monocyte mAbs and esterases have been analyzed in detail by Scott *et al.* (9): they show that many different esterase isoenzymes are contained in normal and leukemic myeloid cells. There is a relationship between the presence of myeloid or monocytic membrane markers and the relative proportion of granulocytic- and monocytic-type of esterase isoenzymes within each cell. This correlation is not as sharp as expected. Moreover, the synthesis of different forms of esterases can be induced *in vitro* without synthesis of messenger RNA (10). As we have used αNBE at pH 5.8 (a cytochemical stain that gives monocyte specificity) (11) we may wrongly interpret this negative reaction as non-monocytic in AUL-1 and HL-60; other factors such as immature stage of differentiation or an abnormal gene dosage may account for a peculiar composition of esterase isoenzymes.

The combined results of OKM1 and LFA-1 mAbs on our cell panel reveal another interesting heterogeneity. Both mAbs precipitate a common β subunit of 95 Kd mol. wt., in addition to the 165-Kd OKM1 α subunit and the 177-Kd LFA-1 α subunit (12). The expression of the alpha subunits varies during leukocyte differentiation: granulocytes express the OKM1 subunit more than the LFA-1 one while monocytes express both equally well (12).

Comparable percentages of OKM1 and LFA-1 are present on both ANLLs, M5b, and AUL-1. HL-60, ML-3, and U937 cell lines have little or no OKM1 but show high levels of LFA-1. The pattern of the fresh M3 leukemia (and of another M3 sample) is completely different: both OKM1 and LFA-1 expressions are depressed. This suggests that the immunological phenotype of the fresh promyelocytic leukemias is quite distinct from monocytic and myelo-monocytic leukemias as well as from HL-60 and ML-3 cell lines.

A large group of Workshop mAbs show an anti-granulocytic reactivity compatible with an anti-LNF III specificity. The immunoreactivity of this epitope depends on the extent of sialic acid substitution (13), on the length of the glycosphingolipids carrying the epitope, or on the presence of proteins and glycolipids close to the antigen (14). This group of mAbs seems indeed myeloid specific; however, the enzymatic removal of sialic acid unmasks LNF III determinant on lymphoblasts (15). We have found that only 60% of ALLs and 75% of previously negative M1 ANLLs acquire LNF III reactivity after desialation: the relationship of this trisaccharide (formerly known as Lewis X or X-hapten) with the blood group carbohydrate antigens may account for this kind of discrete distribution. We have found however no relationship between the ABO and Lewis status of the patients and the LNF III pattern of their leukemic cells (unpublished results). Furthermore ANLL-1, ANLL-2, ML-3, M3, AUL-1, and AUL-2 are negative for blood group-related substances (tested with anti-A and anti-B mAbs from Biotest, Germany; anti-H mAb from Dakopatt, Denmark; anti-Le^a, Le^b and Le^{a+b} mAbs, gift of Dr. M. Herlyn, New York, USA) (data not shown). A reduced reactivity with this cluster of mAbs indicates a monocytic differentiation of the cells; this further suggests that HL-60 bears more monocytic features than ML-3, contrary to other morphological and cytochemical evidence.

Apart from Cluster III and Cluster IV mAbs and their peculiar reactivity, none of the mAbs described in Table 18.5 stains the M3 and AUL-2 cases, except Cluster VI mAbs. If the highly immature status of AUL-2 may account for this, the explanation in the case of M3 is not evident. The refractoriness of promyelocytic leukemia to present on its membrane markers of monocytic or granulocytic differentiation may be explained if we consider M3 as a bipotential progenitor cell still uncommitted to any neutrophilic, basophilic, eosinophilic, or monocytic intermediate precursor cell, but already able to synthesize primary granules and enzymes of both granulocytic and monocytic cells. We know that a normal or near-normal progenitor cell can be induced by appropriate stimuli (phorbol esters *in vitro,* preleukemic syndromes *in vivo*) to undergo maturation by the acquisition of lineage-specific markers, but still retaining its bipotential capacity, even after the promyelocyte stage (7). M3, ANLL-1, and ANLL-2 can be considered examples of the sequential maturation of a bipotential granulo-monocytic cell, along a peculiar differentiative pathway on which the acquisition of granulocytic markers does not prevent the cell from displaying monocytic characteristics and vice versa. The relationship between the normal hematopoietic differentiation and this hypothetical alternative is exemplified in Fig. 18.9.

In addition to diagnostic methods discussed above, we have used the cytogenetic findings to improve the diagnosis of leukemia type. There is a nonrandom association between definite types of chromosomal abnormalities such as deletion of the long arm of chromosome 11 or the t(15;17) translocation and various types of leukemias (16). We also tried to correlate specific chromosomal aberrations with a definite phenotypic pattern. Our attention has been focused on chromosomes 4, 6, 11, and 12, since we know membrane markers and enzymes which are coded by genes localized on these chromosomes (Table 18.7).

No correlation appears to exist between chromosome 12 status and BA-2 expression or between A3D8, F10.44.2, and anti-LNF III mAbs and rearrangement of chromosome 11.

The t(4;11) translocation was reported as diagnostic of a highly undifferentiated type of non-lymphoid leukemia (17). The translocation involves two important sites for the regulation of esterase synthesis (Table 18.7). This fact may be taken into account to explain the lack of esterase positivity in AUL-2. It has, however, been shown that phorbol esters can induce esterase synthesis in the t(4;11) leukemias (18). A rearrangement of chromosome 11 has been reported to occur as a secondary event at the time of overt leukemic transformation of a preleukemic syndrome (19). The phenotypic characteristics of our two ANLL cases are very similar to the previously described myelo-monocytic cells of the preleukemic BM (8), suggesting a clonal expansion of such a type of cells. The detection in only one of our ANLLs of a 11q rearrangement appears to be in agreement with the karyotypic evolution previously described (19).

In conclusion, no unequivocal definition may be advanced, either morphological, immunological, or ultrastructural, about what is monocytic and what is granulocytic, except for the mature peripheral blood cells. A wide range of monocytic- or granulocytic-associated markers exist in leukemia and the balance of the various components gives an idea of the differentiative stage of a leukemic sample. We may postulate that this situation is true for the normal hematopoietic system as well, although we can demonstrate it only under nonphysiological conditions *in vitro*. This balance can be different among various types of leukemic populations. Our analysis of the cell lines compared with the fresh leukemias shows

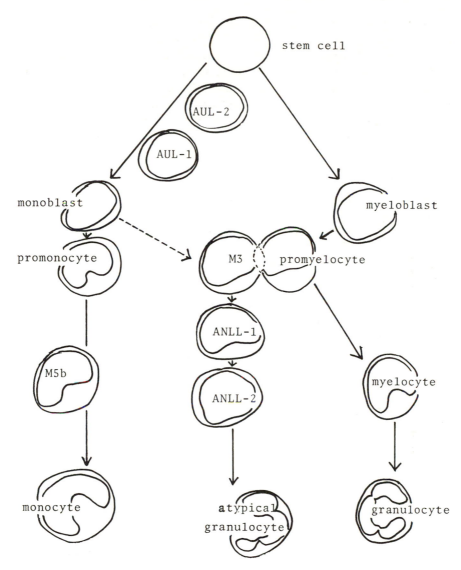

Fig. 18.9. Hypothetical representation of normal and leukemic differentiative pathways. The leukemic cases analyzed are located in the scheme according to their putative maturative stage.

that the former are phenotypically more distant from a normal cell than acute leukemias; in addition they display marked chromosomal aberrations as well as uncoordinated expression of markers. It therefore becomes compulsory to study the fresh leukemias in order to understand the first steps of leukemogenesis.

Acknowledgments. This work was supported by a grant from the Ministero P.I. No. 1202A and by the Progetto Finalizzato Oncologia, C.N.R., 1984. We would like to thank Prof. Giuseppe Masera, Dr. Paolo Paolucci, and Dr. Marino DeTerlizzi for sending some of the leukemias, and the Italian Railway Personnel and the Railway Police for their cooperation in the transport of the samples. G. Cattoretti is supported by a grant from the "Comitato M.L. Verga per lo studio e la cura della leucemia del bambino," Milano.

References

1. Robinson, D., J. Tavares DeCastro, N. Polli, M. O'Brien, and D. Catovsky. 1984. Simultaneous demonstration of membrane antigen and cytochemistry at ultrastructural level: a study with the immunogold method, acid phosphatase and myeloperoxidase. *Brit. J. Haematol.* **56**:617.
2. Standardization in human cytogenetics. Paris conference (1971), supplement. 1975. *Birth Defects Original Art. Ser.* **9**:9.
3. Pryzwansky, K.B., P.G. Rausch, J.K. Spitznagel, and J.C. Herion. 1979. Immunocytochemical distinction between primary and secondary granule formation in developing human neutrophils: correlation with Romanovsky stains. *Blood* **53**:179.
4. Newman, R.A., and M.F. Greaves. 1982. Characterization of HLA-DR antigen on leukemic cells. *Clin. Exp. Immunol.* **50**:41.
5. Peterlin, B.M., T.A. Gonwa, and J.D. Stobo. 1984. Expression of HLA-DR by a human monocyte cell line is under transcriptional control. *J. Mol. Cell. Immunol.* **1**:191.
6. Dayton, E.T., B. Perussia, and G. Trinchieri. 1983. Correlation between differentiation, expression of monocyte-specific antigens and cytotoxic functions in human promyelocytic cell lines treated with leukocyte-conditioned medium. *J. Immunol.* **130**:1120.
7. Perussia, B., E.T. Dayton, V. Fanning, P. Thiagarjan, J. Hoxie, and G. Trinchieri. 1983. Immune interferon and leukocyte-conditioned medium induce normal and leukemic myeloid cells to differentiate along the monocytic pathway. *J. Exp. Med.* **158**:2085.
8. Scott, C.S., A. Cahill, A.G. Bynoe, M.J. Ainley, D. Hough, and B.E. Roberts. 1983. Esterase cytochemistry in primary myelodisplastic syndromes and megaloblastic anemias: demonstration of abnormal staining patterns associated with dysmyelopoiesis. *Brit. J. Haematol.* **55**:411.
9. Scott, C.S., D.C. Linch, A.G. Bynoe, C. Allen, N. Hogg, M.J. Ainley, D. Hough, and D.E. Roberts. 1984. Electrophoretic and cytochemical characterization of alpha-naphthyl acetate esterases in acute myeloid leukemia: relationship with membrane receptor and monocyte-specific antigen expression. *Blood* **63**:579.
10. Yourno, J., J. Walsh, G. Kornatowsky, D. O'Connor, and S. Anand Kumar. 1984. Nonspecific esterases of leukemia cell lines: evidence for activation of myeloid-associated zymogens in HL-60 by phorbol esters. *Blood* **63**:238.
11. Yam, L.T., C.Y. Li, and W.H. Crosby. 1971. Cytochemical identification of monocytes and granulocytes. *Am. J. Clin. Pathol.* **55**:283.

12. Sanchez-Madrid, F., J.A. Nagy, E. Robbins, P. Simon, and T.A. Springer. 1983. A human leukocyte differentiation antigen family with distinct α and a common β subunit. *J. Exp. Med.* **158:**1785.
13. Gooi, H.C., D.J. Thorpe, E.F. Hounsell, H. Rumpold, D. Kraft, O. Förster, and T. Feizi. 1983. Marker of peripheral blood granulocytes and monocytes of man recognized by two monoclonal antibodies VEP8 and VEP9 involves the trisaccharide 3-fucosyl-*N*-acetyllactosamine. *Eur. J. Immunol.* **13:**306.
14. Hakomori, S., and R. Kannagi. 1983. Glycosphingolipids as tumor-associated and differentiation markers. *J. Natl. Cancer Inst.* **71:**231.
15. Tetteroo, P.A.T., M.B. van't Veer, J.F. Tromp, and A.E.G. von dem Borne. 1984. Detection of the granulocyte-specific antigen 3-fucosyl-*N*-acetyllactos-amine on leukemic cells after neuraminidase treatment. *Int. J. Cancer* **33:**355.
16. Fourth International Workshop on chromosomes in leukemia, 1982. 1984. *Cancer Genet. Cytogenet.* **11:**249.
17. Parkin, J.L., D.C. Arthur, C.S. Abramson, R.W. McKenna, J.H. Kersey, R. L. Heideman, and R. D. Brunning. 1982. Acute leukemia associated with the t(4;11) chromosome rearrangement: ultrastructural and immunological characteristics. *Blood* **60:**1321.
18. Nagasaka, M., S. Maeda, H. Maeda, H. Chen, K. Kita, O. Mabuchi, H. Misu, T. Matsuo, and T. Sugiyama. 1983. Four cases of t(4;11) acute leukemia and its myelomonocytic nature in infants. *Blood* **61:**1174.
19. Raskind, W.H., N. Tirumali, R. Jacobson, J. Singer, and P.J. Fialkow. 1984. Evidence for a multistep pathogenesis of a myelodisplastic syndrome. *Blood* **63:**1318.

CHAPTER 19

Immunohistochemical Reactivity of Anti-Myeloid/Stem Cell Workshop Monoclonal Antibodies in Thymus, Lymph Node, Lung, Liver, and Normal Skin

E. Berti, M.G. Paindelli, C. Parravicini, Giorgio Cattoretti, Domenico Delia, and F. de Braud

Introduction

Monoclonal antibodies (mAbs) against human monocytoid cells are "relatively specific" (1,3) and spurious cross-reactivity with other cell types can be detected "*in vitro*" in experimental models (3). Thus, instead of a single mAb it is customary to identify a specific cell type by the use of a panel of mAbs. Among the system of mononuclear phagocytes (MPS) (2,4), however, some cell types show characteristic histological distribution in the various organs and the specificity of a mAb for each single cell type is relatively simple to assess by immunohistochemical methods. In order to test the reactivity of the anti-myeloid stem cell mAbs of the Workshop, various tissues, including most of the cell types belonging in the mononuclear phagocytic system, were studied.* In addition GR positivity was easily detected in the tissues studied.

---

* Abbreviations used in this paper: HC: Hassal corpuscles; TMC: thymic marginal cells; GC: germinal center; DDC: dendritic dermic cells; IDL: interdigitating cells of lymph node; IDT: interdigitating cells of thymus; GR: granulocytes; SH: sinus histiocytes; AM: alveolar macrophages; LC: Langerhans cells; MZ: mantle zone; DRC: dendritic reticulum cells of follicles; GCM: germinal center macrophage—tingible bodies macrophage; KC: Kupffer cells; MPS: mononuclear phagocyte system; MC: monocytoid cells (randomly distributed in vascular and extravascular compartments); Ep: epidermis; TEC: thymic epithelial cells; Nv: nerves; HE: Hepatocytes; Vs: vessels; Mu: muscle; LYGC: scattered lymphoid cells of germinal center; LY: scattered lymphocytes; St: Stroma: reactivity with both collagen and reticulum fibers; EG: Eccrine glands.

Materials and Methods

Three blocks of normal skin, lung, liver, lymph nodes, and thymus removed from different patients have been employed in this study. Specimens were snap-frozen immediately after the surgical removal. 4 μ thick cryostat sections were mounted on the same slide and simultaneously processed. After a brief fixation (5 minutes) with 3% paraformaldehyde in 0.1 M cacodylate buffer, pH 7.2, and washing in phosphate-buffered saline (PBS), pH 7.2, the sections were incubated with 20% normal horse serum in PBS and for 30 minutes with primary mAs (diluted 1/100), biotinylated horse anti–mouse H+L immunoglobulins (Vector Lab), and preformed avidin–biotin–peroxidase complex (ABCPx) (Vector). Reaction was revealed in 3 amino-9 ethylcarbazole and the sections were counterstained with hematoxylin.

For a few selected mAbs immunoelectron microscopy procedures have been performed; for technical details, see Ref. 5.

Results

Group 1: mAbs with a Relative Specificity for Polymorphonuclear Leukocytes Also Reacting with HC and MTC

As shown in Table 19.1, several monoclonal antibodies may be grouped together because of their reactivity with HC and TMC (see Fig. 19.1). In addition mAbs 5, 19, 25, 27, 33, 52, 61, and 77 reacted with sinus histiocytes (SH) while mAbs 28, 70, 81, and 112 detected SH and eccrine sweat glands of the skin. A cytoplasmic pattern of positivity is shown by mAb μ 4.

Group 2: mAbs Reacting with Polymorphonuclear Cells and SH (Table 19.2)

mAbs 12, 15, 31, 41, 42, and 82 reacted with granulocytes and SH only. mAbs 34, 37, 63, 67, 68, 69, 81, and 90 reacted with GC of the lymph node, whereas mAbs 26, 59, 88, and 96 reacted with GR, SH, and giant mononu-

Table 19.1. Reactivities of Group 1 mAbs.[a]

mAbs	Common reactivities	Additional reactivities
4, 13, 80, 91, 110, 111, 116, 117	GR+, HC+, TMC+	—
5, 19, 25, 27, 33, 52, 61, 77	GR+, HC+, TMC+	SH+
28, 70, 81, 112	GR+, HC+, TMC+	SH+, EG+

[a] GR: granulocytes; HC: Hassal corpuscles; TMC: thymic marginal cells; SH: sinus histiocytes; EG: eccrine glands.

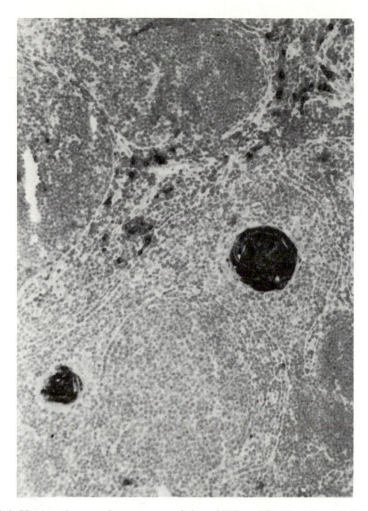

Fig. 19.1. Human thymus. Immunoreactivity of HC and TMC with mAb 5. Hematoxylin counterstain. 160×.

Table 19.2. Reactivities of Group 2 mAbs.[a]

mAbs	Common reactivities	Additional reactivities
12, 15, 31, 41, 42, 82	GR+, SH+	—
34, 37, 63, 68, 69, 90	GR+, SH+	GC+
8, 67	GR+, SH+	GC+, MC+, DDC+
62	GR+, SH+	MZ+
65, 66	GR+, SH+	DDC+, IDL+, IDT+
26, 59, 88	GR+, SH+	MC+
22, 57	GR+, SH+	AM+, KC+, MC+
94	GR+, SH+	MC+, Ep+, EF+

[a] GR: Granulocytes; SH: sinus histiocytes; GC: germinal centers; MZ: mantle zone; DDC: dendritic dermal cells; MC: monocytoid cells; IDL: interdigitating cells of lymph nodes; IDT: interdigitating cells of thymus; AM: alveolar macrophages; KC: Kupffer cells; Ep: epidermis; EF: elastic fibers.

clear cells. mAbs 65 and 66 were positive on GR, SH, DDC, and ID of the thymus. A strong reaction by mAbs 57 and 22 was observed with GR, SH, AM, Kupffer cells, and MC (Figs. 19.2 and 19.3).

Group 3: mAbs Exhibiting a Pan Leukocyte Reactivity

To this group belong mAbs 2, 3, 6, 7, 11, 36, 39, 55, 56, 64, 72, 73, 74, 75, 76, 78, 89, and 106. mAbs 75 and 89 strongly label AM. As shown in Table 19.3 mAbs 2, 3, 6, 7, 11, 64, 74, and 106 displayed a variety of cross-reactivities with non-hematopoietic cells such as epithelial structures, muscle, vessels etc. (Table 19.3).

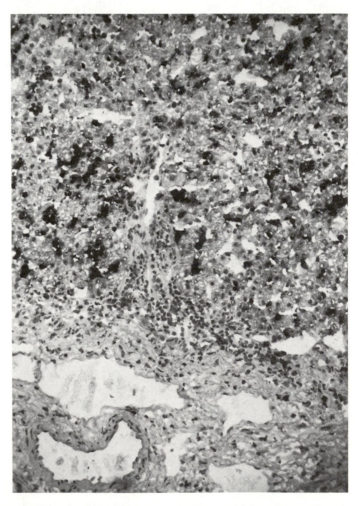

Fig. 19.2. Human liver. Kupffer cells exhibit positivity with mAb 57. Note un-stained lymphocytes. Hematoxylin counterstain. 160×.

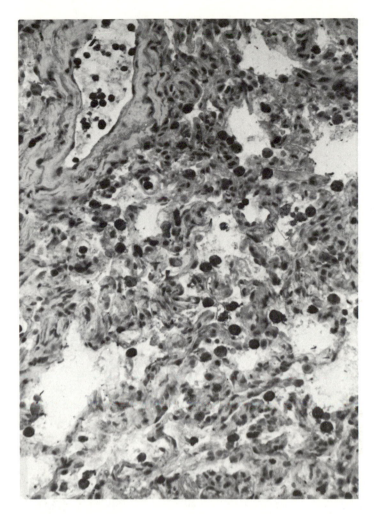

Fig. 19.3. Human lung. Both AM and GR are selectively stained by mAb 22. Hematoxylin counterstain. 250×.

Group 4: mAbs with an Ia-like Reactivity

The Ia-like pattern (positivity of the MZ, GC, IDL, IDT, and vessels) was exhibited by the following mAbs: 38, 40, 60; the latter also showed strongly stained AM.

Group 5: mAbs Reacting with B Lymphocytes

mAbs 51, 83, 84, and 85 strongly labelled B lymphocytes located in the MZ of follicles. mAb 50 stains MZ, GC, and MC.

Table 19.3. Reactivities of Group 3 mAbs in addition to their common pan leukocyte reactivity.[a]

mAbs	Additional reactivities
36, 39, 55, 56, 72, 73, 75, 76, 78, 89	—
2	Ep⁺, EG⁺, Nv⁺, Mu⁺, TEC⁺, Ec⁺, Vs⁺
3	EG⁺, Mu⁺, Ec⁺, Vs⁺, St⁺
6	Ep⁺, EG⁺, Nv⁺, Mu⁺, HE, Vs⁺, St⁺
7	Ep⁺, EG⁺, Mu⁺, Ec⁺, Vs⁺, St⁺
11	Ep⁺, Vs⁺
64	HE, Vs⁺, St⁺
74	St⁺
106	Ep⁺, EG⁺, Vs⁺

[a] Ep: Epidermis; EG: eccrine glands; Nv: nerves; Mu: muscle; TEC: thymic epithelial cells; HE: hepatocytes; Vs: vessels; St: stroma.

Group 6: mAbs Reacting with Platelets

This group comprises mAbs 9, 10, 35, 44, and 101, based on their strong reactivity with megakaryocytes of fetal liver. mAb 9 also reacted with monocytes.

Group 7: mAbs Primary Reacting with Vessels

To this group belong mAbs 14, 18, 23, 32, 79, 87, and 114. Of these, 14 and 114 reacted with vessels only. In addition mAb 18 and mAbs 23 and 87 labelled SH and SH plus GC, respectively. mAbs 32 and 79 stained SH and granulocytes.

Miscellaneous Patterns of Reactivity

Interesting reactivities were observed with the following mAbs. mAb 71 is specific for DRC (Fig. 19.4). mAb 46 strongly reacted with IDL, IDT, DDC, GCM, and AM. mAb 24 displayed positivity for SH, AM, GCM, and GC. mAb 92 gave a strong cytoplasmic staining of all monocytic cells and some lymphocytes in the tested tissues. mAb 98 showed a unique type of reactivity with epithelial cells of the thymus (Fig. 19.5) and basal cells of the epidermis. mAbs 113 and 30 weakly labelled granulocytes. Immunoreactivity of these heterogeneous mAbs is shown in Table 19.4. Finally, mAbs 1, 16, 17, 20, 29, 37, 47, 48, 49, 96, 97 were consistently negative in the tissues tested.

Conclusive Remarks

The heterogeneous immunoreactivity of the MPS cells together with the clustering results obtained by this screening give some suggestions as to

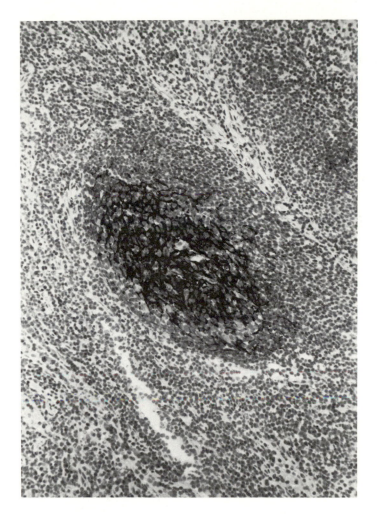

Fig. 19.4. Human lymph node. Selective reactivity of DRC with mAb 71. Hematoxylin counterstain. 160×.

the possible functional and/or differentiative significance associated with these cells.

It is interesting to note that group 1 mAbs reacted with GR, HC, TMC, and often with SH, but lacked reactivity with other cells of the MPS, such as macrophages (AM, KC, GCM) and dendritic cells (LC, ID, DDC, DRC). On the contrary mAbs reacted with cells of the MPS other than SH were frequently unreactive with both HC and TMC.

Antigen-presenting cells (LC, IDL, DDC) were stained by different mAbs in the different organs. IDL reacted with mAbs 38, 40, 46, 60, 64, 65, 66, and 95. Skin LC reacted with mAbs 38 and 40, whereas DDC were

Table 19.4. Reactivities of heterogeneous mAbs.

Type of reactivity	mAbs																							
	24	30	43	45	46	53	54	58	71	86	92	93	95	98	99	100	102	103	104	105	107	113	115	118
GC⁺	+	−	−	−	−	−	−	−	−	−	−	+	−	−	−	−	−	−	−	−	−	−	−	−
DRC⁺	−	+	−	−	−	−	−	−	+	−	−	−	+	−	−	−	−	−	−	−	+	−	+	−
GR⁺	−	−	−	+	−	−	−	−	−	−	−	+	−	−	−	+	−	−	+	−	−	+	−	−
MC⁺	−	−	+	−	+	+	+	+	−	−	+	+	−	+	+	+	+	+	+	−	−	−	−	−
AM⁺	+	−	−	−	+	+	−	−	−	−	+	−	−	−	−	−	−	−	−	−	−	−	−	+
KC⁺	−	−	−	−	−	−	−	−	−	−	+	−	−	−	−	−	+	−	−	−	−	−	−	+
SH⁺	+	−	−	−	+	+	−	−	−	−	+	+	−	−	−	−	+	−	−	−	−	−	−	+
DDC⁺	+	−	+	−	+	−	−	−	−	−	+	+	−	−	−	−	−	−	−	−	−	−	−	+
ID⁺	−	−	−	−	+	−	−	−	−	−	+	−	+	−	−	−	−	−	−	−	−	−	−	−
GCM⁺	+	−	−	−	+	−	−	−	−	−	+	−	−	−	−	−	−	−	−	−	−	−	−	−
LyGC⁺	−	−	−	−	−	+	−	−	−	−	−	−	−	−	−	−	−	−	−	+	−	−	−	−
Ly⁺	+	−	−	−	−	+	−	−	−	−	+	−	−	−	−	+	−	−	+	−	−	−	−	−
St⁺	−	−	−	−	−	−	−	−	−	−	−	−	+	−	−	−	−	−	−	−	−	−	+	−
Nv⁺	−	−	+	−	−	−	+	−	−	+	−	−	−	−	−	−	−	−	−	+	−	−	−	−
EG⁺	−	−	−	−	−	−	+	−	−	−	−	−	−	−	−	+	−	−	−	−	−	−	−	−
Vs⁺	−	−	+	−	−	−	−	−	−	−	−	−	−	−	−	−	−	−	−	−	−	−	−	+
Ep⁺	−	−	−	−	−	−	−	−	−	−	−	−	−	+	−	−	−	−	−	−	−	−	−	−
TEC	−	−	−	−	−	−	−	−	−	−	−	−	−	+	−	−	−	−	−	−	−	−	−	−

Fig. 19.5. Human thymus. Epithelial cells are stained by mAb 98. Hematoxylin counterstain. 160×.

stained with mAbs 37, 41, 43, 67, 68, 69, 92, 93, and 118. All these antibodies displayed miscellaneous cross-reactivities, with the exception of 38, 40, and 60 which exhibit an Ia-like pattern of reactivity in the tested tissues. Typical macrophages (AM, KC, GCM) reacted with many of the Workshop mAbs: AM are stained by 22, 24, 46, 57, 60, 64, 92, 118; KC by 22, 57, 64, 92, 118; GCM of the GC are stained by 24, 46, 64, and 92. mAbs 22, 57, 72, and 118 reacted with both KC and AM. SH are stained by several Workshop mAbs in various groups (see Results). Of the Workshop mAbs, no. 71 is the one which showed a unique reactivity restricted

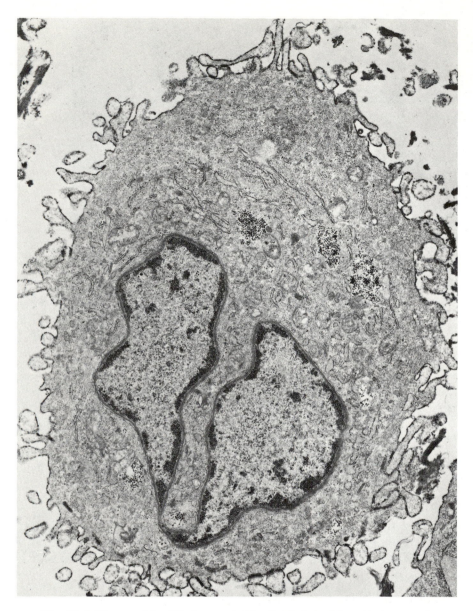

Fig. 19.6. Human skin. Histiocytosis X cell labeled with mAb 46. 6450×.

to DRC only. These cells were also stained by mAbs 2, 3, 7, 95, and 115 which cross-reacted with other cell types.

The broadest spectrum of immunoreactivity with cells of the MPS is shown by mAb 118 which stained SH, KC, AM, and DDC and by mAb 46 which stained AM, SH, ID, GCM, and DDC. Both these mAbs did not display any cross-reaction with other cell types.

In our study paraphormaldehyde fixation was used. Some caution is required in the interpretation of our results due to the possible effects of the fixation procedure on the preservation of antigenic determinants. In addition, in our opinion further studies utilizing immunoelectron micros-copy procedures (Fig. 19.6) would be required to identify *in situ* each individual immunoreactive cell type.

References

1. Hancock, W.W., H. Zola, and R.C. Atkins. 1983. Antigenic heterogeneity of human mononuclear phagocytes: immunohistological analysis using mono-clonal antibodies. *Blood* **62:**1271.
2. Hoefsmit, E.C.M. 1975. Mononuclear phagocytes, reticulum cells and den-dritic cells in lymphoid tissues. In: *Mononuclear phagocytes in immunity infec-tion and pathology,* R. Van Furth, ed. Blackwell Scientific, Oxford, p. 129.
3. Todd, R.F., and S.F. Schlossman. Utilisation of monoclonal antibodies in the characterization of monocyte-macrophage differentiation antigens. In: *Immu-nology of the reticuloendothelial system: a comprehensive treatise,* Vol. VI, J.A. Bellante and H.B. Herscowitz, eds. Plenum, New York (in press).
4. Van Furth, R. 1982. Current view on the mononuclear phagocyte system. *Immunobiol.* **161:**178.
5. Berti, E., M. Monti, S. Cavicchini, and R. Caputo. 1983. The avidin-biotin peroxidase complex (ABPxC) in skin immunoelectron microscopy. *Arch. Dermatol. Res.* **275:**134.

CHAPTER 20

Immunohistological Studies of Anti-Myeloid Monoclonal Antibodies

Wayne W. Hancock, Norbert Kraft, and Robert C. Atkins

Introduction

A considerable number of anti-monocyte monoclonal antibodies (mAb) have now been produced, as discussed in a recent review (1). These antibodies have generally resulted from immunization with peripheral blood mononuclear cells or purified blood monocytes, and have been characterized by analysis of reactivity with blood cells or various cell lines. Few studies have examined the reactivity of such mAbs with cells in tissues, despite knowledge that the main functions of mononuclear phagocytes (MP) involve extravascular tissue macrophages rather than the more functionally immature circulating monocyte precursor forms (2). Indeed, we have found quite variable reactivity of anti-monocyte mAbs with cells in tissue sections (3) and our previous immunohistological evaluation of a panel of established anti-monocyte mAbs showed evidence of variations in antigen expression by macrophages depending upon anatomical site and stage of differentiation (4).

This study further explored the question of monocyte antigen expression *in situ* by analyzing the reactivity of Workshop myeloid panel mAbs with tissues containing resident macrophages, inflammatory macrophages, and related cell types. Though the utility of some monocyte mAbs was confirmed, and a series of granulocyte-specific mAbs were delineated, a considerable number of Workshop mAbs bound non-MP cell types, thereby potentially limiting their usefulness for *in situ* and other studies.

Materials and Methods

Tissues

Fresh surgical specimens of normal tonsil, spleen, liver, and alveolar lavage cells were collected and included as tissues containing resident macrophages. Normal kidney and skin were included as tissues containing wandering macrophages and also resident dendritic cell populations (5). Sections from rejecting renal allografts and sarcoid granulomata were studied, since these contained large numbers of inflammatory macrophages (6) and mature epithelioid and multinucleated macrophages (4), respectively. Freshly obtained tissues and smears of alveolar lavage cells were fixed in periodate–lysine–paraformaldehyde (PLP), washed overnight in phosphate buffer containing 7% sucrose, and stored at −80°C. PLP fixation was used since this step significantly improved cell morphology and, in previous studies (3,7), was shown to have minimal effect on a large series of antigens detected by mAbs.

Immunohistology

PLP-fixed cryostat sections and cell smears were labeled using a two-layer indirect immunoperoxidase technique (7) involving Workshop mAbs (1 : 400), followed by peroxidase-conjugated rabbit anti–mouse immunoglobulins (Dako, Denmark) (1 : 40), incubation with the substrate diaminobenzidine, and counterstaining with hematoxylin. The results of blind evaluation of Workshop mAbs were tabulated for each tissue using relevant criteria such as B cell areas (lymphoid follicles), T cell areas (paracortical and periarteriolar areas), and phagocyte-rich areas such as splenic red pulp.

Results

Ninety-five (83%) of the 115 anti-myeloid mAbs (no samples of M60, M108, or M109 were received) stained various cell types in tissue sections and cell smears. These results are summarized in Table 20.1. mAbs ($N = 20$) which were unreactive with the panel of tissues used were: M17, M20, M31, M44, M45, M48, M49, M58, M62, M68, M69, M78, M82, M97, M98, M101, M102, M103, M113, and M114. Control slides, involving replacement of mAb with either phosphate buffer or mAb unreactive with human tissues, were unstained.

Thirteen mAbs were specific for MP cells (Table 20.2). Within this group, further subclassification was possible based on differential staining patterns of resident or inflammatory macrophages. Only one of these 13

Table 20.1. Summary of tissue reactivity of workshop anti-myeloid mAbs.[a]

Specificity	Workshop designation
MP cells[b]	M24, M29, M34, M37, M43, M46, M67, M93, M96, M99, M100, M105, M107 ($N = 13$)
MP, MNC	M50, M51 ($N = 2$)
MP, endo	M18, M95, M106, M115, M118 ($N = 5$)
MP, epi	M21 ($N = 1$)
MP, epi, endo	M10, M47 ($N = 2$)
MP, P	M8, M22, M26, M41, M42, M53, M55, M56, M57, M59, M63, M88, M104 ($N = 13$)
MP, P, MNC	M72, M73, M75, M76, M89 ($N = 5$)
MP, P, epi	M54, M90, M92 ($N = 3$)
MP, P, epi, endo	M3, M12, M65, M66, M70, M81, M87, M112 ($N = 8$)
P	M15, M28, M94 ($N = 3$)
P, glandular epi	M13, M19, M25, M27, M30, M33, M52, M61, M77, M80, M91, M110, M111, M116, M117 ($N = 15$)
P, glandular and squamous epi	M4, M5 ($N = 2$)
endo	M1, M14, M16, M32, M79 ($N = 5$)
endo, platelets	M35 ($N = 1$)
epi, endo	M9, M23 ($N = 2$)
MNC	M2, M83, M84, M85 ($N = 4$)
MNC, epi	M86 ($N = 1$)
DRC	M71 ($N = 1$)
HLA-ABC-like	M6, M7, M11, M64, M74 ($N = 5$)
HLA-D/DR-like	M38, M40 ($N = 2$)
Leukocyte-common-like	M36, M39 ($N = 2$)

[a] Abbreviations: DRC = dendritic reticulum cells; endo = endothelium; epi = epithelium; MNC = mononuclear cells (including at least some lymphocytes); MP = mononuclear phagocytes; P = polymorphonuclear leukocytes.
[b] For classification here, MP cells include also interstitial dendritic cells of kidney, as discussed in text.

Table 20.2. Subclassification of 13 workshop mAb reactive only with MP cells.[a]

Workshop designation	MP cell reactivity					
	Lymphoid	Liver	Lung	Inflamm.	Granuloma	DC
M24	+	−	+	+	+	−
M29	−	−	−	+	+	−
M34	+	−	+	+	+	IDC
M37	+	−	+	+	+	IDC
M43	+	−	−	−	−	−
M46	+	+	+	+	+	−
M67	−	−	−	+	−	−
M93	−	−	−	+	−	−
M96	+	−	−	+	+	−
M99	−	−	−	+	+	−
M100	+	−	+OCC	+	+	−
M105	+	−	−	+OCC	−	−
M107	−	−	−	+OCC	−	−

[a] Abbreviations: inflamm. = inflammatory MP; DC = dendritic cells; IDC = interstitial dendritic cells of the kidney; OCC = occasional cell positive.

mAbs (M46) stained all categories of MP cells. In particular, all except M46 failed to bind to Kupffer cells of the liver, and only five mAbs bound to alveolar macrophages. Excluding Kupffer cells, a further three anti-monocyte mAbs (M24, M34, M37) labeled all MP cells, though M34 and M37 also stained the closely related interstitial dendritic cells of the kidney but not epidermal Langerhans cells. Interestingly, five mAbs (M29, M67, M93, M99, and M107) showed specificity for macrophages at sites of acute (transplant rejection) or chronic (granulomata) inflammation. Related to this point of MP reactivity, M8 and M22 mAbs bound to all categories of MP, plus polymorphs (and interstitial dendritic cells—M8) but not other cell types. All other shared MP cell/polymorph markers (Table 20.1), like most MP-specific mAbs, failed to stain Kupffer cells.

Three mAbs were granulocyte-specific (Table 20.1), and a further 17 mAbs could be subclassified on the basis of differential additional reactivity with either squamous or glandular epithelium. Other unusual mAbs included a marker specific for dendritic reticulum cells within lymphoid follicles (M71), and a mAb reactive only with platelets and some endothelial cells (M35). mAbs with the tissue distribution characteristic (3) of leukocyte-common antigen (M36, M39), class I (M6, M7, M11, M64, M74), and class II antigens (M38, M40) were also noted.

Discussion

Immunohistological studies with a panel of mAbs directed against a particular cell type can, as shown here, provide a sensitive method for detecting subtle differences between otherwise similar antibodies. In this study of 115 Workshop mAbs directed against myeloid antigens, the majority of mAbs could be subclassified into various small groups within a given Workshop cluster, based on differential patterns of reactivity with MP cells, granulocytes, and parenchymal and other cell types.

Macrophages are most commonly studied *in vitro,* in part because of the availability of suitable criteria for their identification, such as adherence, phagocytic capacity, production and release of various enzymes and other proteins, expression of Fc and complement receptors, and tumoricidal capacity. Attempts to correlate *in vitro* properties of MP cells with properties of MP *in situ* have been limited by the lack of widely accepted criteria for their identification in tissue sections, and by the marked heterogeneity of this cell type (8). The advent of anti-monocyte mAbs, and eventually anti-macrophage mAbs, may provide potent new methods for further studies of MP cell differentiation, activation, and pathophysiology, particularly through hitherto technically difficult studies *in situ.* Although the functional significance, if any, of each of 115 mAbs tested in this study is unknown, several observations pertinent to this research were made.

Firstly, M46 antibody was found to be an operatively useful pan MP cell marker, unique in its lack of reactivity with granulocytes or other cell types despite extensive tissue distribution studies. Secondly, considerable antigenic differences were found between MP cells at different anatomical sites, confirming previous studies with a considerably smaller group of anti-monocyte mAbs (4). In particular, Kupffer cells were found to be devoid of the majority of antigens shared between other MP cells, a finding which may prove significant as the nature of these antigens and their functions are defined. Nevertheless this result again illustrates the need for careful evaluation of studies wherein anti-monocyte mAbs are used to define or detect MP cells in sections or cell suspensions from various tissues. Finally, antigenic differences were noted between MP cells resident in lymphoid and other tissues, and MP involved in immunological or inflammatory processes. The five mAbs involved, M29, M67, M93, M99, and M107, warrant further studies to assess the basis of these antigenic differences and their apparently unique specificities.

References

1. Todd, R.F., and S.F. Schlossman. 1984. Utilization of monoclonal antibodies in the characterization of monocyte-macrophage differentiation antigens. In: *The reticuloendothelial system,* J.A. Bellanti and H.B. Herscowitz, eds. Plenum Publishing Corp, New York, pp. 87–112.
2. Cline, M.J., R.I. Lehrer, M.C. Territo, and D.W. Golde. 1978. Monocytes and macrophages: functions and diseases. *Ann. Intern. Med.* **88**:78.
3. Hancock, W.W., N. Kraft, and R.C. Atkins. 1984. Immunohistochemical studies with monoclonal antibodies to differentiation, cell lineage and site-specific surface antigens of human mononuclear phagocytes. In: *Leucocyte typing,* A. Bernard, L. Boumsell, J. Dausett, C. Milstein, and S.F. Schlossmann, eds. Springer-Verlag, Berlin, Heidelberg, pp. 493–499.
4. Hancock, W.W., H. Zola, and R.C. Atkins. 1983. Antigenic heterogeneity of human mononuclear phagocytes: Immunohistological analysis using monoclonal antibodies. *Blood* **62**:1271.
5. Hancock, W.W., and R.C. Atkins. 1984. Immunohistological studies of the cell surface antigens of human dendritic cells using monoclonal antibodies. *Transplant. Proc.* **16**:963.
6. Hancock, W.W., N.M. Thomson, and R.C. Atkins. 1983. Composition of interstitial cellular infiltrates in renal biopsies of rejecting human renal allografts identified by monoclonal antibodies. *Transplantation* **35**:458.
7. Hancock, W.W., G.J. Becker, and R.C. Atkins. 1982. A comparison of fixatives and immunohistochemical techniques for use with monoclonal antibodies to cell surface antigens. *Am. J. Clin. Pathol.* **78**:825.
8. Dougherty, G.J., and W.H. McBride. 1984. Macrophage heterogeneity. *J. Clin. Lab. Immunol.* **14**:1.

CHAPTER 21

Immunohistological Characterization of Myeloid and Leukemia-Associated Monoclonal Antibodies

Michael A. Horton, Denise Lewis, and Katrina McNulty

The use of monoclonal antibodies has opened up the possibility of studying cell differentiation and tissue heterogeneity in the hematopoietic system in a way which would have been deemed impossible before the advent of hybridoma technology. However, the exquisite specificity of monoclonal antibodies has emphasized the need for detailed studies of the tissue distribution of the antigenic epitopes they recognize. Recent studies have clearly demonstrated the value of immunohistological techniques (as opposed to the study of isolated populations of cells) in evaluating the detailed patterns of reaction of monoclonal antibodies (1). Moreover, these investigations have not only provided further insights to the biology of the cells and their surface antigens recognized by specific antibodies, but have also stressed the need for a thorough analysis of their tissue distribution prior to *in vivo* clinical usage of any particular monoclonal antibody.

In this paper we summarize our findings from immunohistological studies on the tissue distribution of the antigens defined by the myeloid and leukemia-associated monoclonal antibodies of the Second Workshop.

Methods

Standard immunohistological techniques were used throughout this study. Normal or reactive lymphoid tissues (tonsil, lymph node, spleen), kidney, lung, skin, and liver were obtained at operation or postmortem. 12–18 week fetal kidney was obtained from pregnancy terminations. Cryostat sections were fixed in acetone and staining was performed by routine immunofluorescence methods using a polyspecific goat anti–mouse Ig–FITC (Coulter Clone) developing agent.

In the studies of the renal expression of myeloid and leukemia-associated antigens the following control antibodies were used to aid in the identification of the different parts of the nephron (2) [glomerulus + proximal tubule, gp100 CALLA, J5; glomerular capsule + distal tubule, p24, BA2; proximal tubule, mast cell granule component, MCG3 (3); distal tubule, rabbit anti-Tam-Horsfall glycoprotein]; the different developmental stages in fetal kidney (BA2); and non-tubule interstitial and vascular endothelial cells (DrW, DA2.)

Results

Tissue Distribution of Myeloid Antigens

The myeloid monoclonal antibody reactivity in the tissues studied gave complex results (Tables 21.1 and 21.2) and fell into four basic patterns. First, negative or staining only of intravascular myeloid cells. Second, reacting with germinal center and/or interfollicular zone lymphoid cells in tonsil (37 of 116) (Table 21.2). Third, broadly reactive with tissue macrophages in lymphoid and other tissues (18 of 116) (Table 21.2). Fourth, a further 34 antibodies appeared to react with macrophages in a more restricted tissue distribution (Table 21.1). Thus, for example, six antibodies reacted only with Kupffer cells and nine with alveolar macrophages—however, as the number of tissues examined in this study was limited, this conclusion must be taken only as preliminary.

Lymphoid Expression of Leukemia-Associated Antigens

The majority of leukemia-associated antigens (16 of 21) could be demonstrated in sections of normal (reactive) tonsil, spleen, and lymph node. Eight antibodies bound to lymphoid cells within follicles and/or in the interfollicular zones (Table 21.3). A further seven antibodies bound to

Table 21.1. Second Workshop myeloid antibodies giving a restricted tissue distribution in tonsil, liver, lung, and skin.

Kupffer cells	M4, 8, 39, 111, 112, 114
Alveolar macrophages	M11, 26, 42, 54, 60, 69, 70, 73, 106
Langerhans cells	M1, 5, 13, 98
Tonsil interfollicular macrophages	M37, 47, 53, 63, 78, 99, 104
Tonsil follicular macrophages	M24, 43, 71, 93
Follicular + interfollicular macrophages	M45, 79, 100, 105

Table 21.2. Second Workshop myeloid antibodies broadly reactive with tissue macrophages and lymphoid tissues.

Broadly reactive with tissue macrophages: M7, 14, 18, 22, 23, 33, 38, 40, 46, 55, 57, 64, 72, 75, 81, 89, 117, 118

On tonsil sections the following mAbs were:

   Pan-reactive: M1, 2, 6, 7, 11, 23, 35, 36, 39, 50, 55, 56, 60, 66, 72, 73, 74, 75, 76, 78, 89, 106, 114
   Predominantly reactive with germinal centers: M8, 23, 37, 63, 71, 87, 95, 99
   Predominantly reactive with interfollicular zone: M53, 64, 65, 79, 100, 106

Remaining antibodies were either negative or only reacted with intravascular blood cells.

interfollicular dendritic cells; of these five also reacted strongly with tonsillar epithelium. The remainder were negative or pan-reactive.

Renal Expression of Myeloid and Leukemia-Associated Antigens

The distribution of the myeloid and leukemia-associated antigens is summarized in Tables 21.4–21.6.

The majority of myeloid antibodies (85 of 116) failed to bind to either adult of fetal kidney or reacted only with interstitial cells and/or vascular endothelium (Table 21.5). The most common pattern of reactivity with renal parenchyma (seen with 11 antibodies, Table 21.4) was a BA2-like reaction with the mature and immature zones in fetal kidney. Ten antibodies gave restricted binding to proximal and/or distal tubules (but not reacting with glomeruli) (Table 21.5). A further eight antigens were differentially distributed between fetal and adult kidney, demonstrating that their extrahematopoietic expression is regulated during ontogeny.

Of the leukemia-associated monoclonals, seven J5-like and two BA2-like reaction patterns were identified (Table 21.6); the remainder were pan-reactive, bound to interstitial cells or were negative.

Table 21.3. Lymphoid distribution of Second Workshop leukemia antigen panel.

Tonsil reactivity pattern[a]	Antibody number[b]
Follicular lymphocytes	L17
Follicular + interfollicular lymphocytes	L8, 9, 11, 12
Interfollicular lymphocytes	L3, 13
Interfollicular dendritic + tonsil epithelium	L4, 16, 18, 21, 22
Interfollicular dendritic (epithelium neg.)	L19, 20

[a] Similar findings with spleen and reactive lymph node.
[b] L6, pan-reactive; L5, not available; L1, 2, 7, 10, 14, 15, negative.

Table 21.4. Distribution of Second Workshop myeloid antibodies on kidney: ontogenically influenced reactivity patterns.

	Reactivity pattern	
Myeloid antibody number	Fetal kidney	Adult kidney
(A) M23, 87	Mature zone proximal tubules	Pan-reactive
(B) M5, 13, 19, 25, 33, 61, 77, 91, 110, 111, 117	BA2-like distribution	Weak tubular or negative
(C) M98	Negative	Apical zone proximal tubular epithelium
(D) M116	Proximal tubule	Interstitial mononuclear cells
(E) M106	Pan tubular	Pan tubular + glomerular capsule
(F) M79	Pan tubular (mature zone only)	Interstitial cells + endothelium
(G) M105	Glomerular capsule basement membrane	Negative
(H) M14	Negative	Tubular + glomerular basement membrane

Discussion

This study has demonstrated the usefulness of immunohistological techniques in the assessment of the specificity of monoclonal antibodies to hematopoietic cellular antigens; it, thus, allows the general conclusion that specificity defined at the level of isolated, purified populations of hematopoietic cells is, in the main, insufficient when one starts to examine, for example, tissue macrophages *in situ*. Despite these complexities, it is clear that some of the monoclonal antibodies in the Second Workshop

Table 21.5. Distribution of Second Workshop myeloid antibodies on kidney.

	Myeloid antibody number	
Reactivity pattern	Fetal kidney	Adult kidney
Proximal tubules only (with some glomerular capsule staining)	M12, 46, 47, 52, 65, 66	M12, 46, 47, 52, 65, 66
Distal tubules only	M7, 90	M7, 90
Proximal + distal tubules	M2, 21	M2, 21
Interstitial cells only	M9, 10, 28, 35, 38, 39, 40, 56, 72, 75, 89, 100, 112, 118	M1, 4, 6, 9, 10, 26, 27, 28, 30, 41, 42, 55, 57, 69, 70, 72, 75, 76, 82, 88, 89, 104, 116, 118
Interstitial cells + endothelium	M18	M11, 17, 18, 38, 40, 59, 60, 73, 74, 79, 81, 112
Pan-reactive	M3, 64, 95, 115	M3, 64, 95, 115
Negative	The remaining 69 antibodies	The remaining 49 antibodies

Table 21.6. Distribution of the Second Workshop leukemia antigen panel on fetal and adult kidney.

Reactivity pattern	Antibody code number[a]
gp100 CALLA-like (J5 equivalent)	L2, 10, 11, 14, 15, 20, 21
p24 leukemia-associated (BA-2 equivalent)	L13, 19
Vascular + interstitial cells	L4, 8 (fetal and adult); L1, 16 (adult only)
Pan-reactive	L6, 18, 22
Negative	L3, 7, 9, 12, 17 (L1, 16 negative on fetal kidney)

[a] L5 not available.

go some way to identifying both antigens on subpopulations of macrophages and also pan-macrophage determinants.

With the increasing clinical use of monoclonal antibodies it is important to understand the complexity of the extrahematopoietic distribution of cell surface antigens. Thus, the extensive representation of myeloid antigens on renal parenchyma has clinical relevance from at least three standpoints. First, *in vivo* use of antibodies in the treatment of malignancy—clearly the knowledge that some hematopoietic antigens are expressed by kidney will be important in the assessment of toxicological problems which might be encountered with therapeutic use in man. Second, polymorphisms in any of these antigenic determinants either in, or between, renal and hematopoietic tissues might be relevant in the genesis of leukocyte alloantibodies, graft rejection, or immunologically mediated renal dysfunction. Third, as the distribution of some of these antigens is restricted to particular segments of the nephron, they may conceivably be useful as markers of renal damage—measurement of these antigens in urine could thus give useful information about renal function in a variety of clinical situations.

Concluding Remarks

In this paper we summarize our experience on the tissue distribution of myeloid and leukemia-associated antigens. The results emphasize the importance of tissue section studies in the delineation of monoclonal antibody specificity—in particular, the definition of extraneous antibody reactivity will be of considerable importance with the increasing therapeutic use of monoclonal antibodies.

Acknowledgments. The authors acknowledge the support of the Wellcome Trust. M.A. Horton was a Wellcome Trust Senior Research Fellow in Clinical Science.

References

1. Bernard, A., L. Boumsell, J. Dausset, C. Milstein, and S.F. Schlossman, eds. 1984. *Leucocyte typing*. Springer-Verlag, Berlin, Heidelberg.
2. Platt, J.L., T.W. LeBien, and A.F. Michael. 1983. Stages of renal ontogenesis identified by monoclonal antibodies reactive with lymphohaemopoietic differentiation antigens. *J. Exp. Med.* **157**:155.
3. Rimmer, E.F., C. Turberville, and M.A. Horton. 1984. Human mast cells detected by monoclonal antibodies. *J. Clin. Pathol.* **37**:1249.

CHAPTER 22

Differentiation-Associated Stages of Clonogenic Cells in Acute Myeloblastic Leukemia Identified by Monoclonal Antibodies

Kert D. Sabbath and James D. Griffin

Introduction

Acute myeloblastic leukemia (AML) is characterized by defective myeloid proliferation and maturation, accumulation of immature blast cell, and progressive loss of normal bone marrow components (1). Although the circulating leukemic blast cells are uniformly morphologically immature, only a small subset are capable of *in vitro* proliferation in semi-solid media (2–4). This subpopulation of leukemic colony-forming cells (L-CFC) shares certain properties with normal myeloid progenitor cells, including a high ^{3}H-TdR labeling index (5), the ability to divide five or more times *in vitro* (6), and at least a limited capacity for self-renewal (7). It has been proposed that the L-CFC act as progenitor cells *in vivo* and are responsible for the maintenance of the leukemic blast cell population (5,7). Detailed study of these cells has been hindered by low cell numbers and difficulties in the isolation and identification of L-CFC. In particular, the relationship of the L-CFC to normal myeloid stem cells has not been defined. Recently, we described the use of anti-myeloid monoclonal antibodies to identify and enrich L-CFC in patients with AMML (8). It was shown that L-CFC may be distinguished from the majority of leukemic cells, and that L-CFC were less "mature" than their apparent progeny. In this study, we investigated the utility of surface marker analysis in the identification of L-CFC using a panel of monoclonal antibodies identifying discrete stages of myeloid differentiation. Our results show that L-CFC are heterogeneous among different patients and suggest that AML clonogenic cells may arise at multiple levels within the myeloid differentiation pathway.

Materials and Methods

The production and characterization of murine monoclonal antibodies anti-MY3, -MY9, PM-81, AML-2-23, -Mo1, and I2 have been previously described (9–14). Leukemic cells were obtained from peripheral blood or bone marrow of untreated patients and control cells from normal volunteers. The diagnosis of AML was confirmed by standard morphologic and cytochemical criteria (15,16). L-CFC were assayed in semi-solid agar as described (10). The surface antigen phenotype of L-CFC was determined by complement lysis. The percent lysis was evaluated by comparison with a nonbinding control monoclonal antibody. The phenotype of circulating leukemic blast cells was determined by indirect immunofluorescence using flow cytometry (9) or by complement lysis, both techniques giving similar results.

AML Patient Population

The study population was composed of 20 patients with demonstrated *in vitro* colony formation. Twenty percent of patients were classified morphologically as FAB type M1, 10% were M2, 55% were M4, and 10% were M5 with 1 patient not classified by FAB criteria. The surface antigen phenotypes of the L-CFC and blast cell populations are shown in Fig. 22.1. L-CFC expressed Ia and MY9 in all the patients studied while

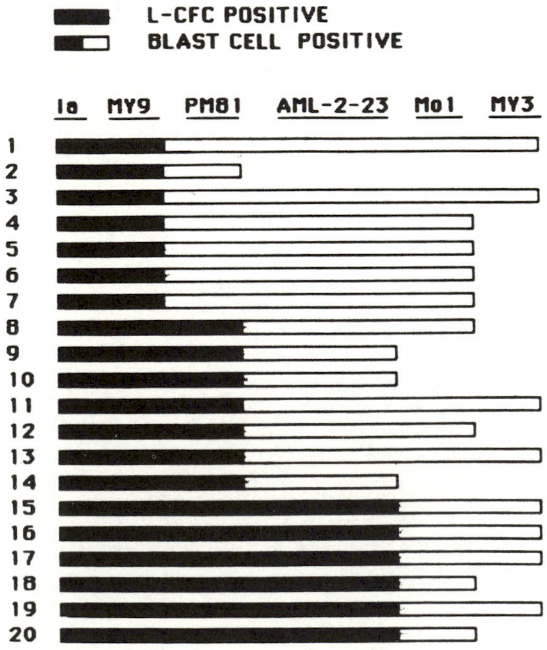

Fig. 22.1. Surface antigen phenotype of L-CFC and total AML cell population.

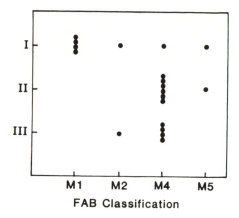

Fig. 22.2. Relationship between L-CFC antigen expression and FAB classification of leukemias.

PM-81 and AML-2-23 were expressed on 65% and 30% of L-CFC, respectively. The antigens Mo1 and MY3 were not observed on L-CFC. In contrast, the antigens Ia, MY9, PM-81, AML-2-23, and Mo1 were detected on the circulating blast cells from more than 85% of the patients tested while MY3 was expressed by 40%. In all cases, the circulating blast cells expressed additional antigens not expressed on L-CFC.

An analysis of the data shown in Fig. 22.1 suggests that three groups can be identified based on the L-CFC antigen expression. Group I L-CFC (7 patients) expressed Ia and MY9; Group II L-CFC (7 patients) expressed Ia, MY9, and PM-81; and Group III L-CFC (6 patients) expressed Ia, MY9, PM-81, and AML-2-23. In all cases the L-CFC phenotype was less mature than the total blast cell population.

The significance of these phenotype groups was tested by examining the relationship between the L-CFC and the FAB classification of each patient (Fig. 22.2). Eighty percent of the FAB M1 and M2 leukemias had Group I L-CFC, while 83% of the more monocytic leukemias (FAB M4 and M5) had Group II or III L-CFC.

The surface antigen phenotypes can be compared to the phenotypes of normal colony-forming cells (Table 22.1). Group I L-CFC are approximately at the maturation level of the normal CFU-GEMM (Ia and MY9

Table 22.1. Surface antigen phenotype of normal myeloid cells.

Cell	Ia	MY9	PM-81	AML-2-23	Mo1	MY3
CFU-GEMM	+++	++	−	−	−	−
CFU-GM (day 14)	+++	++	+	−	−	−
CFU-GM (day 7)	+++	+++	+++	++	−	−
Myeloblast	+++	+++	+++	+++	−	−
Monocyte	+++	+++	++	++	++	+++

positive), Group II L-CFC are the leukemic counterparts of the day 14 CFU-GM (Ia, MY9, and PM-81 positive), and Group III L-CFC the counterparts of the day 7 CFU-GM (Ia, MY9, PM-81, and AML-2-23 positive).

Discussion

The data presented here show that the surface antigen phenotype of L-CFC can be used to investigate L-CFC heterogeneity and to compare L-CFC to normal cellular counterparts. Although multiple L-CFC phenotypes were observed in the 20 patients studied, the L-CFC in individual patients tended to be homogeneous. The L-CFC phenotypes were representative of cells at early, discrete stages of normal myeloid cell development. If the L-CFC cell as assayed is indeed a leukemic "stem" cell, then this suggests that leukemia can arise at multiple identifiable points within the normal differentiation pathway. It was noted that the FAB M1 and M2 leukemias were associated with Group I L-CFC while more "differentiated" M4 and M5 leukemias were associated with Group II and III L-CFC. This suggests that the degree of blast cell maturation is related in part to the degree of leukemic progenitor cell differentiation.

The ability to identify specific normal and leukemic cell fractions raises the possibility of immunologically purging leukemic bone marrow for autologous transplantation (17).

Acknowledgments. This work was supported by PHS Grants CA19389 Project 3, CA36167-01, CA02321, and CA31888 awarded by the National Cancer Institute, DHHS. Dr. Sabbath is a clinical and research fellow at the Massachusetts General Hospital, Boston, MA.

References

1. Cline, M.J., D.W. Golde, R.J. Billing, J.E. Groopman, J. Zighelboim, and R.P. Gale. 1979. Acute leukemia: Biology and treatment. *Ann. Int. Med.* **91:**758.
2. Moore, M.A.S., N. Williams, and D. Metcalf. 1973. *In vitro* colony formation by normal and leukemic human hematopoietic cells: Characterization of the colony-forming cells. *Natl. Cancer Inst.* **50:**603.
3. Aye, M.T., J.E. Till, and E.A. McCulloch. 1972. Growth of leukemic cells in culture. *Blood* **40:**806.
4. Buick, R.N., J.E. Till, and E.A. McCulloch. 1977. Colony assay for proliferative blast cells circulating in myeloblastic leukemia. *Lancet* **1:**862.
5. Minden, M.D., J.E. Till, and E.A. McCulloch. 1978. Proliferative stage of blast cell progenitors in acute myeloblastic leukemia (AML). *Blood* **52:**592.
6. Buick, R.N., M.D. Minden, and E.A. McCulloch. 1979. Self-renewal in culture of proliferative blast progenitor cells in acute myeloblastic leukemia. *Blood* **54:**95.

7. Wouters, R., and B. Lowenberg. 1984. On the maturation order of AML cells: A distinction on the basis of self-renewal properties and immunologic phenotypes. *Blood* **63**:864.

8. Griffin, J.D., P. Larcom, and S.F. Schlossman. 1983. Use of surface markers to identify a subset of acute myeloblastic leukemia cells with progenitor cell properties. *Blood* **62**:1300.

9. Griffin, J.D., J. Ritz, L.M. Nadler, and S.F. Schlossman. 1981. Expression of myeloid differentiation antigens on normal and malignant myeloid cells. *J. Clin. Invest.* **68**:932.

10. Griffin, J.D., D. Linch, K. Sabbath, and S.F. Schlossman. 1984. A monoclonal antibody reactive with normal and leukemic human myeloid progenitor cells. *Leuk. Res.* **8**:521.

11. Ball, E.D., R.F. Graziano, L. Shen, and M.W. Fanger. 1982. Monoclonal antibodies to novel myeloid antigens reveal human neutrophil heterogeneity. *Proc. Natl. Acad. Sci. U.S.A.* **79**:5374.

12. Ball, E.D., R.F. Graziano, and M.W. Fanger. 1981. A unique antigen expressed by myeloid cells and acute blast cells defined by a monoclonal antibody. *J. Immunol.* **130**:2937.

13. Todd, R.F., L.M. Nadler, and S.F. Schlossman. 1981. Antigens on human monocytes identified by monoclonal antibodies. *J. Immunol.* **126**:1435.

14. Nadler, L.M., P. Stashenko, R. Hardy, J.M. Pesando, E.J. Yunis, and S.F. Schlossman. 1980. Monoclonal antibodies defining serologically distinct HLA/DR related Ia-like antigens in man. *Hum. Immunol.* **1**:77.

15. Bennett, J.M., D. Catovsky, M-T Daniel, G. Flandrin, D.A.G. Galton, H. Gralnick, and C. Sultan. 1976. Proposals for the classification of the acute leukemias. *Brit. J. Haematol.* **33**:451.

16. Gralnick, H.R., D.A.G. Galton, D. Catovsky, L. Sultan, and J.M. Bennett. 1977. Classification of acute leukemia. *Ann. Intern. Med.* **87**:740.

17. Ritz, J., S.E. Sallen, R.C. Bast, J.M. Lipton, L.A. Clavell, M. Feeney, T. Hercend, D.G. Nathan, and S.F. Schlossman. 1982. Autologous bone marrow transplantation in CALLA-positive acute lymphoblastic leukemia after *in vitro* treatment with J5 monoclonal antibody and complement. *Lancet* **2**:60.

CHAPTER 23

Antigenic Analysis of Human Malignant Myeloid Cells by Immunoperoxidase

J.G. Levy, P.M. Logan, D. Pearson, S. Whitney, V. Lum, and S. Naiman

Introduction

Antigenic markers associated with subpopulations of hematologic cell types are usually identified by immunofluorescence, either by direct microscopic examination or by flow cytometry. Such methods for identifying cell populations have been applied to the classification of normal peripheral blood leukocyte (PBL) subpopulations or for the identification of monoclonal populations in cases of (mainly) lymphoproliferative disorders (1,2,3). Immunofluorescence is probably the most sensitive procedure at our disposal, at this time, for the identification of cell surface markers on subset populations. However, there are a number of limitations in the use of immunofluorescence procedures for monitoring and classifying cell populations, particularly in defining cells expressing specific markers in neoplastic conditions. This paper describes a procedure which, we feel, may offer advantages over the traditional immunofluorescence procedure.

In this laboratory, using a conventional feedback procedure, we raised a rabbit antiserum which reacted exclusively, in the enzyme-linked immunosorbent assay (ELISA), with membrane extracts of peripheral blood leukocytes of patients with acute nonlymphoblastic leukemia (ANLL) as opposed to equivalent preparations from normal individuals, or those with lymphoproliferative disorders (4). Subsequent studies showed that antisera so raised reacted with an antigen which was present at considerable concentrations in cell extracts from patients with ANLL but was not detectable in equivalent preparations from normal individuals. The antigen was shown to be a protein of molecular weight 68 Kd and pI 7.16 (5). Studies have been carried out with rabbit antisera raised to this 68-Kd protein and results from flow cytometry analyses using this antisera showed the following: (1) PBLs or BM from patients with ANLL or chronic granulocytic leukemia (CGL) reacted strongly with the antiserum whereas (2) cells from normal individuals or those with lymphoproliferative disorders did not and (3) cells from patients in remission from ANLL also reacted strongly with the antiserum (6,7).

When a monoclonal antibody (CAMAL-1) was raised to this 68-Kd common antigen in myelogenous acute leukemia (CAMAL), analogous studies using flow cytometry showed that the monoclonal reagent did not react as reliably as did the rabbit antiserum in immunofluorescence, although both antibodies were directed to the homogeneous 68-Kd protein (8). This observation is not really surprising since the rabbit antiserum to CAMAL would have a variety of epitope specificities whereas the monoclonal antibody, CAMAL-1, would have only one and would therefore be less sensitive on a per molecule basis.

Because of difficulties encountered with CAMAL-1 and other monoclonal antibodies to CAMAL, we sought alternative procedures to immunofluorescence for purposes of examining either BM or PBLs. As a result of these studies, we report here on an indirect immunoperoxidase procedure developed by us which can be applied to the analysis of cell populations for the expression of certain antigenic markers (9).

It must be stated here that we recognize that this procedure may not, in all instances, exhibit the sensitivity of immunofluorescence. However, this immunoperoxidase procedure has a number of advantages: (1) antigen distribution (membrane, cytoplasmic, perinuclear, etc.) can be easily defined; (2) the morphology and type of cells expressing the antigen in question can be relatively easily determined; (3) a permanent record of patient material is easily obtainable; and (4) the procedure is inexpensive and relatively simple. Considering the plethora of monoclonal antibodies available for defining human hematologic subpopulations, it would not be difficult to obtain monoclonal antibodies which, using indirect immunoperoxidase, could define FAB types of myelogenous leukemias or other disorders with relative ease.

This paper covers information on indirect immunoperoxidase staining data using our monoclonal antibody CAMAL-1, as well as results from a panel of monoclonal antibodies raised to myelogenous leukemia antigens and tested by us against a number of cell lines as well as five patient samples.

Materials and Methods

Cell Preparations

PBL or BM samples from patients or normal individuals were separated over Ficoll–Hypaque as previously described (7). For cytospin preparations, cells were washed repeatedly (3 times) in serum-free DME before suspension at a concentration of 2×10^6 per ml. This procedure was found to be extremely important in obtaining cytospin preparations for indirect immunoperoxidase staining, since residual Ficoll or serum in cell preparations can give rise to nonspecific background staining (9). Cytospin preparations were made using a Shandon cytocentrifuge to which approximately 100 μl of cell suspension (3 drops) was applied to each centrifuge cup. Slides were air-dried and stored at room temperature until they were stained. Reports from other laboratories in which similar procedures have been applied have indicated that situations in which ambient temperatures or humidities are excessive ($>25°C$ or 80% humidity) can lead to rupture of cells under these conditions. Therefore, it might be advisable that under such conditions slides should be refrigerated after cytocentrifuging.

Cell lines used in these studies were prepared in essentially the same manner as described above. The cell lines used here were: HL-60, KG1, K562, and WC2. The first three lines are well-known established lines from human myelogenous leukemias. The WC2 line was produced in this laboratory by fusion of the WEHI murine line [provided by Dr. Bootsma and described elsewhere (10)] with PBLs from a CGL patient (11). This fusion product was selected for its ability to produce the CAMAL antigen, currently under study in this laboratory. It was used in this survey to determine whether other monoclonal antibodies raised to human myeloid antigens would also react with it.

Immunoperoxidase Staining

Cytospin preparations were stained according to procedures which have been described in detail elsewhere (9). Briefly, slides were fixed for 30 min in methanol containing 2% (final volume) H_2O_2. Slides were washed in PBS and never allowed to dry throughout the procedure. The test monoclonal antibodies (mAbs) were applied to slides for 30 min, which were then flooded and washed thoroughly with PBS. The second antibody (HRP-labeled rabbit anti–mouse Ig, DAKO) was applied to slides at a 1:100 dilution in PBS with normal human serum at 1:50 for another 30 min. Slides were washed again in PBS and then were developed with DAB for 10 min. They were washed for 5 min in running water prior to staining with hematoxylin.

Results

Summary of Results Obtained with CAMAL-1

We have examined a large number of patient samples using these procedures and our mAb, CAMAL-1, which is directed to an antigen that appears to be ubiquitously expressed in cells of patients with CGL or

Table 23.1. Summary of immunoperoxidase staining results of cells from patients with myelogenous leukemia labeled with CAMAL-1.

Diagnosis	Sample	Number tested	Percent CAMAL-positive ± S.E.M. and range
ANLL	BM	51	14.6 ± 2.1 (0–80.0)
ANLL	PB	34	6.1 ± 2.6 (0–80.0)
CGL (benign)	BM	8	17.6 ± 5.2 (3.0–49.0)
CGL (benign)	PB	16	11.6 ± 2.4 (0–23.0)
CGL (blast crisis)	BM	11	15.4 ± 4.0 (3.0–45.0)
CGL (blast crisis)	PB	9	8.6 ± 3.8 (0.1–37.0)
ANLL remission	BM	66	8.2 ± 1.0 (0.2–35.0)
ANLL remission	PB	59	3.2 ± 0.7 (0–15.0)
ANLL relapse	BM	18	13.4 ± 2.1 (1.0–82.0)

ANLL (regardless of FAB classification). A remarkable aspect of the expression of the CAMAL antigen in these patients is that cells expressing this antigen are also present in BM aspirates of patients in clinical remission from ANLL. A summary of results obtained to date on these patients is shown in Table 23.1. A fairly comprehensive study of cells from patients with lymphoproliferative disorders and normal individuals has also been undertaken. It has been our observation that PBLs of normal individuals contain <0.1% CAMAL-positive cells, whereas BM aspirates of normal individuals may contain a small number of these cells. We have, somewhat arbitrarily, chosen a cutoff of ≤1.0% CAMAL-1 reactive cells in BM and PBLs as delimiting the normal range since no normal samples so far studied have shown levels above 1.0% positive in this test. A summary of results from patients with lymphoproliferative disorders or normal individuals is shown in Table 23.2. It can be seen that there appears to be a difference in % CAMAL-1 reactive cells in the lymphoproli-

Table 23.2. Summary of immunoperoxidase staining results of cells from individuals with preleukemia or lymphoid malignancies, or normals labeled with CAMAL-1.

Diagnosis	Sample	Number tested	Percent CAMAL-positive ± S.E.M. and range
Preleukemia/ myelodysplasia	BM	12	4.7 ± 1.3 (0.3–14.4)
	PB	10	1.9 ± 0.7 (0–7.0)
ALL, primary presentation	BM	34	1.2 ± 0.2 (0–4.2)
ALL, primary presentation	PB	11	0.2 ± 0.09 (0–1.0)
CLL	PB	14	0.2 ± 0.1 (0–1.5)
Lymphoma	BM	67	1.8 ± 0.02 (0–5.0)
Normal	BM	21	0.3 ± 0.07 (0–1.0)
Normal	PB	37	0

ferative patient groups in comparison to normal PBLs or BM in that patients with lymphomas and acute lymphoblastic leukemias frequently have levels of CAMAL-positive cells over the established normal levels. However, it should also be pointed out that, on the average, numbers of positive cells in these patient groups were significantly lower than those seen in patients with myeloproliferative diseases.

Thus, our studies using a mAb (CAMAL-1) directed to a defined antigen expressed apparently at elevated levels and frequencies in cells of patients with myeloproliferative disorders, regardless of disease status (chronic, acute, or remission), and an indirect immunoperoxidase staining procedure, have enabled us to determine the cell types, and, within limits, the concentration and distribution of the antigen within those cells. These findings have been discussed elsewhere (9).

A study that follows a group of ANLL patients over time has been undertaken. It is still premature to draw conclusions of any possible prognostic nature; however, we have observed that the number of PBLs or BM cells which express the CAMAL antigen, as assayed by immunoperoxidase, does not remain static over time. Indeed, these numbers vary considerably in individual patients studied in this manner. Figure 23.1 illustrates a positively labeled promyelocyte from the BM of an ANLL (M3) patient tested post-chemotherapy. Routine hematologic evaluations were unable to determine the presence of residual leukemia at that time and remission status remained difficult to assess. We found, using the

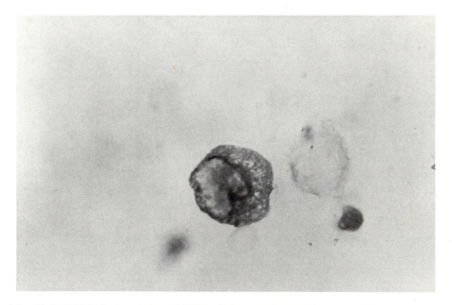

Fig. 23.1. CAMAL-1 staining of BM cell (promyelocyte) from a patient in clinical remission from ANLL (M3).

Table 23.3. Samples used in testing mAb panel.

Patient no.	Diagnosis	Sample	Comments
1	CGL	PB	
2	M5	BM	82% blasts
3	M5	BM	
4	M5	BM	98% blasts
5	M4	BM	68% blasts
6	Normal	PB	

immunoperoxidase slide test, that BM cells from this patient continued to show high numbers (between 15 and 28%) of CAMAL-positive cells on three occasions examined over the next four months during which the patient suffered two relapses.

Results from the Monoclonal Test Panel

The indirect immunoperoxidase procedure used by us has been applied to the examination of a test panel of mAbs directed to leukocyte differentiation markers. In our study, BM cells from five patients and a normal PBL preparation, as well as four cell lines, were prepared and stained with all 116 mAbs in the panel. Descriptions of the patient cell preparations used

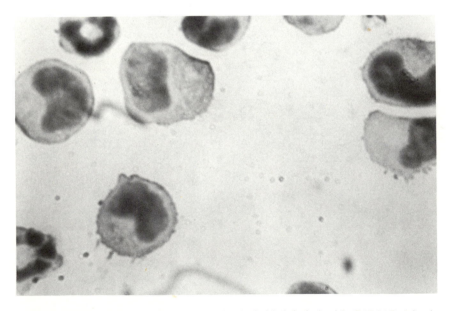

Fig. 23.2. WC2 cell line, a human–murine hybrid, labeled with CAMAL-1 in the immunoperoxidase slide test.

are summarized in Table 23.3. The cell lines included HL-60, KG1, K562, and WC2. The latter cell line was developed in this laboratory and is a fusion product between the murine WEHI cell line and PBLs from a CGL patient. It was selected for CAMAL production and was found to react in immunoperoxidase with CAMAL-1 (Fig. 23.2).

Results of this study are presented in Table 23.4 and summarized in Table 23.5 and representative pictures of results are shown in Fig. 23.3. Although it is difficult to digest all these data, what can be noted from these results is that many of the test mAbs reacted extensively with both patient samples and with the cell lines. Very few of them reacted at significant levels with normal PBLs. In the representative figures shown, we have selected samples to indicate the variations seen between different mAbs with regard to staining patterns (membrane, cytoplasmic, granular cytoplasmic, and/or perinuclear). It was our observation that such staining patterns were characteristic of a given mAb and apparently reflect the distribution and concentration of the antigen with which each reacts.

Table 23.4. Percent positive cells in each cell category when tested with each antibody in test panel.

Antibody number	Patients					Cell lines				Normal blood mononuclear leukocytes
	1 CGL	2 M4	3 M5	4 M5	5 M5	WC2	HL-60	KG1	K562	
1	0	100	0	0	0	0	12	2	0	0
2	22	100	17	<0.1	4	100	100	0	0	0
3	80	90	100	<0.1	40	0	100	0	22	9
4	0	0	0	0.2	7	0	6	0	18	9
5	0	1.7	0	0	8	0	60	0	0	.2
6	25	100	50	100	100	0	100	100	55	<0.1
7	0	1	0	0	6	0	100	0	0	0
8	80	100	54	0.2	9	13	100	0	0	0.2
9	0	100	0	0	8	80	100	100	0	4
10	13	100	90	100	14	80	100	100	0	6
11	0	80	0	0	9	0	5	<0.1	0	6
12	0	18	0.1	100	100	0	100	0	0	2
13	10	0	0	0	18	0	100	0	0	0.2
14	80	0	0	0	59	0	100	0	0	0
15	0	0	0	0	0	0	0	0	0	0
16	0	0	0	0	3	0	100	0	0	0
17	0	0	0	0	40	0	0	0	0	<0.1
18	0	0	0	0	43	0	0	0	0	1.5
19	17	100	16	0.1	44	0	100	0	38	0
20	0	0	0	0	<0.1	0	3	0	0	0
21	0	90	0	0	6	0	100	0	0	0
22	9	90	74	0	10	0	100	0	0	0
23	13	90	100	0	1.5	0	20	0	0	0.2
24	10	90	5.5	0	4.5	0	0	0	0	0
25	33	0	0.1	<0.1	7	0	26	0	0	0

Table 23.4. *Continued.*

Antibody number	Patients					Cell lines				Normal blood mononuclear leukocytes
	1 CGL	2 M4	3 M5	4 M5	5 M5	WC2	HL-60	KG1	K562	
26	34	0	0	0	15	0	0	0	0	2
27	27	88	0	0	13	0	100	0	0	0
28	29	60	85	0	12	0	100	0	0	0
29	0	0.2	0	0	100	0	4	0	0	0
30	27	0	0	0	2	0	35	0	0	0
31	0	0	0.1	0	0	0	0	0	0	0
32	0	80	5.2	0	0	0	100	0	0	0
33	12	4	0	0	0	0	100	0	0	0.1
34	17	0	0.1	0	15	0	0	0	0	0
35	0	0.1	2.5	0	0	0	100	0	0	0
36	12	11	4	0	25	80	10	100	0	80
37	0	0	0	0	0	0	0	0	0	0
38	0	0	0	0	13	0	0	10	0	0
39	14	100	99	0	13	0	25	0	0	80
40	3	50	0	0	0	0	100	0	0	0
41	2	0	0	0	0	0	0	0	0	0
42	0	0	0	0	0	0	0	0	0	0
43	0	<0.1	0	0	0	0	100	0	0	0
44	0	0.2	0	0	0	0	0	0	0	0
45	3	50	0	0	0	0	100	0	0	50
46	39	1	0	0	0	0	100	0	0	<0.2
47	0	0	0	0	0	0	100	0	0	0
48	0	0	0	0	3	0	100	0	0	0
49	0	0	0	0	0.4	0	0	0	0	0
50	17	70	5	100	6	100	100	0.2	0	50
51	29	62	3.5	0	13	0	100	0	0	14
52	0	100	1.5	0	18	0	100	0	0	0.5
53	0	50	50	0	62	0	0	100	0	30
54	19	100	8	0	0	0	100	50	47	10
55	21	0	99	0	0	0	20	0	0	0
56	2	0	0	0	1	0	100	0	0	0
57	0	0	0	0	21	0	0	0	0	0
58	0.5	0	39	0	95	0	5	0	3	0
59	0.5	0	0	0	0	0	1	0	0	0
60	0	100	5	0	21	0	100	0	0	0
61	47	3	0	0	6	0	5	0	0	0
62	0	0	0	0	0	0	0	0	0	0
63	1	0	0	0	0	0	0	0	0	0
64	11	3	0.4	0	7	0	100	0	32	0
65	0	0	0	0	0	0	0	0	0	0
66	5	0.4	0	0	0	0	0	0	0	0
67	5	0	0	0	0	0	0	0	0	0
68	4	0	18	0	0	0	4	0	0	0
69	4	0	0	0	0	0	0	0	0	0
70	84	50	5	0	12	100	50	0	0	10
71	2	0	0	0	0	0	0	0	0	14
72	<0.5	0	1.5	0	0	0	0	0	0	0
73	0	80	3	0	12	60	2	0	0	25
74	0	44	6	0	7	60	100	0	0	0

Table 23.4. *Continued.*

| Antibody number | Patients | | | | | Cell lines | | | | Normal blood mononuclear leukocytes |
	1 CGL	2 M4	3 M5	4 M5	5 M5	WC2	HL-60	KG1	K562	
75	0	0	1	0	7	0	0	0	0	0
76	50	0	32	0	7	0	2	0	0	0
77	15	10	5	0	2	0	100	0	0	0
78	0	1.6	63	0	0	0	0	0	0	25
79	18	0	95	0	6	0	100	0	100	0
80	40	1.5	10	0	13	0	100	0	50	<0.5
81	80	1.5	15	1.1	16	0	20	3	0	0
82	70	0	100	0	5	81	30	0	100	0
83	0	0	67	0	6	0	100	0	0	70
84	0	0	16	0	15	0	40	0	0	36
85	0	1.5	8	0	2	0	40	0	0	80
86	0	0	1	0	0	0	0	0	0	0
87	0	90	11	0	0	0	50	100	0	0
88	0	0	100	0	0	0	100	0	0	25
89	0	50	5	0	0	0	0	0	0	0
90	2	0	<0.1	0	0	0	100	0	0	0
91	7	2	1	0	1.5	0	40	0	0	0
92	30	2	80	<0.1	0.8	0	100	0	0	12
93	0	0	<1	0	<0.1	0	100	0	0	10
94	20	0	0	0	0	0	15	0	0	<1
95	7	0	100	0	0	0	100	0	0	5
96	0	0	0.5	0	0	0	0	0	0	10
97	0	0	6	0	0	0	0	0	0	<1
98	70	0	3	0	70	0	0	0	0	100
99	90	1	8	0	0	0	91	0	0	3
100	0	1.5	9	0	0	0	4	0	0	<1
101	0	0	9	0	0	0	0	0	<1	0
102	0	0	.2	0	0	0	0	0	0	12
103	0	0.1	0.3	0	0	0	0	0	0	0
104	0	0.5	0.5	0	0	0	0	0	0	0
105	0	2	4	0	7	0	4	0	0	5
106	0	0	.3	0	0	0	31	0	0	0
107	0	0	4	0	0	0	4	0	0	0
110	29	0	3	0	0	0	20	0	0	0
111	30	1	7	0	20	0	30	<1	0	10
112	85	<1	5	0	9	0	21	0	0	0
113	0	0	4	0	0	0	8	0	0	2
114	0	0	.6	0	0	0	0	0	0	0
115	3	0	3	0	0	0	0	0	0	0
116	0	0	3	0	0	0	10	0	0	0
117	0	80	85	.4	70	0	71	14	100	12
118	0	0	5	0	0	0	2	0	0	0
Control[a]	0	0	0	0	0	0	0	0	0	01
Rabbit anti-serum[b]	100	100	98	1.8	79	100	51	0	<0.1	0

[a] The control was a monoclonal antibody vs. ferredoxin.
[b] The rabbit antiserum was one prepared in our lab which reacts with cells from AML patients.

Table 23.5. Summary of results obtained with test monoclonals.

Sample	No. reacting positively in immunoperoxidase
PBL-CGL (1)	58/116
BM-ANLL M4 (2)	57/116
BM-ANLL M5 (3)	71/116
BM-ANLL M5 (4)	8/116
BM-ANLL M5 (5)	73/116
Normal PBL	43/116
Cell line HL-60	62/116
Cell line KG1	13/116
Cell line K562	11/116
Cell line WC2	10/116

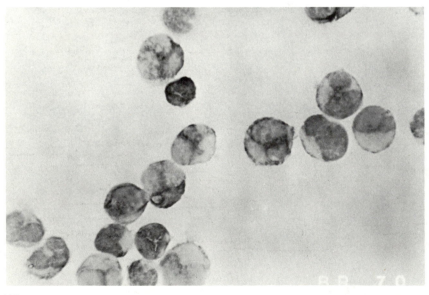

(A)

Fig. 23.3. Representative immunoperoxidase stains from either patient material or cell lines with various monoclonal antibodies. (A) CGL patient's PBLs with test mAb 70; (B), (C) M4 patient's BM stained with CAMAL-1 (arrow shows positive cell) and test mAb 28; (D), (E) M5 patient's BM stained with CAMAL-1 and test mAb 92; (F), (G), (H), (I) HL-60 stained with test mAbs 19, 28, 30, and 110; (J), (K) WC2 stained with test mAbs 23 (showing no reactivity) and 73.

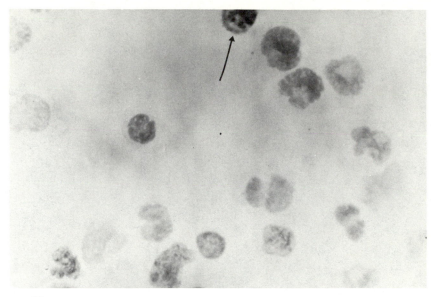

(B)

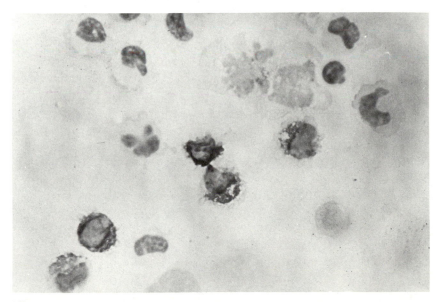

(C)

Fig. 23.3. *Continued.*

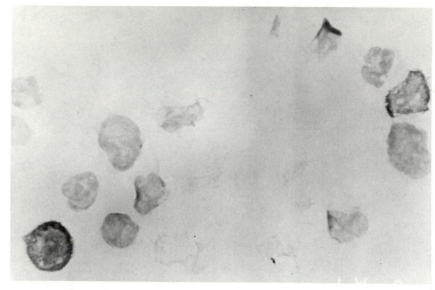

(D)

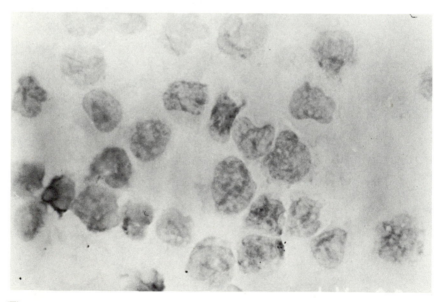

(E)

Fig. 23.3. *Continued.*

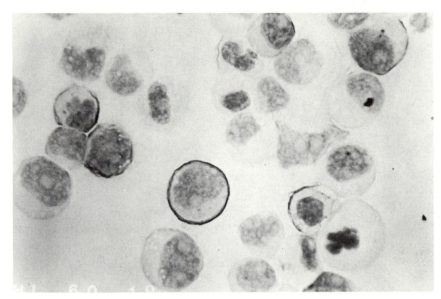

(F)

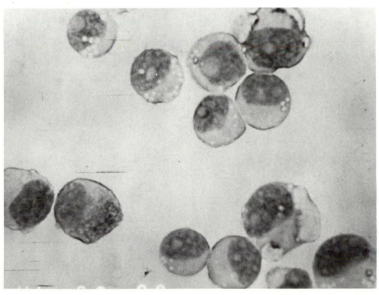

(G)

Fig. 23.3. *Continued.*

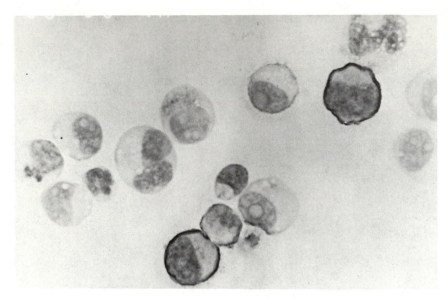

(H)

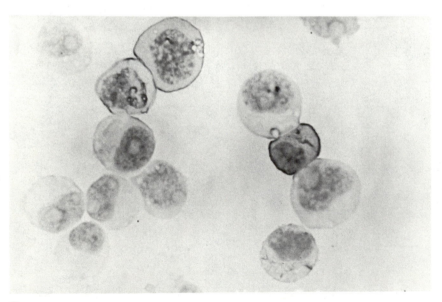

(I)

Fig. 23.3. *Continued.*

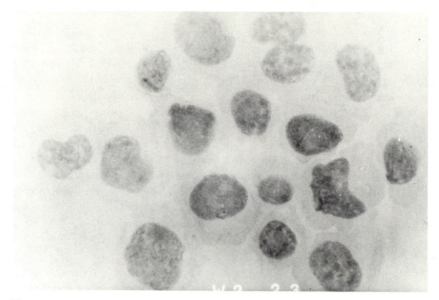

(J)

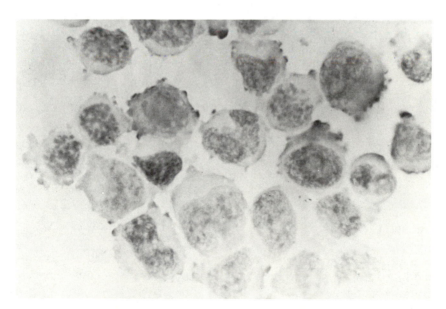

(K)

Fig. 23.3. *Continued.*

Discussion

The immunoperoxidase single cell slide test was originally adapted in our laboratory to investigate the expression of CAMAL in BM and PBL samples of patients with myelogenous leukemia with the CAMAL-1 mAb. The updated summarized results presented here indicate that CAMAL is preferentially expressed on or within cells of patients with myeloproliferative disease. One of the more interesting findings with CAMAL-1 was the observation that reactive cells are frequently present in the BM of ANLL patients who are in clinical remission. Numbers of CAMAL-reactive cells vary extensively between patients in this group. Furthermore, it has become apparent that, for any given remission patient, the number of cells expressing the CAMAL antigen varies over time. These findings explain the wide range of CAMAL-1 reactive cells shown in Table 23.1. The possible use of these observations in terms of patient management and prognosis are currently being investigated.

Many of the 116 mAbs used in this study reacted with a variety of cell preparations in the immunoperoxidase slide test, many of them very strongly. It was of interest to note that there was a lot of variation in staining patterns of the individual mAbs with the patient samples, indicating that few of these mAbs are directed to the same antigen or that epitope specificity, at least, was extremely variable for those mAbs which may react with the same antigen. It was also of interest to note that even when three of the patient samples were from individuals with the same type of ANLL (patients 3, 4, and 5, all M5) staining characteristics varied considerably between patients, indicating marked variability in antigen expression in this subclass.

Although this report is, by necessity, limited, we feel that the study indicates the potential application of this simple technique for routine use in examining antigens in single cell preparations such as BM or PBLs. The advantages that this technique offers in terms of simplicity and information regarding staining patterns and cell morphologies are self-evident.

Summary

A rapid indirect immunoperoxidase slide test has been developed and used in a study of the expression of a common myelogenous leukemia-associated antigen (CAMAL) in or on cells of either bone marrow (BM) aspirates or peripheral blood leukocytes (PBLs), using a monoclonal antibody (CAMAL-1). We have found that both BM and PBLs of patients with either acute non-lymphoblastic leukemia (ANLL) or chronic granulocytic leukemia (CGL) have elevated numbers of CAMAL-positive cells. These elevated levels are also seen in a significant number of ANLL patients in clinical remission. BM from normal individuals has been found to contain <1.0% CAMAL-positive cells (usually <0.5%) and normal PBLs were found to have <0.1% reactive cells. When BM or PBLs from patients with lymphoproliferative disorders were examined, in some instances slightly elevated levels (over normal) of CAMAL-1 reactive cells were seen, although these levels are not comparable to those seen in myeloproliferative disorders.

Using this immunoperoxidase procedure, we have examined a panel of 116 monoclonal antibodies (mAbs), most of which are directed against myeloid-associated antigens. A majority of these monoclonal antibodies (101/116) showed positive staining against at least some of the cell preparations tested, with percent reactivities varying widely. Of the myeloid cell lines tested (HL-60, KG1, and K562), reactivity was most frequent with HL-60 where 62/116 mAbs showed significant staining and 28 of these labeled 100% of the cells. A human–murine hybrid cell line (WC2) derived in this laboratory from a fusion of PBLs from a CGL patient and the WEHI-3B HRPT⁻ cell line was also examined. Ten of the 116 mAbs reacted with this cell line. Patient samples tested included: PBLs from a CGL patient, BM from an ANLL M4 patient, BM from three ANLL M5 patients, and a normal PBL.

References

1. Greaves, M.F. 1981. Analysis of the clinical and biological significance of lymphoid phenotypes in acute leukemia. *Cancer Res.* **41**:4752.
2. Maheu, M., M. Baker, J. Falk, and R.N. Taub. 1981. Immunologic diagnosis and monitoring of human acute leukemias. *Am. J. Pathol.* **103**:139.
3. Foon, K.A., R.W. Schroff, and R.P. Gale. 1982. Surface markers on leukemia and lymphoma cells. Recent advances. *Blood* **60**:1.
4. Al-Rammahy, A.Kh., R. Shipman, A. Jackson, A. Malcolm, and J.G. Levy. 1980. Evidence for a common leukemia-associated antigen in acute myelogenous leukemia. *Cancer Immunol. Immunother.* **9**:181.
5. Shipman, R.C., A. Malcolm, and J.G. Levy. 1983. Partial characterization of a membrane antigen which exhibits specificity for cells of patients with acute myelogenous leukemia. *Brit. J. Cancer* **47**:849.
6. Malcolm, A.J., R.C. Shipman, and J.G. Levy. 1982. Detection of a tumor associated antigen on the surface of human myelogenous leukemia cells. *J. Immunol.* **128**:2599.
7. Malcolm, A.J., P.M. Logan, R.C. Shipman, R. Kurth, and J.G. Levy. 1983. Analysis of human myelogenous leukemia cells in the fluorescence-activated cell sorter using tumor specific antiserum. *Blood* **61**:858.
8. Malcolm, A.J., R.C. Shipman, P.M. Logan, and J.G. Levy. 1984. A monoclonal antibody to myelogenous leukemia: Isolation and characterization. *Exp. Hematol.* **12**:539.
9. Logan, P.M., A.J. Malcolm, and J.G. Levy. 1984. Detection of a common antigen in human myelogenous leukemia using immunoperoxidase. *Diagn. Immunol.* **2**:86.

10. Van Kessel, A.H.M.G., P.A.T. Tetteroo, A.E.G. Kr. von dem Borne, A. Hagemeijer, and D. Bootsma. 1983. Expression of human myeloid associated surface antigens in human myeloid cell hybrids. *Proc. Natl. Acad. Sci. U.S.A.* **80:**3748.
11. Mew, D., V. Lum, C.-K. Wat, G.H.N. Towers, C.-H.C. Sun, R.J. Walter, W. Wright, M.W. Berns, J.G. Levy. 1985. The ability of specific monoclonal antibodies and conventional antisera conjugated to hematoporphyrin to label and kill selected cell lines subsequent to light activation. *Cancer Res.* in press.

CHAPTER 24

Expression of Lymphocyte Antigens on Blast Cells from Patients with Chronic Granulocytic Leukemia

P.M. Lansdorp, J.G.J. Bauman, and W.P. Zeijlemaker

Introduction

Chronic granulocytic leukemia (CGL) is a myeloproliferative disorder and is associated, in more than 90% of cases, with a karyotypic marker, the Philadelphia chromosome (Ph[1]) (1,2). Although during the chronic phase of the disease the blood picture is dominated by leukemic granulocytes of various maturation stages, the disease is not confined to the granulocyte lineage. Isoenzyme (G6PD) studies have revealed that in most patients non-lymphoid cells in the peripheral blood are in majority derived from the malignant clone (3,4). Such studies have also shown involvement of B lymphoid cells in CGL (5), and together with additional evidence this has led to the notion that the target cell for malignant transformation in CGL is a common progenitor for myeloid cells and B lymphocytes (6). Based on cell membrane marker studies during CGL blast crisis, the involvement of T cell progenitors in CGL is controversial (6–8,12). Clonal evolution of CGL cells in the blast crisis of the disease (2) and/or the inability of CGL cells to mature to well-defined T cells (6,12) may be at the basis of this controversy.

In studies with (cryopreserved) low-density blast cells from the circulation of patients with chronic phase CGL, we observed a striking heterogeneity in functional (9) and antigenic (10) properties. These results were interpreted as indicative of the broad differentiation potential of the target cells for malignant transformation in CGL, which is in agreement with published evidence (reviewed in Ref. 6). To detect possible differentiation of the low-density CGL blasts along the lymphoid lineage, their reactivity with a panel of monoclonal antibodies directed against lymphoid-specific antigens was analyzed on a FACS. Subpopulations that were identified were sorted and analyzed with respect to functional properties, i.e., the proliferative response to different hematopoietic growth factors.

Materials and Methods

Cells

The two patients with Ph^1-positive CGL and the isolation of buffy coat cell suspensions from their peripheral blood were described previously (9). Leukapheresis of patient 1 (♀, 1961) was performed after a three-month period without chemotherapy in 1981, several years after the diagnosis of CGL had been made. Buffy coat cells from patient 2 (♀, 1967) were obtained prior to other forms of treatment, shortly after diagnosis in 1983. Low-density cell fractions were obtained by centrifugation of buffy coat cells over Ficoll–Hypaque (d = 1.062 g/cm³). Cells were cryopreserved in aliquots and for each experiment individual samples were thawed.

Labeling Procedure

Cells were labeled by indirect immunofluorescence as described (10). Briefly, cells (10×10^6/ml) were incubated for 45 min with monoclonal antibody, washed twice with a buffered salt solution containing 5% (v/v) fetal calf serum, incubated for 45 min with goat anti–mouse Ig–FITC (CLB, Amsterdam), washed again, and resuspended in Iscove's medium (Gibco, Paisley, Scotland) containing 10% (v/v) fetal calf serum and 2 μg/ml propidium iodide (PI). Throughout the experiment, cells were kept at 0°C. As a negative control cells were labeled with goat anti–mouse Ig–FITC only. A positive control was obtained by incubation of cells with a monoclonal antibody (4D2D11) against β2-microglobulin (prepared by P. van Mourik in our laboratory). The other monoclonal antibodies were obtained from Dr. Haynes (3A1), from commercial sources (Leu 2a, Leu 3a, and Leu 4, Becton-Dickinson, B-D, Sunnyvale, CA, USA, and J5, Ortho Pharmaceuticals) or from the Second Workshop T cell panel. The latter were used at a dilution of 1:400, whereas optimal concentrations of the other antibodies were selected by titration experiments.

Cell Sorting and Analysis

Suspensions of labeled cells were analyzed and cells sorted using a FACS II (B-D). The instrument was set to allow three-parameter analysis and sorting of cells as described by Visser et al. (11), with the laser at 488 nm (0.6 W). The parameters measured were forward light scatter (FLS), red fluorescence (P1), and green fluorescence (FITC). An FLS window was used to exclude both small (e.g., platelets and cellular debris) and large particles (clumps of cells) from analysis and sorting. A PI window was used to restrict measurements and sort procedures to viable (PI-negative) cells. FITC fluorescence was recorded using a logarithmic amplifier.

Analysis of Sorted Cells

The proliferative response of sorted cells to some hematopoietic growth factors was measured as described (9). Briefly, 10^3 sorted cells were cultured in round-bottomed microtiter plates in 100 μl Iscove's medium containing 10% (v/v) fetal calf serum and either a) no further addition; b) 2 U/ml erythropoietin (EPO, Connaught, step IV, Toronto, Canada) to stimulate terminal erythroid differentiation; c) giant cell tumor conditioned medium (GCT, Gibco) 10% (v/v), which contains stimulators for myeloid progenitor cells; d) a mixture of EPO (2 U/ml) and human mixed lymphocyte culture conditioned medium (MLC-CM) 10% (v/v), which stimulates early erythroid differentiation as well as myelopoiesis. MLC-CM does not contain measurable IL-2 activity.

[^{3}H]Thymidine incorporation of the cells over the last four hours of a four-day culture period was measured to quantitate proliferation.

Results

Expression of Lymphocyte Antigens on Low-density CGL Cells

Low-density cells from two patients with CGL were recovered from liquid nitrogen storage and labeled with monoclonal antibodies and goat anti–mouse Ig–FITC. In general more than 90% of the cells were vital as judged by PI exclusion. Examples of FACS analysis of such suspensions are shown in Figs. 24.1 and 24.2. All antibodies shown yield a characteris-

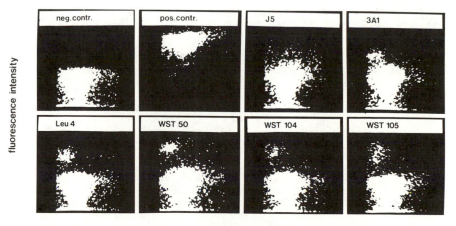

Fig. 24.1. Dual-parameter dot plots of low-density blast cells from patient 1 labeled with the indicated antibodies and goat anti–mouse Ig–FITC. Analysis of 4 × 10^3 propidium iodide-negative cells in the forward light-scatter window. Fluoresence intensity (FITC) is plotted on a logarithmic scale.

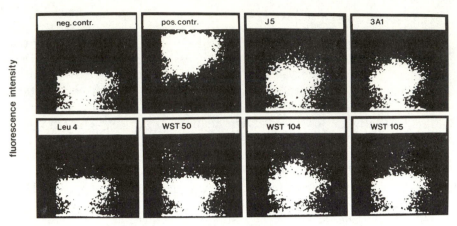

forward light scatter

Fig. 24.2. As Fig. 24.1, patient 2.

tic staining pattern that can be distinguished from the negative control. Most of the cells are stained with the positive control antibody, anti-β2-microglobulin. Different lymphocyte-specific antibodies stain a low percentage of cells with variable intensity. Note that most of the antibodies (except J5) stain a small percentage of small cells (low FLS) in patient 1. This is not seen with CGL-2. The antibody 3A1 shows weak staining of some larger cells and strong staining with a small percentage of small cells. J5 does not stain these small cells but shows (very) weak staining of some average sized cells. Leu 4 antibodies stain a small percentage of small-sized cells in patient 1, whereas the T cell Workshop antibodies T50, T104, and T105 each yield different staining patterns.

The cells from patient 1 were used to screen a number of antibodies from the Workshop T cell panel (see this series, Vol. 1, Ch.1). From the

Table 24.1. Reactivity of Workshop T cell monoclonal antibodies with low-density cells from CGL patient 1.

Staining pattern	Antibodies from workshop T cell panel
Leu 4-like[a]	29, 48, 49, 51, 55, 56, 57, 58, 59, 60, 61, 101, 107, 108, 111, 112, 113, 142, 144, 150, 155
3A1-like[a]	30, 31, 32, 33, 34, 37, 38, 39, 40, 41, 43, 44, 45
Other	50, 52, 53, 62, 63, 104, 105, 106, 109, 134, 139, 152, 156
Less than 1% positive cells	35, 36, 42, 55, 64, 65, 66, 102, 103, 110, 116, 128, 129, 130, 131, 132, 133, 135, 136, 137, 138, 140, 141, 143, 145, 146, 147, 148, 149, 151, 153, 154, 159

[a] Similar dot-plot patterns as indicated antibodies (see Fig. 24.1). Different cells and/or cell numbers may have been labeled.

Table 24.2. Reactivity[a] of lymphocyte antibodies with low-density cells from two patients with CGL.

Antibody	CGL-1	CGL-2
3A1	13	12
J5	9	8
T11	6	<1
Leu 1	6	<1
Leu 2a	3	<1
Leu 3a	3	<1
Leu 4	6	<1
50	9	2
104	4	14
105	5	4
152	12	11

[a] Expressed as the percentage of cells with fluorescence above background (which was set at the FACS channel number enclosing 99% of the cells from the negative control). Mean value of two experiments. For each experiment 10^4 cells were analyzed.

results (Table 24.1) we selected a number of antibodies, showing unexpected or very strong reactivity, for further study. The percentage of cells from the two CGL patients labeled by these and some other well-known lymphocyte-specific antibodies is given in Table 24.2. Less than 1% of the cells from patient 2 expressed mature T cell markers, whereas 6% of the cells from patient 1 were stained with Leu 4 or Leu 1. The percentage of cells stained with Leu 2a and Leu 3a antibodies (both 3%) is in agreement with a fully mature T cell marker phenotype of the Leu 4-positive cells from patient 1. From these results we concluded that these cells were most likely normal lymphocytes, not derived from the malignant clone. The reactivity of 3A1 and the shown Workshop monoclonal antibodies with larger cells was unexpected (Figs. 24.1, 24.2, and 24.4). The functional properties of these cell suspensions were, therefore, further investigated.

Proliferative Response of Sorted Cells

Cell suspensions from the two CGL patients labeled with the indicated antibodies (Fig. 24.3) were analyzed on a FACS and fixed numbers of cells were sorted into sterile test tubes on the basis of their fluorescence intensity ($-$, $+$, and all). The proliferative response of sorted cells to different stimuli was quantitated after four days of culture (Fig. 24.3). Without addition to the culture medium 10^3 sorted cells incorporated approximately 10^2 cpm ^{3}H-TdR. The cells from patient 1 showed a weak

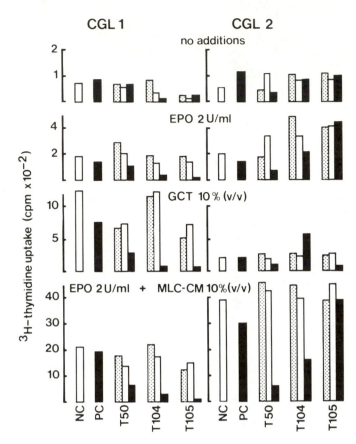

Fig. 24.3. Proliferative response of 10^3 sorted CGL cells to indicated stimuli. Cells were sorted on the basis of fluorescence intensity after labeling with the indicated antibodies. The percentage fluorescence-positive cells with these antibodies is given in Table 24.2. NC = negative control, PC = positive control (anti-β2-microglobulin). Cells sorted: ⊞ all cells; □ fluorescence-negative cells; ■ fluorescence-positive cells. Mean of duplicate reactions.

response to EPO, and the cells from patient 2 showed a stronger response to this stimulator. EPO-responsive cells from patient 1 were T50, T104, and T105 negative, whereas responsive cells from patient 2 were also present in T104- and T105-positive cell fractions. The response to stimulatory substances present in giant cell tumor (GCT) conditioned medium of patient 1 was much higher than that of patient 2. Reactivity was restricted to unlabeled cell fractions with the exception of T104 positive cells from patient 2. The highest proliferative response was observed if erythropoietin was added together with mixed lymphocyte culture conditioned medium (MLC-CM). The reaction pattern resembled that of EPO alone.

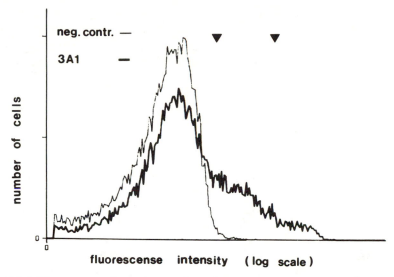

Fig. 24.4. Fluorescence intensity profile of blast cells from CGL patient 1 labeled with 3A1. Arrows indicate fluorescence intensity of sorted cell fractions (Table 24.3). Cells after identical staining procedures without 3A1 served as a negative control (neg. contr.)

The cells from both patients were sorted with respect to fluorescence intensity after labeling with 3A1. The cells from patient 1 were sorted into three fractions (Fig. 24.4), whereas the cells from patient 2 were sorted into a positive and a negative fraction (Fig. 24.2, Table 24.3). Sorted cells were cultured with or without (human) MLC-CM as a source of stimula-

Table 24.3. 3A1-positive CGL cells are responsive to factors present in MLC-CM.

| | | | Stimulator[b] | | | |
| | Sort[a] | | None | | MLC-CM | |
Cells	%	f.i.	Mean	SD	Mean	SD
CGL-1 3A1 all	100	0–300*	221	86	1755	212
CGL-1 3A1 fluor. neg.	87	0–100	187	16	1532	259
CGL-1 3A1 large pos.	10	100–120	177	57	2879	926
CGL-1 3A1 small pos.	3	120–300	20	22	426	121
CGL-2 3A1 all	100	0–300	41	17	393	29
CGL-2 3A1 neg.	88	0–100	22	20	287	56
CGL-2 3A1 pos.	12	100–300	35	20	291	86

[a] Cells were sorted according to 3A1 fluorescence intensity (= f.i. on a logarithmic scale; * FACS channel number).
[b] The proliferative response of low-density CGL cells to stimulatory factors present in mixed lymphocyte culture conditioned medium (MLC-CM) compared to the response in medium alone. Mean and standard deviation of [³H]thymidine incorporation of triplicate reactions (cpm of 1000 cells/well) after 4 days of culture.

tory factors for early erythropoiesis (i.e., burst-promoting activity) as well as myelopoiesis (i.e., colony-stimulating activity). The results (Table 24.3) indicate that CGL cells responsive to factors present in MLC-CM are enriched (CGL-1) or present (CGL-2) in sorted populations of 3A1 (moderate) positive cells.

Discussion

Isoenzyme studies have revealed that the progeny of Ph[1]-positive cells in chronic granulocytic leukemia may include erythrocytes, platelets, monocytes, granulocytes (including eosinophils), and B lymphoid cells (3–5). The involvement of the T cell lymphoid lineage in CGL is less clear (6,12). Expression of T-lymphoid markers on Ph[1]-positive cells is rare but T-lymphoblastic transformation of CGL has been described by several authors (7,8,12). The rarity of T-lymphoid lineage transformation in CGL may be explained by assuming that T cell progenitors derived from the Ph[1]-positive multipotential stem cells are at a low risk of undergoing secondary oncogenic events associated with blastic transformation (2,12). If this assumption is correct, the study of cells from patients in the chronic phase of the disease may offer a better chance to detect the possible involvement of T cell progenitors. We therefore analyzed the expression of T-lymphoid markers on (cryopreserved) low-density blast cells from patients with CGL in the chronic phase of the disease. In previous studies we had found that these blast cells showed only little variation in morphology but a striking heterogeneity in functional and antigenic properties (9,10). Staining of these cells with a panel of monoclonal antibodies specific for lymphoid markers was accurately quantitated using a FACS (Figs. 24.1, 24.2, and 24.4). This technology allowed detection of small numbers of lymphoid cells with a fully mature membrane marker phenotype, probably representing normal lymphocytes not derived from the malignant clone (Table 24.2). In addition, some of the blast cells were stained with J5, 3A1, and the Workshop antibodies T50, T104, and T105 (Figs. 24.1, 24.2, and 24.4). When these cells were sorted and analyzed with respect to their capacity to elicit a proliferative response to some hematopoietic stimulators a heterogeneous picture emerged (Fig. 24.3 and Table 24.3). The reactivity of 3A1 with CGL cells capable of a proliferative response to stimulatory factors present in MLC-CM (Fig. 24.4, Table 24.3) was unexpected in view of the reported lymphocyte specificity of this antibody (13). Further characterization of the 3A1 positive CGL cells as well as studies with normal hematopoietic cells and this antibody are required in view of the possible use of 3A1 or similar antibodies (Table 24.1) in clinical studies.

The results of this study further emphasize the heterogeneity of functional and antigenic properties of low-density blast cells from CGL patients in the chronic phase of the disease (9,10). In addition, the analytical power of monoclonal antibodies, flow cytometry, and cell sorting procedures to characterize small subpopulations of cells is clearly illustrated. In combination with functional assays of sorted cells, such as the proliferation assay used in this study, the differentiation potential of small subsets of cells can be evaluated. This approach may be required to assay generation of pre–T cells by Ph¹-positive stem cells.

Summary

Peripheral blast cells from two patients with chronic granulocytic leukemia (CGL) were obtained by density centrifugation. Up to 20% of the (cryopreserved) low-density cells formed hematopoietic colonies. Subpopulations of the cells from both patients expressed erythrocytic, platelet, and myelo-monocytic membrane markers. To detect possible differentiation of CGL cells along the lymphoid lineage, antibodies specific for lymphocyte antigens, including antibodies from the Workshop T cell panel, were used for immunofluorescence and FACS separation. Small subpopulations (6% and less than 1% for CGL-1 and CGL-2, respectively) of viable cells were stained with antibodies recognizing mature T cell markers. These cells were most likely not derived from the malignant clone. More cells were stained by 3A1, J5, and the Workshop monoclonals T50, T104, T105, and T152. Fluorescence-positive and -negative cells were sorted and cultured in microtiter plates. After four days the proliferative response to various stimulators was measured. Some erythropoietic responsive cells of CGL-2 bound the antibodies T50, T104, and T105; responsive cells from CGL-1 did not stain with these antibodies. Moreover, the cells from CGL-2 that responded to giant cell tumor (GCT) conditioned media were enriched in the T104-positive population. 3A1-positive (non-lymphoid) cells from both patients were responsive to stimulators present in mixed lymphocyte culture conditioned medium. Our data illustrate the heterogeneity of CGL blast cells and indicate that to establish the differentiation potential of these cells, membrane marker analysis has to be combined with functional assays. The demonstration of T cell progenitor involvement in CGL may, therefore, require application of appropriate assays for such cells.

Acknowledgments. The authors thank Dr. P.C. Huijgens (Hematology Dept., Free University, Amsterdam), Dr. B. Löwenberg and Dr. I. Touw (Rotterdam Radio Therapeutical Institute), and Mrs. M.J.G.J. Wijngaarden-Dubois (CLB, Amsterdam) for provision of clinical specimens. Dr. B. Haynes (Duke University Medical Center, Durham, N.C.) kindly provided 3A1 monoclonal antibody. This work was supported in part by the Foundation for Medical Research FUNGO, which is subsidized by the Netherlands Organization for the Advancement of Pure Research (ZWO).

References

1. Nowell, P.C., and D.A. Hungerford. 1960. Chromosome studies on normal and leukemic human leukocytes. *J. Natl. Cancer Inst.* **25**:85.
2. Rowley, J.D. 1980. Ph¹-positive leukaemia, including chronic myelogenous leukaemia. *Clinics in Haematology* **9**:55.
3. Fialkow, P.J., R.J. Jacobson, and T. Panayannopoulou. 1977. Chronic myelocytic leukemia: clonal origin in a stem cell common to the granulocyte, erythrocyte, platelet and monocyte/macrophage. *Am. J. Med.* **63**:125.
4. Koeffler, H.P., A.M. Levine, M. Sparkes, and R.S. Sparkes. 1980. Chronic myelocytic leukemia: Eosinophils involved in the malignant clone. *Blood* **55**:1063.
5. Martin, P.J., V. Najfeld, and P.J. Fialkow. 1982. B-lymphoid cell involvement in chronic myelogenous leukemia: implications for the pathogenesis of the disease. *Cancer Genet. Cytogenet.* **6**:359.
6. Greaves, M.F. 1982. "Target" cells, differentiation and clonal evolution in chronic granulocytic leukaemia: a "model" for understanding the biology of malignancy. In: *Chronic granulocytic leukaemia,* M.T. Shaw, ed. Praeger Publishers, New York, pp. 15–47.
7. Hermann, F., W.F. Ludwig, and P. Kolecki. 1984. Blastic crisis in CML showing early T-lymphoblastic transformation. *Brit. J. Haematol.* **56**:75.
8. Griffin, J.D., R. Tantravahi, G.P. Canellos, J.S. Wisch, E.L. Reinherz, G. Sherwood, R.P. Beveridge, J.F. Daley, H. Lane, and S.F. Schlossman. 1983. T-cell surface antigens in a patient with blast crisis of chronic myeloid leukemia. *Blood* **61**:640.
9. Lansdorp, P.M., F.O. Oosterhof, and W.P. Zeijlemaker. 1985. Colony-forming cells in chronic granulocytic leukemia. I. Proliferative responses to growth factors. *Leuk. Res.* In press.
10. Lansdorp, P.M., J.G.J. Bauman, M.J.E. Bos, A.E.G. Kr. von dem Borne, F.O. Oosterhof, P.v. Mourik, P.A.T. Tetteroo, and W. P. Zeijlemaker. 1985. Colony-forming cells in chronic granulocytic leukemia. II. Analysis of membrane markers. *Leuk. Res.* In press.
11. Visser, J.W.M., J.G.J. Bauman, A.H. Mulder, J.F. Eliason, and A.M. de Leeuw. 1984. Isolation of murine pluripotent hemopoietic stem cells. *J. Exp. Med.* **59**:1576.
12. Jacobs, P., and M. Greaves. 1984. Ph¹-positive T lymphoblastic transformation. *Leuk. Res.* **8**:737.
13. Haynes, B.F., G.S. Eisenbarth, and A.S. Fauci. 1979. Human lymphocyte antigens: production of a monoclonal antibody that defines functional thymus-derived lymphocyte subsets. *Proc. Natl. Acad. Sci. U.S.A.* **76**:5821.

CHAPTER 25

Frequent Detection of Erythroid and Platelet Antigens in Myeloid Blast Crisis of CML

P. Bettelheim, D. Lutz, O. Majdic, O. Haas, and
W. Knapp

Introduction

The terminal phase of chronic myeloid leukemia (CML) is often characterized by an accumulation of immature cells. The differentiation stage of these immature cells can vary considerably, however.

By immunological means we were recently able to discern a high degree of heterogeneity of blast populations in patients with CML blast crisis (CML-BC) (1). Meanwhile we extended our study and present here the results from 62 CML patients in blast crisis.

Materials and Methods

Peripheral blood and bone marrow specimens were obtained from 62 patients with CML-BC. Twenty-three of the CML-BC patients were repeatedly studied in the course of the disease.

The diagnosis of blast crisis was based on the presence of >30% blasts in peripheral blood or bone marrow. Fifty-three patients were in blastic transformation following the typical chronic phase. Nine patients presented with a Ph¹-positive AL but reverted to a chronic phase after treatment. Fifty-five patients were adults (>16 years) and seven were pediatric patients.

Bone marrow and peripheral blood smears were stained with May-Grünwald-Giemsa. The blast populations were morphologically classified as granulocytic, granulo-monocytic, granulo-megakaryocytic, granulo-erythrocytic, granulo-erythro-megakaryocytic, lymphoid, mixed, or undifferentiated.

For the immunological studies mononuclear cells were isolated as described previously (2).

The monoclonal antibodies used in this study and their major specificities are listed in Table 25.1.

The binding of the various antibodies to isolated mononuclear cells was assessed as previously described (2) by indirect immunofluorescence with fluoresceinated goat F(ab′)₂ anti–mouse IgG + IgM antibodies, or in case of the VIM-D5 antigen, by direct immunofluorescence with rhodamine-labeled VIM-D5 hybridoma antibody. It was essential for our studies that the anti–mouse IgG + Igm conjugate did not cross-react with human or rabbit immunoglobulins. Double fluorescent stainings were performed as described (2).

Table 25.1. Monoclonal antibodies used in study of blast crisis.

Designation	Specificity	Reference
Granulo-monocytic markers		
VIM-D5	Granulocytic cells, monoblasts	Majdic *et al.* (3)
VIM2	Granulocytic cells, monocytes	Majdic *et al.* (4)
VIM8	Granulocytic cells, platelets, mono-blasts	Knapp *et al.* (5)
MCS2	Granulocytic cells, monocytes	Sugimoto *et al.* (6)
MY9	Myeloid precursors, monocytes	Griffin and Schlossman (7)
63D3	Monocytes, granulocytes	Ugolini *et al.* (8)
VIM-D2	Monocytes	Majdic *et al.* (9)
Erythroid markers		
VIE-G4	Glycophorin A	Liszka *et al.* (10)
LICR LON R10	Glycophorin A	Anstee and Edwards (11)
LICR LON R18	Glycophorin A	Anstee and Edwards (11)
Platelet markers		
AN51	Platelet glycoprotein Ib	McMichael *et al.* (12)
J15	Platelet glycoprotein IIb/IIIa com-plex	Vainchenker *et al.* (13)
C17-27	Platelet glycoprotein IIIa	Tetteroo *et al.* (14)
C8-13	Platelet glycoprotein IIa	Tetteroo *et al.* (15)
T cell markers		
WT-1	Thymocytes, peripheral T cells, some myeloid blasts	Vodinelich *et al.* (16)
OKT1	T cells, B-CLL	Reinherz *et al.* (17)
OKT11	T cells (E⁺)	Kung *et al.* (18)
9.6	T cells (E⁺)	Kamoun *et al.* (19)
Na1/34	Thymocytes	McMichael *et al.* (20)
Leu 4	Peripheral T cells	Ledbetter *et al.* (21)
Leu 3a	T helper cells	Evans *et al.* (22)
Leu 2a	T suppressor cells	Evans *et al.* (22)
B cell markers		
VIL-A1	Common ALL antigen	Knapp *et al.* (23)
VIB-C5	(Pre-)B cells, granulocytes	Majdic *et al.* (9)
Y29/55	B cells	Forster *et al.* (24)
Anti-IgM	IgM	Lansdorp unpubl.
HLA-DR markers		
VID-1	HLA-DR	Majdic unpubl.

Terminal deoxynucleotidyl transferase (TdT) was determined bio-chemically and by immunofluorescence techniques as described previously (2).

Fluorescence reactivity of the cells was evaluated using a Leitz microscope with incident illumination and equipped for the dual-wavelength method.

Results

Our study on the typing of blast populations revealed a highly heterogeneous reaction pattern of blast cells in CML-BC and we were able to distinguish four major groups:

1. myeloid blast crisis ($n = 38$)
2. lymphoid blast crisis ($n = 18$)
3. mixed myeloid and lymphoid populations ($n = 5$)
4. one unclassifiable blast sample.

Myeloid Blast Crisis

Blast crises were considered myeloid when at least one of the anti-myeloid antibodies listed in Table 25.1 was positive and when no reactivity with anti-lymphoid antibodies could be detected. Within this group the blast cells of 12 patients expressed granulo-monocytic cell surface characteristics only, whereas in 26 other blast populations additional reactivities with monoclonal antibodies directed against erythroid and/or platelet antigens could be detected.

Within this subcategory, granulomono-megakaryocytic ($n = 20$), granulomono-erythrocytic ($n = 3$), and granulomono-erythromegakaryocytic ($n = 3$) reaction patterns could be distinguished.

Table 25.2 demonstrates the marker profiles of individual blast populations of the blast crises with erythrocytic and/or megakaryocytic involvement.

A considerable number (10–80%) of blast cells which expressed platelet determinants were found in the group with megakaryocytic involvement. In 17 of the 23 cases these thrombocyte antigen-positive presumptive platelet precursors remained morphologically unidentified. A remarkable overlap of granulo-monocytic and platelet markers could be found in two samples (patients 1 and 20). In the samples taken from five patients, blasts reactive with glycophorin A antibodies were observed. A clear-cut overlap of glycophorin A (LICR LON R10, LICR LON R18) and of thrombocyte GPIII (C17-27) positive cells was detectable in patient 25. In addition, this patient's blasts reacted with the myeloid markers MCS2 and MY9 (40 and 90%, respectively) but did not express the HLA-DR antigen.

Table 25.2. Immunological blast cell characteristics in patients with

Patient no.	Morph. classif.[a]	Immun. type[a]	VIM-D5	VIM2	VIM8	MCS2	MY9	63D3
1	G–MK	GM–MK	$-^c$	+++	+	++	nt	–
2	GM	GM–MK	+++	+++	+++	+++	nt	+++
3	U	GM–MK	–	+	++	+	nt	–
4	G	GM–MK	–	+	–	+++	nt	–
5	G–MK	GM–MK	++	++	++	+	nt	–
6	G	GM–MK	–	++	–	+++	nt	–
7	G	GM–MK	+	++	++	+++	nt	+
8	G–MK	GM–MK	++	++	nt	+	nt	–
9	G	GM–MK	–	–	–	+++	nt	–
10	G	GM–MK	–	+	–	+	nt	–
11	G	GM–MK	+	+	+	++	nt	–
12	G	GM–MK	+	+	+	++	+++	nt
13	G	GM–MK	–	++	+	+++	+++	+
14	GM	GM–MK	++	+++	++	+++	+++	++
15	U	GM–MK	–	–	–	–	+++	–
16	G–MK	GM–MK	–	+	nt	+++	nt	–
17	GM–MK	GM–MK	+	+++	++	+++	+++	++
18	MK	GM–MK	–	–	–	++	–	–
19	G	GM–MK	++	+++	++	++	nt	–
20	U	GM–MK	–	–	–	++	+++	–
21	GE	GM–E	+	++	++	+	nt	–
22	U	GM–E	–	–	–	–	+++	–
23	GE	GM–E	–	+	+	++	nt	–
24	GE	GM–E–MK	–	+	–	+	nt	–
25	U	GM–E–MK	–	–	+++	++	+++	–
26	GE–MK	GM–E–MK	+	+	+	+	+++	–

[a] G = Granulocytic; GE = granulo-Erythrocytic; GM–MK = granulomono-megakaryocytic; GM–E–MK = granulomono-erythro-megakaryocytic; GM = granulo-monocytic; G–MK = granulo-megakaryocytic; GM–E = granulomono-erithrocytic; U = undifferentiated.
[b] "L" = TdT and all T and B cell antigens listed in Table 25.1.

Lymphoid Blast Crisis

The marker profile in lymphoid blast crisis corresponded immunologically to the phenotype of acute lymphoblastic leukemia.

In 14 cases the blasts displayed the typical cALL type (VIL-A1[+], VIB-C5[+], HLA-DR[+], TdT[+], S/c IgM[−], "T"[−]), in three cases the blast populations were of pre-B type (VIL-A1[+], VIB-C5[+], HLA-DR[+], TdT[−], SIgM[−], $c\mu^+$, "T"[−]), and in one case of a "Null" ALL-type (VIL-A1[−], VIB-C5[+], HLA-DR[+], TdT[+], SIgM[−], $c\mu^-$, "T"[−]). In no case could myeloid antigens be detected on blast cells.

Blast Crisis with Mixed Myeloid and Lymphoid Populations

A mixed myeloid–lymphoid blast population could be demonstrated in five cases by double fluorescence staining, whereby the distinct popula-

myeloid blast crisis with megakaryocytic and/or erythropoietic involvement.

VIM-D2	HLA-DR	VIE-G4	R10 R18	AN51	J15	C8-13	C17-27	"L"[b]
−	++	−	−	−	+++	+++	+++	−
++	+++	−	−	+	+	nt	nt	−
−	++	−	−	+	+++	nt	nt	−
−	+++	−	−	−	+	+	+	−
−	++	−	−	++	++	++	++	−
−	+++	−	−	−	+	−	+	−
nt	+	−	−	−	+	+	nt	−
−	+	−	−	−	+	nt	nt	−
−	+++	−	nt	−	+	nt	nt	−
−	−	−	−	+	+	+	+	−
−	+++	−	−	−	+	nt	nt	−
−	+++	−	−	+	+	+	+	−
nt	+++	nt	−	++	++	+	++	−
nt	+++	nt	−	−	+	−	+	−
nt	+++	nt	−	−	+	−	+	−
nt	++	nt	−	nt	nt	nt	+	−
nt	+++	nt	−	+	+	+	+	−
−	−	−	−	−	++	+	+++	+++[d]
nt	+++	−	−	−	+	−	++	−
−	−	nt	−	+	+	+	+++	+++[e]
−	−	+	++	−	−	nt	nt	−
nt	−	−	+	−	−	−	−	+++[e]
−	+++	+	+	−	−	−	−	−
−	+	+	+	−	+	+	+	−
−	−	+	++	+	++	++	+++	+++[e]
nt	++	+	++	−	+	+	+	−

[c] Blasts staining: − = 0–10%; + = 10–20%; ++ = 20–50%; +++ = >50%; nt = not tested.
[d] Blasts staining with 9.6.
[e] Blasts staining with VIB-C5.

tion of the myeloid type did not overlap with that of the lymphoid type.

In four of these cases the lymphoid blasts displayed the typical cALL type (VIL-A1$^+$, VIB-C5$^+$, HLA-DR$^+$, TdT$^+$, S/cμ^-, "T"$^-$) and in one case the blasts were of c/T type (VIL-A1$^+$, VIB-C5$^-$, HLA-DR$^-$, TdT$^+$, S/cμ^-, WT-1$^-$).

Unclassifiable Blast Population

The blast cells of one patient presented a highly unusual marker profile: 80% of the blasts were HLA-DR positive, 90% reacted with the antibodies OKT11 and 9.6.

All other lymphoid markers tested were negative. Surprisingly, 20% of the blasts reacted with the platelet-specific antibodies J15, C8-13, and C17-27.

Phenotype Changes in the Course of the Disease

The blast cells of three of 23 patients examined showed phenotypic changes. In one patient the blasts changed from the lymphoid to the granulomono-erythro-megakaryocytic type while remaining TdT-positive; in the second patient the blasts switched from granulo-monocytic TdT-negative to granulo-monocytic TdT-positive; in the third patient glycophorin A-positive blasts appeared in the original granulomono-megakaryocytic blast population as the disease progressed.

Discussion

As established in acute leukemia, immunological cell typing can be useful in characterizing blast populations of CML patients in blast crisis. In our present study all 62 blast samples, with one exception, were classified with the monoclonal antibodies used.

The most intriguing result of this study was the unexpectedly frequent occurrence of blast cells with platelet and/or erythrocytic surface determinants in non-lymphoid blast crises. In 23 of 38 (61%) myeloid blast populations we found a considerable proportion (10–80%) of blast cells with platelet surface markers and in 6 of 38 blast populations (16%) we observed a substantial number of blast cells with the erythrocytic marker glycophorin A.

These data are in sharp contrast to the situation in acute leukemia. With the exception of two erythroleukemias, we have never observed blast cell reactivity with the glycophorin A antibody VIE-G4 (11). Similar results have also been reported by Greaves *et al.* (25) using the LICR LON R10 antibody. Blast cells with thrombocytic surface characteristics are also rarely seen in acute leukemias (2,4). At present we have observed only one acute leukemia with exclusive thrombocytic differentiation. In a second case, we found a minor admixture (10%) of megakaryocytic blasts with a majority of myeloid cells in the peripheral blood (unpublished observations).

Four different antibodies were employed to detect platelet determinants bearing blast cells (AN51, J15, C8-13, C17-27). As has recently been reported, platelet complexes GP Ib and 6P IIb/IIIa can also be demonstrated on monocytes (26). In our study, however, we find a positive reaction of blast cells with monocyte-specific markers (63D3, VIM-D2) only in 5 of 23 cases where the blasts reacted with the monoclonal antibodies mentioned above.

The difference in the encountered degree of involvement of the metakaryocytic and erythrocytic systems could probably be traced back to the fact that the platelet markers recognize also highly immature megakaryocytic precursor cells, whereas monoclonal antibodies directed against glycophorin A only detect erythroid precursors at a relatively late stage of differentiation (27).

The results of our investigations also suggest that there is no specific difference between the marker profile in lymphoid blast crisis and in acute lymphoblastic leukemia. The lymphoid blasts exclusively display lymphoid markers and express neither granulo-monocytic nor megakaryocytic or erythrocytic antigens.

Yet, double populations frequently occurred as compared to acute leukemia. In these five cases approximately equal proportions of the immature cells showed either myeloid or lymphoid characteristics. One possible explanation of this phenomenon is the partial maturation of pluripotential stem cells into these two different cell systems.

We also could observe the phenomenon of phenotypic transition in the course of the disease, a fact previously stated in other studies (28,29). Out of 23 blast crises tested, we found a change of the immunological phenotype in three cases as the disease progressed. This phenomenon, probably generated by therapeutic agents, also indicates the increased capacity of this particular process to differentiate along several pathways.

Summary

Blast cells from 62 patients with chronic myeloid leukemia in blast crisis (CML-BC) were immunologically phenotyped with a panel of 27 monoclonal antibodies and studied for terminal deoxynucleotidyl transferase (TdT) content. Out of 62 blast populations, 38 showed a myeloid, 18 a lymphoid, 5 a mixed, and 1 an unclassifiable marker profile.

By using monoclonal antibodies reactive with glycophorin A and platelet antigens, respectively, we encountered, in contrast to acute myeloid leukemia, a frequent involvement of erythropoietic (16%) as well as megakaryocytic (61%) systems in myeloid blast crisis. According to standard criteria platelet determinants bearing blasts were not identified morphologically in 17 of 23 cases.

The lymphoid marker profile of blast cells in lymphoid blast crisis of CML remained the same as in acute lymphoblastic leukemia.

In five cases the blast samples consisted of two distinct types of blast populations, one with myeloid features and the other with lymphoid ones. In the course of disease, three out of 23 tested cases showed a phenotypic change.

The data confirm the assumed involvement of the pluripotential stem cell in the process of CML with the capacity to differentiate along several pathways.

Acknowledgments. Monoclonal antibodies were kindly provided by D.J. Anstee (LICR LON R10, LICR LON R18), K. Forster (Y29/55), P.M.

Landsdorp (C8-13, C17-27, anti-IgM), J.D. Griffin (MY9), J. Minowada (MCS2), P.C. Kung (OKT antibodies), R.L. Evans (Leu antibodies), W.I.M. Tax (WT-1), and A.J. McMichael (Na1/34, AN51, J15). We wish to thank Mrs. Brigitte Fischer-Colbrie, Mrs. Susanne Beranek, and Mrs. Dagmar Mandl for their skillful technical assistance, Mag. Susanne Klöcker and Alexandra M. Karlhuber for their help in translating, and Mrs. Andrea Jager for typing the manuscript. This work was supported by Fonds zur Förderung der wissenschaftlichen Forschung in Österreich and Kamillo Eisner Stiftung, Switzerland.

References

1. Bettelheim, P., D. Lutz, O. Majdic, E. Paietta, O. Haas, W. Linkesch, E. Neumann, L. Lechner, and W. Knapp. 1985. Cell lineage heterogeneity in blast crisis of chronic myeloid leukaemia. *Brit. J. Haematol.,* **59**:395.
2. Bettelheim, P., E. Paietta, O. Majdic, H. Gadner, J. Schwarzmeier, and W. Knapp. 1982. Expression of a myeloid marker on TdT-positive acute lymphocytic leukemic cells: evidence by double-fluorescence staining. *Blood* **60**:1392.
3. Majdic, O., K. Liszka, D. Lutz, and W. Knapp. 1981. Myeloid differentiation antigen defined by a monoclonal antibody. *Blood* **58**:1127.
4. Majdic, O., P. Bettelheim, H. Stockinger, W. Aberer, K. Liszka, D. Lutz, and W. Knapp. 1984. M2, a novel myelomonocytic cell surface antigen and its distribution on leukemic cells. *Int. J. Cancer* **33**:617.
5. Knapp, W., O. Majdic, H. Stockinger, P. Bettelheim, K. Liszka, U. Köller, and C. Peschel. 1984. Monoclonal antibodies to human myelomonocytic differentiation antigens in the diagnosis of acute myeloid leukemia. *Med. Oncol. Tum. Pharmacother.,* **1**:257.
6. Sugimoto, T., E. Tatsumi, K. Takeda, K. Minato, K. Sagawa, and I. Minowada. 1984. Modulation of cell surface antigens induced by 12-*O*-tetradecanoylphorbol 13-acetate in two myeloblastic cell lines, a promyelocytic cell line, and a monoblastic cell line: Detection with five monoclonal antibodies. *J. Natl. Cancer Inst.* **72**:923.
7. Griffin, J.D., and S.F. Schlossman. 1984. Expression of myeloid differentiation antigens in acute myeloblastic leukemia. In: *Leucocyte typing,* A. Bernard, L. Boumsell, J. Dausset, C. Milstein, and S.F. Schlossman, eds. Springer-Verlag, Berlin, Heidelberg, p. 404–410.
8. Ugolini, V., G. Nunez, R.G. Smith, P. Stastny, and J.D. Capra. 1980. Initial characterization of monoclonal antibodies against human monocytes. *Proc. Natl. Acad. Sci. U.S.A.* **77**:6764.
9. Majdic, O., P. Bettelheim, K. Liszka, and D. Lutz. 1982. Leukämiediagnostik mit monoklonalen Antikörpern. *Wr. Klin. Wochenschr.* **94**:387.
10. Liszka, K., O. Majdic, P. Bettelheim, and W. Knapp. 1983. Glycophorin A expression in malignant hematopoiesis. *Am. J. Haematol.* **15**:219.
11. Anstee, D.J., and P.A.W. Edwards. 1982. Monoclonal antibodies to human erythrocytes. *Eur. J. Immunol.* **12**:228.
12. McMichael, A.J., N.A. Rust, J.R. Pilch, R. Sochynsky, J. Morton, D.Y.

Mason, C. Ruan, G. Tobelem, and J. Caen. 1981. Monoclonal antibody to human platelet glycoprotein I. *Brit. J. Haematol.* **49:**501.

13. Vainchenker, W., J.F. Deschamps, J.M. Bastin, J. Guichard, M. Titeux, J. Breton-Gorius, and A.J. McMichael. 1982. Two monoclonal antiplatelet antibodies as markers of human megakaryocyte maturation: immunofluorescent staining and platelet peroxidase detection in megakaryocyte colonies and *in vivo* cells from normal and leukemic patients. *Blood* **59:**514.

14. Tetteroo, P.A.T., P.M. Lansdorp, O.C. Leeksma, and A.E.G.K. von dem Borne. 1982. Monoclonal antibodies against human platelet glycoprotein IIIa. *Brit. J. Haematol.* **55:**509.

15. Tetteroo, P.A.T., F. Massaro, A. Muder, R. Schreuder-van Gelder, and A.E.G.K. von dem Borne. 1984. Megakaryoblastic differentiation of pro-erythroblystic K562 cell-line cells. *Leuk. Res.,* **8:**197.

16. Vodinelich, L., W. Tax, Y. Bai, S. Pegram, P. Capel, and M.F. Greaves. 1983. A monoclonal antibody (WT1) for detecting leukemias of T-cell precursors (T-ALL). *Blood* **62:**1108.

17. Reinherz, E.L., P.C. Kung, G. Goldstein, and S.F. Schlossman. 1979. A monoclonal antibody with selective reactivity with functionally mature human thymocytes and all peripheral human T cells. *J. Immunol.* **123:**1312.

18. Kung, P.C., M.A. Talle, M. DeMaria, M. Butler, J. Lifter, and G. Goldstein. 1980. Strategies for generating monoclonal antibodies defining human T lymphocyte differentiation antigens. *Transplant. Proc.* **12**(suppl.):141.

19. Kamoun, M., P.J. Martin, J.A. Hansen, M.A. Brown, A.W. Siedak, and R.C. Nowinski. 1981. Identification of a human T lymphocyte surface protein associated with the E-rosette receptor. *J. Exp. Med.* **153:**207.

20. McMichael, A.J., J.R. Rilch, G. Galfre, D.Y. Mason, J.W. Fabre, and C. Milstein. 1979. A human thymocyte antigen defined by a hybrid myeloma monoclonal antibody. *Eur. J. Immunol.* **9:**205.

21. Ledbetter, J.A., R.L. Evans, M. Lipinski, C. Cunningham-Rundles, R.A. Good, and L.A. Herzenberg. 1981. Evolutionary conservation of surface molecules that distinguish T lymphocyte helper/inducer and T cytotoxic/suppressor subpopulations in mouse and man. *J. Exp. Med.* **153:**310.

22. Evans, R.L., D.W. Wall, C.D. Platsoucas, F.P. Siegal, S.M. Fikrig, C.M. Testa, and R.A. Good. 1981. Thymus-dependent membrane antigens in man: Inhibition of cell-mediated lympholysis by monoclonal antibodies to Th_2 antigens. *Proc. Natl. Acad. Sci. U.S.A.* **78:**544.

23. Knapp, W., O. Majdic, P. Bettelheim, and K. Liszka. 1982. VIL-A1, a monoclonal antibody reactive with common acute lymphatic leukemia cells. *Leuk. Res.* **6:**137.

24. Forster, K.K., F.G. Gudat, Marie F. Girard, Renate Albrecht, J. Schmidt, C. Ludwig, and J.-P. Obrecht. 1982. Monoclonal antibody against a membrane antigen characterizing leukemic human B-lymphocytes. *Cancer Res.* **42:**1927.

25. Greaves, M.F., C. Sieff, and P.A.W. Edwards. 1983. Monoclonal anti-glycophorin as a probe for erythroleukemias. *Blood* **61:**645.

26. Bai, Y., H. Durbin, and N. Hogg. 1984. Monoclonal antibodies specific for platelet glycoproteins react with human monocytes. *Blood* **64:**139.

27. Robinson, J., C. Sieff, D. Delia, P.A.W. Edwards, and M. Greaves. 1981. Expression of cell-surface HLA-DR, HLA-ABC and glycophorin during erythroid differentiation. *Nature* **289:**68.

28. Hughes, A., B.A. McVerry, H. Walker, K.F. Bradstock, A.V. Hoffbrand, and G. Janossy. 1981. Heterogeneous blast cell crisis in Philadelphia negative chronic granulocytic leukaemia. *Brit. J. Haematol.* **47**:563.
29. Greaves, M.F. 1982. Target cells differentiation and clonal evolution in chronic granulocytic leukemia: a 'model' for understanding the biology of malignancy. In: *Chronic granulocytic leukemia.* M.T. Shaw, ed. Praeger, New York, p. 15–47.

CHAPTER 26

Reactivity of a Monoclonal Antibody Defining Human Hematopoietic Multipotential Progenitors

Fay E. Katz, Robert W. Tindle, Robert Sutherland, and Melvyn F. Greaves

The cell surface antigenic phenotype of human hematopoietic progenitor cells can be mapped using a combination of monoclonal antibodies and either cell sorting (1–3) or C'-mediated cytotoxicity (2,4). A number of investigators have shown that all types of human progenitor cells carry the HLA-DR (6–8) and HLA-ABC (6) determinants, and the leukocyte common antigen (5), whereas the majority of these cells appear to lack the A and B blood group antigens (9), the common ALL antigen (10), and the 40-Kd glycoprotein recognized by the anti–T cell monoclonal antibody, WT1 (11). The multipotential or clonal progenitor cells constitute only 1–2% of normal bone marrow cells and are therefore difficult to identify either morphologically or by functional assays. Attempts to enrich for them have relied on methods such as density gradient separation (12) or binding to lectins (13). Clonal assays and long term *in vitro* culture systems are now available for the identification and quantitation of unipotent and multipotent hematopoietic progenitors (14–17), and recently several monoclonal antibodies have been described which react with all measurable types of human progenitor cells (2,18). One of these antibodies, RFB-1, has been used successfully in multi-parameter cell sorting to enrich the progenitor cell population in human bone marrow 90–150-fold (19). There is, however, still no antibody available which is restricted in its reactivity to the multipotential progenitor or stem cell alone.

In this report, we describe a monoclonal antibody B1.3.C5 raised against the acute myelogenous leukemia cell line KG-1, which identifies a glycoprotein of 110–120 Kd on SDS–PAGE, and which by its reactivity appears to recognize an antigen expressed on early hematopoietic progenitors.

Materials and Methods

Cell Lines, Leukemia Samples, and Normal Cells

Established hematopoietic and other cell lines were obtained from a variety of sources and cultured as described previously (20). Cells from patients with leukemia or other hematological disorders were obtained as part of a general diagnostic screening program operating in this laboratory for hospitals throughout the U.K. Normal bone marrow was obtained from adult volunteers or bone marrow transplant donors, and was aspirated into Iscove's medium containing preservative-free heparin. Following separation on 60% Percoll, the mononuclear cells were washed twice and depleted of the majority of monocytes and macrophages by adherence to plastic overnight in Iscove's medium and 20% FCS. T cell depletion was performed by E-rosetting using neuraminidase or AET-treated sheep red blood cells (SRBC) and separation on 60% Percoll.

Hybridoma Production

The monoclonal antibody B1.3.C5 was raised in mice immunized with the acute myelogenous leukemia cell line KG1, as described previously (21). It is of the IgG_1 subclass and does not fix complement.

Cell Sorting and Morphological Identification

$30-80 \times 10^6$ bone marrow mononuclear cells were labeled with saturating concentrations of B1.3.C5 ascites and incubated for 30 min on ice. The cells were washed twice and counterstained with a FITC $F(ab')_2$ goat anti–mouse Ig for a further 30 min at 4°C, washed twice, and separated on a FACS I (Becton Dickenson) under sterile conditions into positive, gap (weakly positive), and negatively staining cells. Control bone marrow cells received second antibody alone.

Cytospin preparations were made from control and sorted fractions, and stained with May-Grunwald-Giemsa. Differential counts were performed on 100 cells per smear.

Colony Assays—CFU-GEMM Assay*

The fractionated cells plus the stained unseparated control were resuspended at 3×10^6/ml in Iscove's medium and cultured in a mixture of Iscove's medium, 10% PHA-LCM, 10% FCS, 5×10^{-5} M 2ME,

---

* Abbreviations used in this paper: CFU-GEMM, colony-forming unit, granulocyte, erythroid, megakaryocyte, monocyte; CFU-GM, colony-forming unit granulocyte/macrophage; BFU-E, burst-forming unit erythroid; CFU-meg, colony-forming unit megakaryocyte; 2ME, 2-mercaptoethanol; PHA-LCM, phytohemagglutinin–leukocyte conditioned medium.

20% human plasma, and methylcellulose at a final concentration of 0.8%. 1 u/ml erythropoietin (Terry Fox Labs., Vancouver) was added and duplicate 1-ml aliquots of culture mixture plated. The plates were incubated for 12–14 days at 37°C in a humidified atmosphere with 5% CO_2, 95% air, and colonies were scored on day 13–14.

Lymphoid Colony Assay

The fractionated cells plus control were resuspended in Iscove's medium at 6×10^6/ml, and cultured in Iscove's medium supplemented with 20% FCS, 5×10^{-5} M 2ME, 20% T cell conditioned medium (PHA-TCM), and methylcellulose at a final concentration of 0.8%. Normal T (E^+) cells (mitomycin C-treated) were added as feeders at a concentration of 9×10^6/ml. Replicate 0.1 ml aliquots were plated in 96-well flat-bottomed Linbro plates and incubated at 37°C in high humidity and 5% CO_2, 95% air for 7 days. After 7 days of culture, colonies containing more than 20 cells were counted. 20–200 colonies from each fraction were pooled, the cells washed three times, and stained with a selection of monoclonal antibodies followed by FITC F(ab')$_2$ goat anti–mouse Ig. Stained cells were fixed in 95% ethanol and examined using a fluorescence microscope.

Biochemical Analysis

Surface labeling of KG1 cells with ^{125}I and lactoperoxidase was carried out using standard methods as described previously (22). Immune complexes were analyzed using standard SDS–PAGE methods, and visualized by autoradiography (23).

Affinity Purification and Blocking Experiments

B1.3.C5 ascites fluid was purified by affinity chromatography on a column of goat anti–mouse Ig covalently bound to Sepharose 4B. The purified antibody had a protein concentration of 0.8 mg/ml. To test the hypothesis that B1.3.C5 might recognize a hematopoietic growth factor or its receptor, mononuclear bone marrow cells were treated with an excess of affinity-purified antibody over a 5-point concentration range (0.35–6 μg/ml) for 30 min at 4°C, and then plated immediately without washing in the CFU-GEMM assay. Affinity-purified W6/32 antibody (anti-HLA-ABC monomorphic)-treated bone marrow cells were used as a control.

Results

Antibody Reactivity and Characterization

Monoclonal antibody B1.3.C5 showed very restricted reactivity, reacting with only 3–8% cells in normal bone marrow, but not with PBLs or lymphoid tissue. It was reactive with the majority of acute myeloblastic leukemias, null ALLs, and acute lymphoblastic leukemias, but unreactive with the more "mature" leukemias and lymphomas, e.g., CLL, B-ALL, T-ALL, or CGL. With the exception of the immunizing cell line KG1 and its variant KG1-a (to which it bound strongly), B1.3.C5 was unreactive with all other hematopoietic cell lines tested (both myeloid and lymphoid); nor did it show any reactivity on a variety of other malignant non-hematopoietic cell lines.

Cell Sorting and Reactivity with Hematopoietic Progenitors

The reactivity of B1.3.C5 was assessed by separating on the FACS normal bone marrow mononuclear cells stained with antibody into positive (5–14%), weakly positive (30–35%), and negatively (50–60%) staining cells (Fig. 26.1). Morphologically, the positive fraction consisted mainly of large and small lymphocytes and blasts. The more mature myeloid precursors were found in the weak positive or negative fraction, as were the normoblasts (Table 26.1).

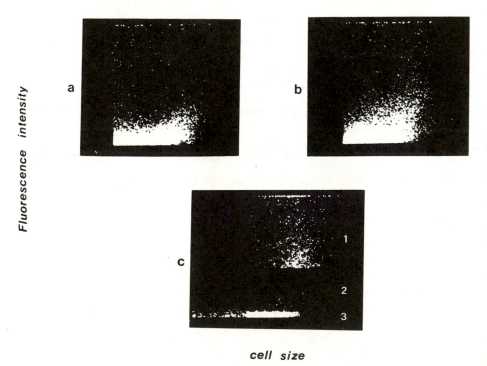

Fig. 26.1. FACS profile of B1.3.C5-labeled normal bone marrow mononuclear cells (b) separated on a FACS I into positive (1c), weakly positive (2c), and negative (3c) fractions compared to the control (a).

Table 26.1. Morphological composition of FACS-sorted B1.3.C.5 positive, weakly positive, and negative cells, compared to the control population.

	Normo-blasts	Eosinophils	Promyelocytes, metamyelocytes	Lymphocytes	Blasts
Positive[a]	8%	9%	27%	26%	30%
Weak positive	7%	13%	71%	8%	1%
Negative	66%	nc	18%	2%	18%
Control (stained/ unseparated)	25%	14%	40%	20%	2%

[a] TdT: 3% in positive fraction, 0% in other fractions.

By far the greater majority of colony-forming cells, that is, CFU-GEMM, BFU-E, CFU-GM, and CFU-meg were found in the positive fraction, with a variable percentage of erythroid colonies in the weakly positive fraction. Table 26.2 lists the results of three experiments, where the mean colony counts per 10^5 cells plated are expressed as a percentage of the control, calculated as exampled below:

$$\frac{\%\text{B1.3.C5 progenitors in original population}} = \frac{\begin{array}{c}\text{colonies}/10^5 \text{ cells} \\ \text{in positive fraction}\end{array} \times \begin{array}{c}\% \text{ cells sorted} \\ \text{into positive fraction}\end{array}}{\begin{array}{c}\text{Colonies}/10^5 \text{ cells in stained} \\ \text{unseparated control}\end{array}}$$

In contrast, virtually all the lymphoid colony-forming cells, both T and B cells, were found in the negative fraction (Table 26.3). However, this assay probably does not detect or permit the growth of lymphoid precursor cells which are present in levels of less than 1–3% in the B1.3.C5 positive fraction, and can be identified by the nuclear enzyme terminal deoxynucleotidyl transferase (TdT).

Biochemical Characterization of the Structure Recognized by B1.3.C5 and Blocking Experiments

KG1 cells labeled with ^{125}I and lactoperoxidase were incubated with B1.3.C5 ascites. Immunoprecipitation with *S. aureus* was carried out, and the immune complexes analyzed on SDS–PAGE and by autoradiography. Under both reducing and nonreducing conditions, the antigen detected by B1.3.C5 appeared to be a single polypeptide of 110–120 Kd (Fig. 26.2).

Blocking Experiments

These showed that, even using a 10-fold excess (6 μg/ml) of purified antibody, there was no significant effect of antibody on colony growth compared to the control (data not shown).

Table 26.2. Reactivity of B1.3.C.5 with hematopoietic colony-forming cells.

Fraction	% Cells in fraction	Colony-forming cells recovered/10^5 cells plated[a]			
		CFUc	BRU-E	CFU-meg	CFU-GEMM
Bone marrow					
Positive	8	401	125	18	11
		(22%)	(41%)	(26%)	(39%)
Weak positive	30	106	78	3	0.7
		(13%)	(27%)	(8%)	(3%)
Negative	57	8	4	0	0
		(3%)	(7%)	(0%)	(0%)
Control		136	25	4	3.5
Fetal liver					
Positive	7	256	38	8	6.5
		(24%)	(38%)	(28%)	(91%)
Weak positive	30	35	2	0	0
		(7%)	(4%)	(0%)	(0%)
Negative	63	1.5	0	0	0
		(1.2%)	(0%)	(0%)	(0%)
Control		76	7	2	0.3

[a] The results of three experiments are shown, where the mean colony counts per 10^5 cells plated are expressed as a percentage of the control, calculated as shown below:

$$\frac{\% \text{ B1.3.C.5 positive progenitors in original population}}{} = \frac{\frac{\text{Colonies}/10^5 \text{ cells}}{\text{in positive fraction}} \times \frac{\% \text{ cells sorted}}{\text{into positive fraction}}}{\frac{\text{Colonies}/10^5 \text{ cells in stained}}{\text{unseparated control}}}.$$

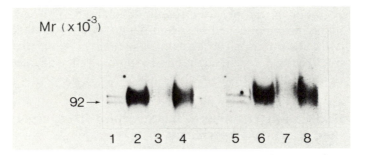

Fig. 26.2. Autoradiograph of SDS–PAGE analysis of human immune complexes isolated with monoclonal antibody B1.3.C5. Tracks 1 and 5, ^{14}C-labeled standard molecular weight markers. Tracks 2 and 6, immune complexes from KG1 lysates. Tracks 3 and 7, immune complexes from Ph' chromosome-positive leukemic blast cells (Patient 1). Tracks 4 and 8, immune complexes from Ph' chromosome-positive leukemic blast cells (Patient 2). Tracks 1–4 run under reducing conditions, and tracks 5–8 under nonreducing conditions.

Table 26.3. Reactivity of B1.3.C.5 with lymphoid colony-forming cells.[a]

| Fraction | Lymphoid colonies/ 10⁵ cells plated | Colony phenotype | |
		OKTIIA positive[b]	SmIg positive[c]
Control			
(stained/unseparated)	70	61%	41%
Positive fraction	89	44%	57%
	(6.3%)		
Negative fraction	267	73%	16%
	(100%)		

[a] The results of two experiments are shown, where the mean colony counts per 10⁵ cells plated are expressed as a percentage of the control, calculated as exampled in the legend to Table 26.2.
[b] Pan T cell marker.
[c] Pan (mature) B cell marker.

Discussion

Our data show that a monoclonal antibody, B1.3.C5, raised against the myeloid leukemia line KG1 identifies a 110–120-Kd glycoprotein which appears to be preferentially expressed on leukemias of very immature phenotype and 3–8% of normal bone marrow cells. Morphologically, the cells in the bone marrow which carry the B1.3.C5 antigen are lymphoid or blast cell-like; the more mature cells of the myeloid lineage and the red cell precursors are unreactive. Following FACS separation of B1.3.C5 positive from negative cells (either bone marrow or fetal liver), the majority of colony-forming cells of all types (with the exception of lymphoid) were found in the positive fraction. The similarity of these data to those recently published by Civin et al. (18) for their antibody MY10 is very evident. Both antibodies react with 3–8% of cells in normal bone marrow, but are unreactive with PBLs. All colony-forming progenitor cells are found in the MY10 positive fraction. Both antibodies detect a glycoprotein of 110–120 Kd, but whether or not they both recognize the same structure remains to be determined.

A number of other monoclonal antibodies have been described which also show preferential reactivity for bone marrow hematopoietic progenitors. A series of antibodies produced against the erythroleukemia line K562 were able to effectively inhibit CFU-GM colony growth, but only one of these also inhibited the growth of BFU-E and was reactive with approximately 3% of bone marrow cells (24).

Papayannopoulou et al. (25) recently described the properties of another series of antibodies produced against a second erythroleukemia cell line, HEL, which also displays multi-lineage surface markers. Their most specific anti–progenitor cell antibody reacted with 4–8% of bone marrow cells and was present on the bulk of the myeloid and erythroid progeni-

tors. However, as it was unreactive with the KG1 cell line and precipitated a protein of 24 Kd, it probably recognizes another surface structure
which is also preferentially expressed on hematopoietic progenitors. Antibody RFB-1 binds to all colony-forming cells (2), but also reacts with
marrow T cells and T cell precursors (26). However, by labeling bone
marrow (depleted of both T cells and mature myeloid cells) with RFB-1,
followed by two-parameter FACS sorting, Bodger *et al.* (19) have
achieved a 150-fold enrichment of CFU-GEMM. Using the B1.3.C5 antibody and sorting on the basis of fluorescence intensity alone, we can
achieve a similar considerable enrichment of CFU-GEMM, as optimal
numbers of multipotential colonies are seen if the B1.3.C5 positive cells
are plated at $1-2 \times 10^4$ cells/ml, at which level the assay is thought to be
truly clonal (27).

In summary, the B1.3.C5 antibody could be used to produce enriched
populations of human pluripotent progenitor cells either as potential immunogens to generate further monoclonal antibodies to stem cell-related
antigens, or to investigate the role of growth factors which regulate progenitor cell development. Yet another possibility would be its use clinically to transplant lymphocyte-depleted, progenitor cell-enriched bone
marrow in allogeneic transplants.

Summary

The development of *in vitro* clonal assays and long-term culture systems
for human unipotent and pluripotent hematopoietic progenitor cells has
permitted the study of these infrequent cell types in both normal and
abnormal hematopoiesis. The cell surface antigenic phenotype of these
cells can be mapped using antibodies combined with cell sorting, and
purified populations of progenitor cells can be obtained in this way. The
monoclonal antibody B1.3.C5 was raised by immunizing mice with the
acute myelogenous leukemia cell line KG1. It is of the IgG_1 subclass, does
not fix complement, and recognizes a glycoprotein of 110–120 Kd.
B1.3.C5 reacted only with the leukemias of immature phenotype, and was
unreactive with CLL, B-ALL, T-ALL, lymphomas, or normal PBLs.
Only 3–8% of normal bone marrow cells reacted with the antibody, and it
was this small B1.3.C5 positive population which contained all the measurable hematopoietic colony-forming cells (excluding lymphoid). The
results suggest that B1.3.C5 could be successfully used to isolate purified
populations of progenitor cells to study the role of growth factors in
progenitor cell development, or provide stem cell-enriched fractions for
allogeneic bone marrow transplantation.

Acknowledgments. We are very grateful to Grace Lam and Laura Davis
for operating the FACS, to Dr. Jennifer Treleaven and Tracy Wallis for

typing the manuscript, and Dr. Hilary Blacklock for providing bone marrow samples.

References

1. Sieff, C., D. Bicknell, G. Caine, J. Robinson, G. Lam, and M.F. Greaves. 1982. Changes in cell surface antigen expression during haemopoietic differentiation. *Blood* **60**:703.

2. Bodger, M.P., C.A. Izaguirre, H.A. Blacklock, and A.V. Hoffbrand. 1982. Surface antigenic determinants on human pluripotent and unipotent haemopoietic progenitor cells. *Blood* **61**:1006.

3. Janossy, G., G.E. Francis, D. Capellaro, A.H. Goldstone, and M.F. Greaves. 1978. Cell sorter analysis of leukaemia associated antigens on human myeloid progenitors. *Nature* **276**:176.

4. Andrews, R.G., B. Torok-Storb, and I.D. Bernstein. 1983. Myeloid-associated differentiation antigens on stem cells and their progeny identified by monoclonal antibodies. *Blood* **62**:124.

5. Beverley, P.L.C., D. Linch, and D. Delia. 1980. Isolation of human haemopoietic progenitor cells using monoclonal antibodies. *Nature* **287**:332.

6. Robinson, J., C. Sieff, D. Delia, P.A.W. Edwards, and M.F. Greaves. 1981. Expression of cell surface HLA-DR, HLA-ABC and glycoprotein during erythroid differentiation. *Nature* **289**:68.

7. Falkenberg, J.H.F., J. Jansen, N. van der Vaast-Duinkerken, W.J.F. Veerhof, J. Blotkamp, H.M. Goselink, J. Parlevliet, and J.J. van Rood. 1984. Polymorphic and monomorphic HLA-DR determinants on human haemopoietic progenitor cells. *Blood* **63**:1125.

8. Linch, D., L.M. Nadler, E.A. Luther, and J.M. Lipton. 1984. Discordant expression of human Ia-like antigens on haemopoietic progenitor cells. *J. Immunol.* **132**:2324.

9. Blacklock, H.A., F. Katz, R. Michalevicz, G.P. Hazelhurst, H.G. Prentice, and A.V. Hoffbrand. 1984. A and B blood group antigen expression on mixed colony cells and erythroid precursors: relevance for human allogeneic bone marrow transplantation. *Brit. J. Haematol.* **58**:267.

10. Clavell, L.A., J.M. Lipton, R.C. Bast, M. Kudish, J. Pesando, S.F. Schlossman, and J. Ritz. 1981. Absence of common ALL antigen on normal bipotent myeloid, erythroid and granulocyte precursors. *Blood* **58**:333.

11. Myers, C.D., P.E. Thorpe, W.J.C. Ross, A. Cumber, F.E. Katz, W.J. Tax, and M.F. Greaves. 1984. An immunotoxin with therapeutic potential in T cell leukaemia: WT1-Ricin A. *Blood* **63**:1178.

12. Francis, G.E., M.A. Wing, J.J. Berney, and J.E.T. Guimaraes. 1983. CFU-GEMM, CFU$_{Gm}$, CFU$_{Mk}$: analysis by equilibrium density centrifugation. *Exp. Haematol.* **11**:481.

13. Morstyn, G., N.A. Nicola, and D. Metcalf. 1980. Purification of haemopoietic progenitor cells from human marrow using a fucose-binding lectin and cell sorting. *Blood* **36**:798.

14. Fauser, A.A., and H.A. Messner. 1978. Granuloerythropoietic colonies in human bone marrow, peripheral blood and cord blood. *Blood* **52**:1243.

15. Fauser, A.A., and H.A. Messner. 1979. Identification of megakaryocytes,

macrophages, and eosinophils in colonies of human bone marrow containing neutrophilic granulocytes and erythroblasts. *Blood* **53**:1023.

16. Izaguirre, C.A., M.D. Minden, A.F. Howatson, and E.A. McCulloch. 1980. Colony formation by normal and malignant human B-lymphocytes. *Brit. J. Cancer* **42**:430.

17. Messner, H.A., N. Jamal, and C.A. Izaguirre. 1982. The growth of large megakaryocyte colonies from human bone marrow. *J. Cell. Physiol.* (Suppl.) **1**:45.

18. Civin, C.I., L.C. Strauss, C. Brovall, M.J. Fackler, J.F. Schwartz, and J.H. Shaper. 1984. Antigenic analysis of haematopoiesis, III. A haemopoietic progenitor cell surface antigen defined by a monoclonal antibody raised against KG1-a cells. *J. Immunol.* **133**:157.

19. Bodger, M.P., I.M. Hann, R.F. Maclean, and M.J. Beard. 1984. Enrichment of pluripotent haemopoietic progenitor cells from human bone marrow. *Blood* **64**:774.

20. Greaves, M.F., D. Delia, R. Sutherland, J. Rao, W. Verbi, J. Kemshead, G. Hariri, G. Goldstein, and P. Kung. 1981. Expression of the OKT monoclonal antibody defined antigenic determinants in malignancy. *Int. J. Immunopharmacol.* **3**:283.

21. Tindle, R.W., R.A.B. Nichols, L.C. Chan, D. Campana, D. Catovsky, and G.D. Birnie. 1985. A monoclonal antibody which recognises lymphoid/myeloid precursors of potential use in the characterisation of 'mixed phenotype' acute leukaemias. *Leuk. Res.* **9**:1.

22. Sutherland, D.R., J. Smart, P. Niaudet, and M. Greaves. 1978. Acute lymphoblastic leukaemia associated antigens II. Isolation and partial characterisation. *Leuk. Res.* **2**:115.

23. Sutherland, D.R., C.E. Rudd, and M.F. Greaves. 1984. Isolation and biochemical characterisation of a human T cell associated glycoprotein. *J. Immunol.* **133**:327.

24. Young, N.S., and S.P. Hwang-Chen. 1981. Anti-K562 cell monoclonal antibodies recognise haemopoietic progenitors. *Proc. Natl. Acad. Sci. U.S.A.* **78**:7073.

25. Papayannopoulou, Th., M. Brice, T. Yokochi, P.S. Rabinovitch, D. Lindsley, and G. Stamatayannopoulos. 1984. Anti–HEL cell monoclonal antibodies recognise determinants that are also present in haemopoietic progenitors. *Blood* **63**:326.

26. Bodger, M.P., G.E. Francis, D. Delia, S.M. Granger, and G. Janossy. 1981. A monoclonal antibody specific for immature human haemopoietic cells and T lineage cells. *J. Immunol.* **127**:2269.

27. Powell, J.S., P.J. Fialkow, and J.W. Adamsŏn. 1984. Human mixed colonies: unicellular or multicellular origin-analysis by G-6-PD. *Brit. J. Haematol.* **57**:89.

CHAPTER 27

Expression of an Acute Myelogenous Leukemia-Associated Antigen (NHL-30.5) on Immature Leukemic Cells

David Askew, Allen C. Eaves, and Fumio Takei

Established human acute myelogenous leukemia cell lines are generally believed to retain the overall phenotype of the leukemic blast cells from which they were derived. Most of these lines can be induced to express a variety of differentiated markers using a number of inducing agents. These observations have been used to support the interpretation that leukemic blast cells are arrested in normal differentiation programs (1), although in some cases the leukemic cells have been reported to differ phenotypically from their normal counterparts (2,3).

We have previously reported the initial characterization of a monoclonal antibody (NHL-30.5) reactive with cells from patients with acute myelogenous leukemia (4). In this report we summarize the reactivity of the antibody with various patient categories and with myeloid leukemic cell lines induced to differentiate *in vitro*.

Materials and Methods

Monoclonal Antibodies

The production of the NHL-30.5 monoclonal antibody has been described elsewhere (4). NHL-62.14 is an anti–transferrin receptor antibody. The specificity of NHL-62.14 was confirmed by immunoprecipitation of the antigen and by its reactivity with proliferating cells.

Preparation of Human Cells

Heparinized peripheral blood and marrow aspirates were obtained with informed consent from patients with various hematologic malignancies. Marrow cells from transplant donors and lymphoma or Hodgkins patients (with noninfiltrated marrows) were used as normal controls. Marrow cells were spun at 800 × g for 4 min and the buffy coat treated with ammonium chloride to remove residual red cells. Peripheral blood mononuclear cells were obtained by centrifugation of 1.077 g/ml Ficoll–Hypaque (LSM, Litton Bionetics, Kensington, MD) or 1.077 g/ml Percoll (Pharmacia, Uppsala, Sweden).

Cell Fractionation

Granulocytes, lymphocytes, monocytes, and erythrocytes were purified from the peripheral blood of healthy volunteers as previously described (4).

Cell Lines

The promyelocytic leukemia cell line HL-60 (5) was supplied by Dr. J. Levy (Dept. of Microbiology, University of British Columbia, Vancouver, B.C.). The Ph^l-positive cell line K562 (6) was obtained from Dr. B.B. Lozzio (Dept. of Medical Biology, University of Tennessee, Knoxville, TN), and the myeloblastic leukemia line KG-1 (7) was obtained from Dr. H.P. Koeffler (Dept. of Medicine, University of California, Los Angeles). The B cell lymphoma cell lines SU-DHL-1, 4, 6, 8, and 10 (8) were obtained from Dr. D.R. Howard (Terry Fox Laboratory, B.C. Cancer Research Centre) and the T-cell line Jurkat from Dr. D. Kilburn (Dept. of Microbiology, University of British Columbia). All lines were incubated in an atmosphere of 5% CO_2 at 37°C in medium containing 10% FCS.

FACS Analysis

Cells were stained for FACS analysis as previously described (4) and analyzed on a FACS IV (Becton Dickinson, Sunnyvale, CA). The instrument was routinely standardized using glutaraldehyde-fixed chicken red blood cells and fluorescent microspheres. Debris was gated out on the basis of light scatter.

Differentiation of HL-60/KG-1 Cells Induced by DMSO (9) or Retinoic Acid (10)

Cells were cultured in plastic tissue culture flasks at a seeding concentration of 2 × 10^5 cells/ml in DMEM (HL-60) or MEM (KG-1) containing 10% FCS and either 1.25% DMSO or 1 × 10^{-6} M retinoic acid (RA). The cultures were incubated at 37°C for 5 days. Treated and control cultures showed >80% viability by Trypan Blue exclusion. On days 1, 3, and 5, the cells were labeled for FACS analysis with NHL-30.5 and FITC-conjugated rabbit F(ab')$_2$ anti–mouse Ig, followed by an incubation for 5 min with 25 μ/ml of propidium iodide. Dead cells were then gated out from the analysis on the basis of propidium iodide staining. Morphological and cytochemical assessment of the extent of differentiation was conducted independently.

Differentiation of HL-60/KG-1 Cells Induced by TPA (11)

Cells were cultured in 100-mm diameter tissue culture dishes at a seeding concentration of 2×10^5 cells/ml in medium containing 10% FCS and 1.6 $\times 10^{-8}$ M TPA. Cells were harvested on days 1 and 2 for FACS analysis. Nonadherent cells were removed, and the adherent cells treated with saline containing 0.2% EDTA and 0.1% BSA. The strongly adherent cells were gently removed with a rubber policeman.

Immunoprecipitation

The iodogen procedure was used to label cell surfaces, and the antigen precipitated as previously described (4). The immunoprecipitate was then analyzed by SDS–PAGE.

Results

Reactivity of NHL-30.5 with Human Cells

The reactivity of NHL-30.5 with a variety of normal and malignant hematopoietic cells was tested by fluorescent staining. Samples were reacted with the antibody and FITC-conjugated rabbit $F(ab')_2$ anti–mouse Ig and then analyzed on the fluorescence-activated cell sorter (FACS). Table 27.1 shows the percent positive cells determined by FACS analysis for all AML patients tested to date, together with the percent blasts in each sample and the FAB classification for each patient. A leukemia was classified as antigen positive if >10% of the cells stained with antibody. Thus, 21/25 AML patients were considered NHL-30.5 positive. All patient samples that reacted with the antibody had at least 10% blasts, although there was no correlation between the percent blasts and the percent positive cells.

Table 27.2 indicates the reactivity of the antibody with normal hematopoietic cell populations: bone marrow, granulocytes (>90% granulocytes by morphology), monocytes (>80% positive for nonspecific esterase), lymphocytes, PHA-stimulated lymphocytes, erythrocytes (ABO), splenocytes, and platelets. None of these showed any staining above 2% positive.

Marrow cells from patients with various other hematologic malignancies were also examined for their reactivity with NHL-30.5. The results of these tests are summarized together with the AML data in Table 27.3.

Table 27.1. Reactivity of NHL-30.5 monoclonal antibody with cells from patients with AML.

AML patient number	FAB classification	% Blasts		FACS analysis % Positive	
		PB	BM	PB	BM
1	M1	98	97	49	51
2	M1	82	94	26	55
3	M1	63	71	16	NT
4	M1	58	81	66	NT
5	M1	21	35	15	NT
6	M2	51	60	53	46
7	M2	67	32	21	21
8	M2	6	20	53	11
9	M2	29	71	NT	12
10	M2	38	41	20	46
11	M2	86	89	42[a]	NT
12	M4	72	71	36	38
13	M4	42	47	64	26
14	M4	38	49	36	29
15	M4	8	33	26	21
16	M4	22	82	27	17
17	M4	57	72	58	NT
18	M4	88	NT	70	NT
19	M4	94	79	70[a]	NT
20	M4	77	78	64	51
21	M5	84	82	58	29
22[b]	M2	9	32	2	NT
23	M3	Occ.	9	<1	NT
24	M4	67	91	<1[a]	NT
25	M5	86	87	<1	NT

[a] Leukapheresis samples.
[b] Relapsed 11 months later with 46% and 18% positive cells in peripheral blood and bone marrow, respectively.

Table 27.2. Reactivity of NHL-30.5 with normal hematopoietic cells.

Cells	% Positive (FACS)
Normal marrow	<2
Granulocytes	<1
Monocytes	<1
Lymphocytes	<1
PHA-stimulated lymphocytes	<1
Erythrocytes	<1
Spleen cells	<1
Platelets	<2

Table 27.3. Summary of results of testing blood and/or marrow cells from various patient categories with NHL-30.5 monoclonal antibody.

Diagnosis	Number tested	Number positive
AML M1	5	5
AML M2	7	6
AML M3	1	0
AML M4	10	9
AML M5	2	1
Chronic myelo-monocytic leukemia	1	1
Myelofibrosis[a]	1	1
Acute lymphoblastic leukemia	10	1
Chronic myeloid leukemia (blast crisis)	3	1
AML in remission	5	0
Acute lymphoblastic leukemia in remission	3	0
Chronic myelogenous leukemia (chronic phase)	10	0
Polycythemia vera	1	0
Multiple myeloma	1	0
Myelodysplasia	3	0
Chronic lymphocytic leukemia	2	0
Normal bone marrow	10	0

[a] A 66-year-old male who presented with myelofibrosis and a peripheral leukocyte count of 2,700 with 19% blasts. Six months after testing, his peripheral counts rose to 123,000 with 50% blasts, and 2 months later he developed myeloid skin infiltrates and died unresponsive to chemotherapy.

Four non-AML samples were found to be positive: a chronic myelo-monocytic leukemia (CMML), a myelofibrosis (MF) (the patient subsequently developed AML), one acute lymphoblastic leukemia (ALL) (of 10 studied), and a patient with chronic myelogenous leukemia (CML) in blast crisis. Significant numbers of blast cells were present in each case.

Differentiation of HL-60 Cells with DMSO or RA

Approximately 70–90% of cultured HL-60 cells showed reactivity with NHL-30.5. Binding of ^{125}I-labeled purified NHL-30.5 antibody to HL-60 cells gave an estimation of 4,000 molecules/cell. When the cells were induced to differentiate along the granulocyte lineage by incubation in the presence of DMSO (Fig. 27.1) or retinoic acid (Fig. 27.2), the number of cells expressing the antigen decreased. This decrease began on day 1 and continued steadily until the fluorescence profiles of the induced cells became identical to those of the negative control. On day 5 only 3% of the DMSO-induced cells and <1% of the RA-induced cells were positive. Less than 250 NHL-30.5 molecules/cell could be detected on the induced cells using iodinated antibody. Following exposure to DMSO, >80% of the cells showed morphological evidence of differentiation beyond the promyelocyte stage. NHL-62.14, a monoclonal antibody with apparent

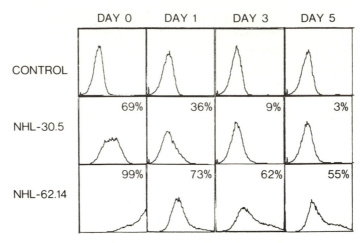

Fig. 27.1. FACS analysis of the reactivity of NHL-30.5 monoclonal antibody with HL-60 cells induced to differentiate with DMSO. Cells were incubated in the presence of 1.25% DMSO and analyzed for reactivity on days 1, 3, and 5 following the addition of DMSO. The fluorescence profiles of untreated HL-60 cells are shown on day 0. The control antibody is an unrelated monoclonal antibody raised against mouse lymphocytes and NHL-62.14 is an anti–transferrin receptor antibody. Percentages refer to the % positive cells. The horizontal axis represents fluorescence intensity (log) and the vertical axis represents cell number.

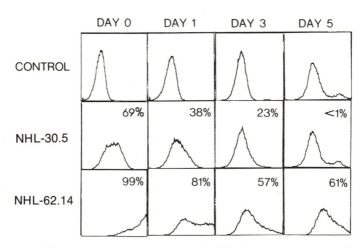

Fig. 27.2. FACS analysis of the reactivity of NHL-30.5 monoclonal antibody with HL-60 cells induced to differentiate with retinoic acid. Cells were treated with 1×10^{-6} M retinoic acid on day 0, and analyzed for reactivity on days 1, 3, and 5 following induction as in Fig. 27.1.

specificity for the transferrin receptor (produced in the same fusion as NHL-30.5), was used as a positive control. Greater than 95% of cultured HL-60 cells were positive for NHL-62.14, but if induced to differentiate only 60% of the cells were positive (fluorescence intensity was also much weaker).

The NHL-30.5 and NHL-62.14 antigens were immunoprecipitated from the surface of iodinated HL-60 cells before and after induction with DMSO. Figure 27.3 shows that NHL-30.5 could only be precipitated from the uninduced cells. The NHL-62.14 antigen was precipitated before and after induction, although there was less precipitated from the induced cells.

The KG-1 myeloblastic leukemia cell line also expressed the NHL-30.5 antigen, although the number of molecules/cell appeared to vary between 500–2,300. Under normal culture conditions 30–50% of the cells were positive and this did not change if the cells were incubated in the presence of DMSO.

Differentiation of HL-60/KG-1 Cell Lines with TPA

The HL-60 and KG-1 cell lines have previously been shown to acquire monocytoid characteristics following incubation in the presence of TPA

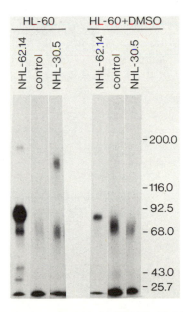

Fig. 27.3. Immunoprecipitation of NHL-30.5 NHL-62.14 antigens on HL-60 cells before and after induction with DMSO (day 5). Target cells were labeled with ^{125}I and the antigens were immunoprecipitated from the cell lysates with NHL-30.5 and NHL-62.14 monoclonal antibodies. The analysis was carried out on SDS–PAGE (5%) under reducing conditions.

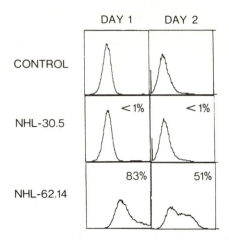

Fig. 27.4. FACS analysis of the reactivity of NHL-30.5 monoclonal antibody with HL-60 cells induced to differentiate with TPA. Cells were harvested on days 1 and 2 following addition of the TPA and analysis was carried out as in Fig. 27.1.

(11). Figure 27.4 shows the fluorescence profiles of TPA-induced HL-60 cells stained with NHL-30.5 and NHL-62.14 antibodies. Within one day following induction, 85% of the cells were adherent and had to be removed with a rubber policeman. On days 1 and 2 following addition of the TPA, <1% of the cells (adherent and nonadherent) expressed the NHL-30.5 antigen. Half of the adherent cells were also negative for NHL-62.14 on day 2. Sixty percent of the TPA-induced KG-1 cells were adherent on

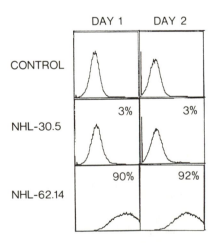

Fig. 27.5. FACS analysis of the reactivity of NHL-30.5 monoclonal antibody with KG-1 cells induced to differentiate with TPA. Cells were harvested on days 1 and 2 following addition of the TPA and analysis was carried out as in Fig. 27.1.

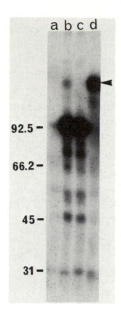

Fig. 27.6. Immunoprecipitation of NHL-30.5 antigen from peripheral blood cells from a patient with AML. The sample contained 77% blast cells (M4 classification). The lysate from the iodinated cells was immunoprecipitated with a control antibody raised against mouse lymphocytes (lane a), with NHL-62.14 (lane b), with OKT9 (lane c), and with NHL-30.5 (lane d).

days 1 and 2, but they easily went into suspension following a 5-min incubation in saline containing 0.2% EDTA and 0.1% BSA. On each day only 3% of the induced KG-1 cells were NHL-30.5 positive, but >90% still reacted with NHL-62.14. (Fig. 27.5).

Immunoprecipitation

Immunoprecipitation of the antigen from iodinated HL-60 cell surfaces has previously revealed a M.W. of approximately 180,000 under both reducing and nonreducing conditions (4). The antigen has subsequently been immunoprecipitated from an AML sample (M4) containing 77% blast cells (Fig. 27.6).

Discussion

This report summarizes the characterization of a murine IgG1 monoclonal antibody that reacts with a significant proportion of hematopoietic cells from newly diagnosed or relapsed AML patients. Twenty-one of twenty-five AML samples tested to date have been found to react. Since the antibody was raised against a promyelocytic leukemia cell line it is somewhat surprising that the one promyelocytic leukemia tested did not react. The peripheral blood from this patient contained only the occasional blast cell, however, and bone marrow cells were not tested. The negative result may therefore have been due to limitations in detection. Three other patients with significant blast counts also did not show detectable numbers of NHL-30.5-positive cells. One of these, patient number 22 (Table 27.1), relapsed 11 months after testing with 46% NHL-30.5-positive cells in the peripheral blood and 18% positive in the marrow. Among the 21 positive patients, there was no correlation between blast count in the specimen and the percent positive cells determined by FACS.

In addition to reacting with cells from patients with a diagnosis of AML, the antibody was found to react with cells from four other patients not considered to have AML at the time of study (Table 27.3). One of these was a patient with myelofibrosis who had 19% blasts in his blood and whose blood and marrow cells produced numerous abnormal colonies in methylcellulose assays. Eight months later this patient developed AML. The second was a patient diagnosed as chronic myelo-monocytic leukemia with 28% blasts in his marrow. This patient also produced numerous abnormal colonies in methylcellulose, but remains alive two years later with a moderate anemia, thrombocytopenia, and white count of 6,000 of which 60% are atypical mononuclear cells. The third patient was an adult ALL with 92% blasts. Since NHL-30.5 appears to be a myeloid antigen, then this might be interpreted as a form of lineage infidelity (3). Alternatively, it is possible that NHL-30.5 may be an antigen expressed on some primitive lymphoid as well as myeloid cells. This was the only ALL of ten analyzed that demonstrated reactivity, however, and five B cell lymphoma cell lines (SU-DHL-1, 4, 6, 8, and 10) and the T cell line Jurkat were also negative, suggesting that lymphoid reactivity is a rare event. A fourth patient appeared to be an ALL on initial examination but cytochemical studies and cell surface phenotyping showed a lymphoid, as well as a myeloid, blast population. Subsequently, the patient was shown to be Ph^1-positive indicating early blast crisis in a patient whose primary disorder was chronic myelogenous leukemia.

Two cell lines (HL-60 and KG-1) initially derived from patients with AML express the NHL-30.5 antigen. The apparent M.W. of the antigen on HL-60 cells is 180,000, where it is expressed on 70–90% of the cells with approximately 4,000 molecules/cell. The antigen does not appear to be phosphorylated (data not shown). Approximately 30–50% of KG-1 cells express the same M.W. antigen, although the number of molecules/cell appears to be variable (500–2,300). The antigen can also be immunoprecipitated from AML cells (Fig. 27.6). The Ph^1-positive cell line K562 expresses the antigen very weakly (<10% of the cells).

If the HL-60 cell line is induced to differentiate using DMSO or RA, the cells acquire properties characteristic of mature granulocytic cells (9,10). This process takes 5 days and, as demonstrated by fluorescent staining (Fig. 27.1) and immunoprecipitation (Fig. 27.3), involves the loss of the NHL-30.5 molecule beginning on the first day. HL-60 cells can also be induced to differentiate into cells with markers of mature monocytic cells by incubating in the presence of TPA (11). Cells treated in this manner are also negative for NHL-30.5 antigen expression. The KG-1 cell line, considered to be less mature than HL-60, can also be induced with TPA to become adherent but reports suggest that its differentiation capacity is less extensive (12). Nevertheless, both the adherent and nonadherent cells from TPA-treated KG-1 cultures lose the NHL-30.5 antigen.

These observations lend support to the idea that the NHL-30.5 antigen is a differentiation antigen that may be present on immature myeloid cells, and gradually lost as the cells differentiate. This would provide an explanation for the expansion of NHL-30.5-positive cells seen in AML and other hematologic disorders characterized by the presence of immature hematopoietic cells. Current efforts are directed at determining if a population in normal marrow, particularly hematopoietic progenitors, expresses the antigen.

Summary

The number of patients tested to date with the NHL-30.5 monoclonal antibody is reported. Hematopoietic cells from 21/25 patients with acute myelogenous leukemia have been found to react with the antibody, in contrast to cells from normal individuals which demonstrate no reactivity. The HL-60 and KG-1 AML cell lines both express the antigen, but lose it when they are induced to differentiate. The antigen can be immunoprecipitated from the surface of HL-60, KG-1, and AML cells, where it has an apparent molecular weight of 180,000 under both reducing and nonreducing conditions.

Acknowledgments. F. Takei is a Research Scholar of the Medical Research Council of Canada. This research is supported by the National Cancer Institute of Canada, the Cancer Control Agency of British Columbia, and the B.C. Cancer Foundation.

References

1. Sachs, L. 1980. Constitutive uncoupling of pathways of gene expression that control growth and differentiation in myeloid leukemia. A model for the origin and progression of malignancy. *Proc. Natl. Acad. Sci. U.S.A.* **77:**6152.
2. Bettelheim, P., E. Paietta, O. Majdic, M. Gadner, J. Schwarzmeier, and W. Knapp. 1982. Evidence of a myeloid marker on TdT-positive acute lympho-

cytic leukemia cells: Evidence of double-fluorescence staining. *Blood* **60:**1392.

3. McCulloch, E.A. 1984. The blast cells of AML. *Clinics in Haematology* **13:**503.

4. Askew, D.S., A.C. Eaves, and F. Takei. 1985. NHL-30.5: a monoclonal antibody reactive with an acute myelogenous leukemia (AML)-associated antigen. *Leuk. Res.* **9:**135.

5. Collins, S.J., R.C. Gallo, and R.E. Gallagher. 1977. Continuous growth and differentiation of human promyelocytic leukemia cells in suspension culture. *Nature* **270:**347.

6. Lozzio, C.B., and B.B. Lozzio. 1975. Human chronic myelogenous leukemia cell-line with positive Philadelphia chromosome. *Blood* **45:**321.

7. Koeffler, H.P., and D.W. Golde. 1978. Acute myelogenous leukemia: a human cell line responsive to colony stimulating activity. *Science* **200:**1153.

8. Epstein, A.L., R. Levy, and H. Kim. 1978. Functional characterization of 10 diffuse histiocytic lymphoma cell lines. *Cancer* **42:**2379.

9. Collins, S.J., F.W. Ruscetti, R.E. Gallagher, and R.C. Gallo. 1978. Terminal differentiation of human promyelocytic leukemia cells induced by dimethyl-sulfoxide and other polar compounds. *Proc. Natl. Acad. Sci. U.S.A.* **75:**2458.

10. Breitman, T.R., S.E. Selonick, and S.J. Collins. 1980. Induction of differentiation of the human promyelocytic leukemia cell (HL-60) by retinoic acid. *Proc. Natl. Acad. Sci. U.S.A.* **77:**2936.

11. Rovera, G., D. Santoli, and C. Damsky. 1979. Human promyelocytic leukemia cells in culture differentiate into macrophage-like cells when treated with a phorbol diester. *Proc. Natl. Acad. Sci. U.S.A.* **76:**2779.

12. Ferrero, D., S. Pessano, G.L. Pagliardi, and G. Rovera. 1983. Induction of differentiation of human myeloid leukemias: surface changes probed with monoclonal antibodies. *Blood* **61:**171.

CHAPTER 28

Down-Regulation of Promyelocytic Cell Transferrin Receptor Expression by Cholera Toxin and Cyclic Adenosine Monophosphate

J.B. Trepel, R.D. Klausner, O.R. Colamonici, S. Pittaluga, and L.M. Neckers

The human promyelocytic cell line HL-60 has been studied as an *in vitro* model of impaired differentiation characteristic of acute nonlymphocytic leukemia (1–3). Because of the plasticity of this cell line, we used HL-60 cells to examine the biochemical events controlling expression of the transferrin receptor (TF_R), a cell-surface glycoprotein which appears to play a pivotal role in cell regulation.

The Role of the Transferrin Receptor

The TF_R is a 180-Kd homodimer of 90-Kd subunits. It binds the iron-transport protein transferrin with very high affinity, and translocates the metal–protein complex to an acidic intracellular compartment where low pH releases iron from the transferrin–receptor complex, following which the receptor and apotransferrin recycle to the cell surface (4–8). The role of the TF_R in iron transport and the mechanisms controlling receptor recycling have been intensively studied and will not be addressed here. In the experiments presented here we will be examining the biochemical regulation of TF_R expression. It should be noted that it is not entirely clear that the only role of the TF_R is in iron transport. Although there is evidence connecting transferrin receptors with the control of cell proliferation, not enough is known about the role of iron in the control of eukaryotic cell growth to evaluate the need for TF_R-mediated increases in intracellular iron to facilitate such events as total genomic replication. Certain trends, however, are clear. The TF_R is not found, or is weakly expressed, on most nonproliferating cells. Events that move resting cells toward cell division, such as mitogenic stimulation or malignant transformation, also induce TF_R expression (9–11). In addition, anti-TF_R antibodies have been shown to inhibit cell proliferation in a number of systems (12–14).

Biochemical Regulation of Transferrin Receptor Expression

Although there is a considerable amount of phenomenologic data describing increases or decreases in TF_R number, until recently little was known of the underlying biochemical events. It has been reported that phorbol ester treatment of HL-60 cells causes a decrease in surface TF_R number (15,16). Very recently, Cuatrecasas and coworkers found that this decrease is associated with a rapid TF_R internalization, and increase in the level of TF_R phosphorylation (17). In the experiments reported here we found that the cholera exotoxin (CT) also induces a decline in surface TF_R number, and that this effect is mimicked by dibutyryl cyclic adenosine monophosphate (db cAMP). In contrast, low doses of dibutyryl cyclic guanosine monophosphate (db cGMP) did not decrease TF_R number, but rather induced a modest surface TF_R increase. CT and db cAMP both induced a redistribution of TF_R from the cell surface to the cytoplasm, and CT stimulated TF_R phosphorylation. These data suggest that cyclic nucleotide balance *in vivo* may modulate TF_R expression, which in turn has an impact on the control of cell growth and maturation.

Cholera Toxin Decreases HL-60 Surface TF_R Number

Figure 28.1 is a flow microfluorometry histogram of TF_R expression comparing untreated HL-60 cells to HL-60 cells following treatment for 24 hr with 100 nM cholera toxin. The histogram was generated on a Becton-Dickinson FACS II using OKT9 (monoclonal anti–transferrin receptor antibody, Ortho Diagnostics), followed by fluoresceinated goat anti-mouse antibody, as described previously (13). CT has been reported to contribute to HL-60 cell differentiation (18). CT caused a marked decrease in surface TF_R levels. Both treated and untreated populations show a relatively Gaussian distribution of TF_R number, suggesting that CT is causing a fairly uniform decrease in TF_R expression, on a per cell basis. As shown in Fig. 28.2, CT is a very potent stimulus inducing down-regulation of surface TF_R expression. This effect of CT is first seen in the femtomolar range, with a 45% decrease in surface TF_R expression at 125 fM CT. Interestingly, although at the lowest doses TF_R expression is very sensitive to CT, there is a long plateau region before CT again induces a dose-dependent TF_R decrease. This result is not inconsistent with two classes of receptor, or two mechanisms of receptor down-regulation. The time course of CT-stimulated surface TF_R decline is shown in Fig. 28.3. Following a latency of approximately 7 hr, CT induces a decline in surface TF_R which is long-lasting, and perhaps irreversible. In a time course of 6 days (data not shown) CT treatment resulted in depression of surface TF_R values for 6 days, but control receptor levels also began declining as the cells became confluent. CT is an 84-Kd protein composed of an A subunit,

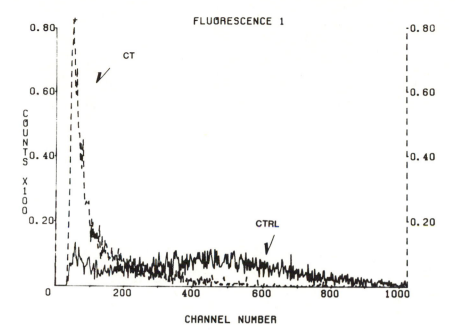

Fig. 28.1. Cholera toxin inhibits surface TF_R expression. A flow microfluorometry histogram was generated comparing surface TF_R expression on untreated HL-60 cells to that on HL-60 cells treated with 100 nM CT for 24 hr. Cells were initiated in culture at 0.5×10^6/ml. TF_R levels were detected on a Becton Dickinson FACS II using the anti–transferrin receptor antibody OKT9.

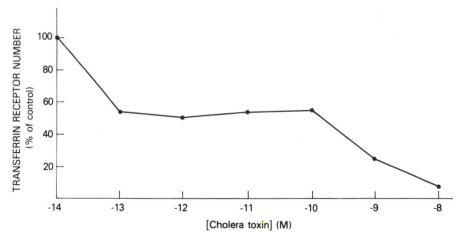

Fig. 28.2. Cholera toxin inhibition of surface TF_R expression: Dose response. HL-60 cells were cultured at 0.5×10^6/ml for 24 hr with the final concentration of CT as indicated. Histograms of surface TF_R number expression were generated as in Fig. 28.1. The level of TF_R expression was calculated from the median channel number of the fluorescence peak using a PDP-11 minicomputer (Digital Equipment Corp.) and a program provided by the Division of Computer Research and Technology, NIH.

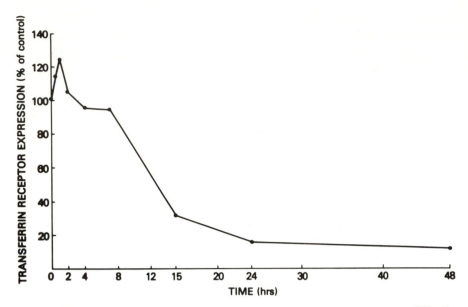

Fig. 28.3. Cholera toxin inhibition of surface TF_R expression: Time course. HL-60 cells were cultured at $0.5 \times 10^6/ml$ in the presence of 10 nM CT. Levels of TF_R expression were calculated as in Fig. 28.2.

which acts as an ADP-ribosyltransferase (19–21) and is responsible for the remarkable ability of CT to elevate cyclic adenosine monophosphate (cAMP) in all eukaryotic cells tested, coupled to a B subunit, which is the cell-binding moiety of the CT molecule (22). The B subunit *per se* does not affect adenylate cyclase activity. Table 28.1 shows a comparison between the ability of the intact CT molecule and the CT B subunit to modulate cell

Table 28.1. Effect of intact CT versus CT B subunit on HL-60 surface TF_R expression.[a]

Concentration (µg/ml)	% of Control TF_R	
	Intact CT	CT B subunit
0.1	6	91
0.25	6	84
0.50	7	94
1.0	8	97
2.0	6	72
3.0	5	70
4.0	3	72

[a] HL-60 cells were cultured at an initial concentration of 0.5×10^6 cells/ml for 24 hr with the concentrations of intact CT or CT B subunit indicated. TF_R levels were calculated using the median channel number from flow microfluorometry histograms of surface TF_R expression.

surface transferrin receptors. The B subunit at 10 nM induced a 3% decline in surface TF$_R$, while the same concentration of intact CT caused a 92% decrease in receptor expression. As these data suggest that elevated intracellular cAMP is mediating the CT effect, we tested the ability of db cAMP to modulate surface TF$_R$.

Dibutyryl cAMP Stimulates Decreased Surface TF$_R$ Expression

Figure 28.4 demonstrates the marked inhibition of surface TF$_R$ induced at 24 hr by 1 mM db cAMP. As was evident for CT in Fig. 28.1, db cAMP induced an apparently uniform decrease in per cell surface TF$_R$ number. Figure 28.5 demonstrates the dose-dependent inhibition of HL-60 surface TF$_R$ expression by treatment for 24 hr with db cAMP. Although the effect of db cAMP is similar to that induced by CT, db cAMP is less potent. This is not surprising as CT covalently modifies the guanyl nucleotide binding site of adenylate cyclase (23), resulting in activation and continuous generation of cAMP. As a specificity control we tested the effect of db cGMP on surface TF$_R$ expression. In contrast to db cAMP, db cGMP in the

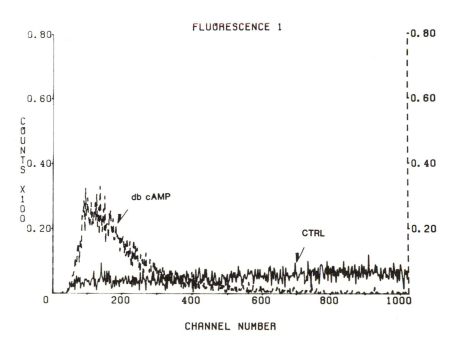

Fig. 28.4. Inhibition of surface TF$_R$ expression by 1 mM dibutyryl cAMP. The histogram of surface TF$_R$ expression was generated as described in Fig. 28.2. HL-60 cells were cultured for 24 hr in the presence of db cAMP.

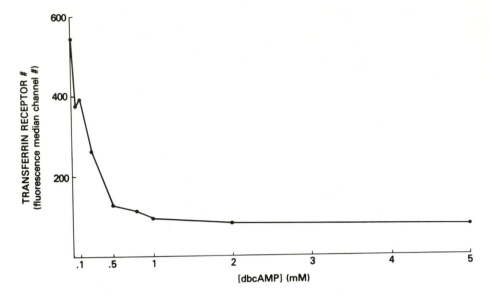

Fig. 28.5. Inhibition of HL-60 TF_R expression by dibutyryl cAMP: Dose response. Cells were cultured and TF_R levels determined as described in Fig. 28.2.

range of 5 μM through 250 μM caused a modest increase in surface TF_R expression, which is reversed at the highest doses. This effect is demonstrated in Table 28.2. Figure 28.6 compares the DNA histograms of untreated versus CT-treated HL-60 cells, showing that after 24 hr 10 nM CT decreased the number of S phase cells and increased the number of cells in G_1. Treatment with 100 nM CT for 48 hr resulted in almost complete synchronization in G_1 (data not shown).

Table 28.2. Dibutyryl cGMP: Effect on surface TF_R and comparison with dibutyryl cAMP.[a]

Concentration (μM)	% of Control TF_R	
	db cGMP	db cAMP
5	110	68
10	122	71
50	109	48
100	112	23
250	110	21
500	53	18
1	68	15

[a] HL-60 cells were cultured at an initial concentration of 0.5 × 10⁶ cells/ml for 24 hr with the concentrations of cyclic nucleotide indicated. TF_R levels were calculated using the median channel number from flow microfluorometry histograms of surface TF_R expression.

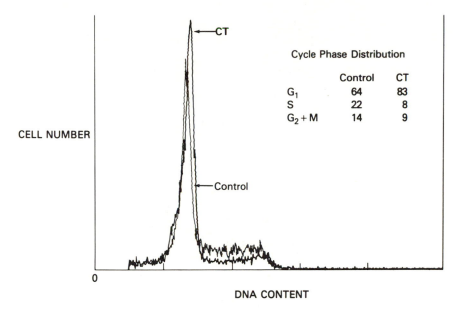

Fig. 28.6. DNA histogram of untreated HL-60 cells versus HL-60 cells treated for 24 hr with 10 nM cholera toxin. Following the 24-hr incubation time the cells were fixed in 50% ethanol, treated with 500 units per ml ribonuclease for 30 min at 37°C, and incubated for 30 min with the fluorescent DNA-intercalating dye propidium iodide.

Effect of CT and db cAMP on the Distribution of Transferrin Receptors

Since CT and db cAMP decreased surface TF$_R$ it was of interest to see if these treatments affected the ratio of surface TF$_R$ to total cellular transferrin receptor levels, as had been reported following phorbol ester treatment. We performed a preliminary experiment in which this ratio was determined in untreated, CT-treated, and db cAMP-treated HL-60 cells. Total TF$_R$ levels were measured by an iodinated transferrin binding assay using lysed and solubilized cells. Surface TF$_R$ levels were quantified by an iodinated transferrin binding assay using intact cells (see Ref. 24). Table 28.3 shows that the ratio of surface to total TF$_R$ was decreased by both CT and db cAMP treatment.

Cholera Toxin Increases the Level of Transferrin Receptor Phosphorylation

As CT and db cAMP induced TF$_R$ redistribution, and phorbol ester had been reported to induce receptor redistribution via TF$_R$ phosphorylation, we tested the level of TF$_R$ phosphorylation after CT treatment. HL-60

Table 28.3. CT and db cAMP decrease the ratio of surface TF_R to total TF_R.[a]

Treatment	Ratio
Control	0.55
CT	0.34
db cAMP	0.28

[a] HL-60 cells were cultured at 0.5 × 10^6 cells/ml for 24 hr. The CT concentration was 100 nM, and the concentration of db cAMP was 1 mM. Surface TF_R levels were quantified by an iodinated transferrin binding assay using intact cells. Total cellular TF_R levels were determined by lysing the cells, incubating with iodinated transferrin, filtering the suspension, and counting the iodinated material trapped on the filter.

cells were treated for 24 hr with 100 nM CT, metabolically labeled with [^{32}P]orthophosphoric acid. The TF_R were then immunoprecipitated with OKT9, and the immunoprecipitated material was analyzed by gel electrophoresis under reducing conditions, followed by autoradiography. After CT treatment and immunoprecipitation with anti-TF_R antibody, there was an increase in phosphorylation of a band migrating at 90-Kd. The gel autoradiogram was scanned on a densitometer and Table 28.4 shows that

Table 28.4. Cholera toxin increases the level of transferrin receptor phosphorylation.[a]

	Phosphorylation level (arbitrary units)
Control	7
Cholera Toxin	67

[a] HL-60 cells were treated for 24 hr with 100 nM CT and metabolically labeled for 3 hr with [^{32}P]orthophosphoric acid. The cells were incubated with OKT9, solubilized with Triton X-100 and sodium deoxycholate, centrifuged, and a suspension of protein A–Sepharose was added to the supernatant. The sample was tumbled for 1 hr at 40°C, washed with the solubilization mix, and boiled for 5 min in Laemmli sample buffer. Following centrifugation to remove the protein A–Sepharose, the samples were electrophoresed through a 12% gel under reducing conditions, the gel was fixed and dried, and exposed to X-Omat film at −70°C. Quantification of degree of phosphorylation was made via densitometry.

CT induced a 9.5-fold increase in TF_R phosphorylation. This result is consistent with the idea that CT-mediated surface TF_R down-regulation occurs subsequent to CT stimulation of TF_R phosphorylation.

Conclusions

We conclude that intracellular cyclic nucleotide levels may play a role in regulating the level of cell-surface transferrin receptor expression, and that this control may be exerted via receptor redistribution subsequent to cyclic nucleotide-stimulated TF_R phosphorylation. There are many reports in the literature suggesting that cyclic nucleotides play an important regulatory role in the control of cell proliferation and differentiation. Specifically, it has been observed that decreased cAMP levels have been associated with dedifferentiation and unregulated cell growth (for review, see Ref. 25). It has also been observed that in some rapidly growing cells cAMP levels are actually elevated, and it has been suggested that cyclic nucleotides may exert both positive and negative modulatory roles in controlling cell division (26). One may speculate that a mechanism of cyclic nucleotide regulation of cell proliferation in normal cells may be cyclic nucleotide control of surface TF_R expression, and that loss of cyclic nucleotide regulation of surface TF_R expression may play a role in malignant transformation.

References

1. Gallagher, R., S. Collins, J. Trujillo, K. McCredie, M. Ahern, S. Tsai, R. Metzgar, G. Aulakh, R. Ting, F. Ruscetti, and R. Gallo. 1979. Characterization of the continuously differentiating myeloid cell line (HL-60) from a patient with acute promyelocytic leukemia. *Blood* **54**:713.
2. Collins, S.J., F.W. Ruscetti, R.E. Gallagher, and R.C. Gallo. 1978. Terminal differentiation of human promyelocytic leukemia cells induced by dimethyl sulfoxide and other polar compounds. *Proc. Natl. Acad. Sci. U.S.A.* **75**:2458.
3. Rovera, G., D. Santoli, and C. Damsky. 1979. Human promyelocytic leukemia cells in culture differentiate into macrophage-like cells when treated with a phorbol diester. *Proc. Natl. Acad. Sci. U.S.A.* **76**:2779.
4. Karin, M., and B. Mintz. 1981. Receptor-mediated endocytosis of transferrin in developmentally totipotent mouse teratocarcinoma stem cells. *J. Biol. Chem.* **256**:3245.
5. Schneider, C., R. Sutherland, R. Newman, and M. Greaves. 1982. Structural features of the cell surface receptor for transferrin that is recognized by the monoclonal antibody OKT 9. *J. Biol. Chem.* **257**:8516.
6. van Renswoude, J.K. Bridges, J. Harford, and R.D. Klausner. 1982. Receptor-mediated endocytosis of transferrin and the uptake of Fe in K 562 cells: identification of a nonlysosomal acidic compartment. *Proc. Natl. Acad. Sci. U.S.A.* **79**:6186.

7. Dautry-Varsat, A., A. Ciechanover, and H.F. Lodish. 1983. pH and the recycling of transferrin during receptor-mediated endocytosis. *Proc. Natl. Acad. Sci. U.S.A.* **80:**2258.

8. Klausner, R.D., G. Ashwell, J. van Renswoude, J.B. Harford, and K.R. Bridges. 1983. Binding of apotransferrin to K 562 cells: Explanation of the transferrin cycle. *Proc. Natl. Acad. Sci. U.S.A.* **80:**2263.

9. Larrick, J.W., and P. Cresswell. 1979. Modulation of cell surface iron transferrin receptors by cellular density and state of activation. *J. Supramol. Struct.* **11:**579.

10. Trowbridge, I.S., and M.B. Omary. 1981. Human cell surface glycoprotein related to cell proliferation is the receptor for transferrin. *Proc. Natl. Acad. Sci. U.S.A.* **78:**3039.

11. Sutherland, D.R., D. Delia, C. Schneider, R.A. Newman, J. Kemshead, and M.F. Greaves. 1981. Ubiquitous cell surface glycoprotein on tumor cells is proliferation-associated receptor for transferrin. *Proc. Natl. Acad. Sci. U.S.A.* **78:**4515.

12. Trowbridge, I.S., and F. Lopez. 1892. Monoclonal antibody to transferrin receptor blocks transferrin binding and inhibits human tumor cell growth *in vitro. Proc. Natl. Acad. Sci. U.S.A.* **79:**1179.

13. Neckers, L.M., and J. Cossman. 1983. Transferrin receptor induction in mitogen-stimulated human T lymphocytes is required for DNA synthesis and cell division and is regulated by interleukin 2. *Proc. Natl. Acad. Sci. U.S.A.* **80:**3494.

14. Neckers, L.M., S.P. James, and G. Yenokida. 1984. Role of transferrin receptors in B cell activation. *J. Immunol.* **133:**2437.

15. Yeh, C.-J.G., M. Papamichael, and W.P. Faulk. 1982. Loss of transferrin receptors following induced differentiation of HL-60 promyelocytic leukemia cells. *Exp. Cell Res.* **138:**429.

16. Rovera, G., D. Ferreo, G.L. Pagliardi, J. Vartikar, S. Pessano, L. Bottero, S. Abraham, and D. Lebman. 1982. Induction of differentiation of human myeloid leukemias by phorbol diester: phenotypic changes and modes of action. *Ann. N.Y. Acad. Sci.* **397:**211.

17. May, W.S., S. Jacobs, and P. Cuatrecasas. 1984. Association of phorbol ester-induced hyperphosphorylation and reversible regulation of transferrin membrane receptors in HL-60 cells. *Proc. Natl. Acad. Sci. U.S.A.* **81:**2016.

18. Olsson, I.L., T.R. Breitman, and R.C. Gallo. 1982. Priming of human myeloid leukemic cell lines HL-60 and U 937 with retinoic acid for differentiation effects of cyclic adenosine 3′:5′-monophosphate-inducing agents and a T-lymphocyte-derived differentiation factor. *Cancer Res.* **42:**3928.

19. Moss, J., and M. Vaughan. 1977. Mechanism of action of choleragen. Evidence for ADP-ribosylating activity with arginine as an acceptor. *J. Biol. Chem.* **252:**2455.

20. Trepel, J.B., D.-M. Chuang, and N.H. Neff. 1977. Transfer of ADP-ribose from NAD to choleragen: A subunit acts as catalyst and acceptor protein. *Proc. Natl. Acad. Sci. U.S.A.* **74:**5440.

21. Trepel, J.B., D.-M. Chuang, and N.H. Neff. 1981. Polypeptide hormones and chromatin-associated proteins act as acceptors for cholera-toxin catalyzed ADP-ribosylation. *J. Neurochem.* **36:**538.

22. Gill, D.M., and R. Meren. 1978. ADP-ribosylation of membrane proteins

catalyzed by cholera toxin: Basis of the activation of adenylate cyclase. *Proc. Natl. Acad. Sci. U.S.A.* **75:**3050.

23. Cassel, D., and T. Pfeuffer. 1978. Mechanism of cholera toxin: Covalent modification of the guanylnucleotide-binding protein of the adenylate cyclase system. *Proc. Natl. Acad. Sci. U.S.A.* **75:**2669.

24. Klausner, R.D., J. Harford, and J. van Renswoude. 1984. Rapid internalization of the transferrin receptor in K 562 cells is triggered by ligand binding or treatment with a phorbol ester. *Proc. Natl. Acad. Sci. U.S.A.* **81:**3005.

25. Pastan, I. 1975. Cyclic AMP and the malignant transformation of cells. In: *Advances in metabolic disorders,* Vol. 8, *Somatomedians and some other growth factors.* Academic Press, New York, pp. 377–383.

26. Parker, C.W. 1976. Control of lymphocyte function. *New England J. Med.* **295:**1180.

CHAPTER 29

Expression of Myeloid and B Cell-Associated Antigens on T Lineage Cells

P. Mannoni, P. Dubreuil, B. Winkler-Lowen, D. Olive, L. Linklater, and C. Mawas

A variety of monoclonal antibodies against differentiation antigens which appear to define antigens expressed mainly in one cell lineage have now been established. However, it has already been recognized that some of these antigens are also expressed on cells from other lineages. One of the best examples is the CALLA antigen originally described as specific for early stages of B cell differentiation and subsequently described on granulocytes and other tissues (1). Monoclonal antibodies recognizing the oligosaccharide structure alpha-fucosyl-N-acetyllactosamine (IG10, My 1, 80H5, etc.) (2) have been published originally as specific for the definition of the myeloid lineage. Thereafter it was shown by ourselves (this series, Vol. 1, Ch. 26) and others (2,3) that this structure was expressed also on cells originating in the non-myeloid pathway. On the other hand, it has been shown that leukemic cells, when tested with mAbs, could express differentiation antigens belonging to different lineages (4–6). This "lineage infidelity" can be seen as a specific pattern for leukemic differentiation, or contrarily as the counterpart of a normal stage observed only on progenitor cells (4). For the above reasons it was interesting to test the monoclonal antibodies submitted to the myeloid/stem cell Workshop and to the B cell and leukemia Workshop for their cross-reactivity with cells from other lineages. In this study we report the results obtained on cells belonging to the T cell differentiation pathway.

Materials and Methods

The whole panel of monoclonal antibodies was tested by immunofluorescence, using a flow cytometer cell sorter (EPICS V, Coulter-Coultronics, Hialeh, FL) to determine the percentage of cells recognized by each mAb. FITC-coupled F(ab')$_2$ fragment of goat anti–mouse and anti–rat Ig (Cappel) was used as secondary antibody. At least 10,000 cells were counted for each test, and the percentage of positive cells was determined by comparison with negative controls. A result was considered positive when at least 20% of cells were found positive. Different calculation programs were used (INTGRA, SUBTRACTION, or IMMUNO) to compare the number of fluorescent cells with the controls (Coulter-Coultronics).

In order to study stages of the T cell differentiation pathway, we selected the following cells and cell lines: Reh (null cell line), 1301, CCRF-CEM, MOLT-4, ICHIGAWA, (T-ALL cell lines). Thymocytes were obtained with parental consent from a 5-year-old child undergoing cardiac surgery. T cells were prepared from normal donors by rosetting with AET-treated sheep red blood cells. Activated T cells were obtained by PHA stimulation (1% PHA-M in RPMI 10% FCS, Gibco Laboratory) and tested at day 4. Kinetic studies were also performed with selected mAbs by studying mAb binding at day 0, 1, 4, and 7. Activated T cells were purified by incubation in, and elution from, a nylon wool column. Leukemic cells isolated from two patients with T-ALL were also tested. As part of the general Workshop study, the same panel was also tested on myelomonocytic cell lines (KG1, ML1, HL-60, U 937), erythroid cell lines (K562, HEL), a pre-B cell line (Nalm-6), and on leukemic cells isolated from patients with chronic or acute myeloid leukemia.

Results and Discussion

Figure 29.1 summarizes the overall results of tests of the Workshop myeloid mAbs on T-derived cells. Only 30 of the 118 mAbs submitted were completely negative when tested on either T differentiated cells, activated T cells, or leukemic T cells. Sixty-one mAbs were found to react with some T cell lineage cells, among which 49 gave consistent and clear positive results (Fig. 29.1). "X" represents a group of mAbs which recognize the oligosaccharide alpha-fucosyl-*N*-acetyllactosamine, or X-hapten (M5, 13, 19, 25, 33, and 91 and 80H5 produced in our laboratory) (2,7). Black corresponds to more than 80% positive cells, dark gray between 50 and 80%, and light gray from 20 to 50%. All cell lines were tested in exponential growth phase. Antibodies reacting weakly with one or two suspensions of leukemic cells only were considered as nonreactive on the T cell lineage.

Some clusters could be defined according to the serological pattern observed on T lineage cells: a first mAb group (M24, 21, 6, 31, 32, 38, 58, 11, 74, 39, 72, 73, 75, 89, 41, 55, 36, 53, 56, 50) can be identified by its strong reactivity on several T leukemia cell lines and/or thymocytes, and T and activated T cells. Among them are included all LFA or LFA-like mAbs submitted to the Workshop except M26 which was found to react weakly or not at all. All but one (M11) reacted on thymocytes and on T

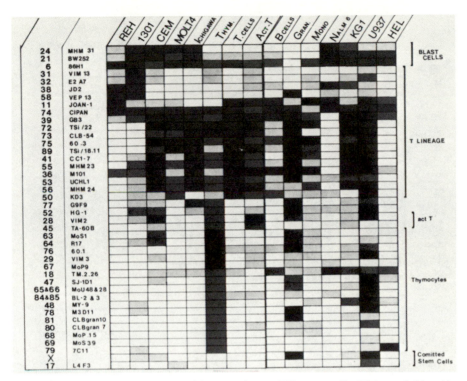

Fig. 29.1. Myeloid Workshop mAbs reactive on T lineage cells. The myeloid mAb expressed on cells from the T lineage: The X includes the 6 mAbs with reactivity against X-hapten. The black squares represent reactivity of the corresponding mAb on more than 80% of the target cells, gray squares 50 to 80%, and light gray 20 to 50%.

and activated T cells. This group however showed heterogeneity: antigens corresponding to three mAbs (M24, 21, and 6) were expressed mainly on immature T cells (1301, CCRF-CEM) and on poorly differentiated cell lines (Reh, Nalm-6). These were all negative when tested on mature granulocytes and M24 was also negative on mature T cells. mAbs M31, 32, 38, 58, and 11 (VIM13, E2A7, JD2, VEP13, JOAN-1) identified antigens mainly reactive on blast or poorly differentiated T cells. Only 86H1 and JOAN-1 reacted strongly with mature T cells. All of these antibodies recognized leukemic cells isolated from two patients with T-ALL (data not shown).

Two antibodies, M28 and M45 (VIM2, TA-60B), negative on resting T cells were clearly positive at day 4 and day 7, respectively, on PHA-activated T cells (Fig. 29.3). Unlike antibodies against the interleukin-2 receptor (TAC and others) they were not reactive on thymocytes.

A third group was identified by its main reactivity on thymocytes (M63,

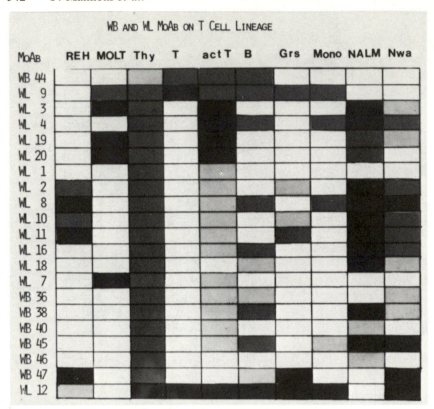

Fig. 29.2. B and leukemia-associated mAbs reactive on T lineage cells.

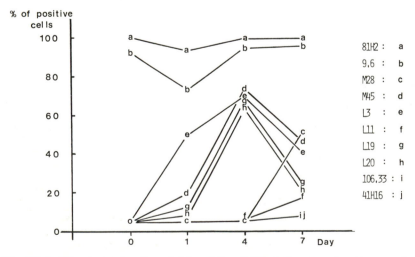

Fig. 29.3. Kinetic study of myeloid and B cell Workshop panel mAbs reactive on purified activated T cells. Days 0, 1, 4, and 7.

64, 76, 29, 67, 18, 47, 65, 66, 84, 85, 48, 78, 81, 80, 68, 69, 79). Some members of this group were also moderately positive on T or activated T cells.

Antibodies recognizing the oligosaccharide structure X (WM5, 13, 19, 25 and probably 33, 91) showed a weak expression on early leukemic T cell lines but were completely negative on mature T cells.

Very few mAbs from the B cell Workshop panel were found to be clearly reactive on T cells or T-derived cell lines (Fig. 29.2). On the contrary when mAbs from the leukemic kit were tested, 14 out of 20 bound to the membrane of T lineage cells. Two of them, WL9 and 12, showed a broad reactivity but were negative on Nalm-6. The most common pattern of this leukemic mAb series was its strong reactivity with thymocytes since all were found positive (Fig. 29.2).

Figure 29.3 summarizes the kinetics of mAbs which were reactive with PHA-activated T cells and not with resting T cells. Most of them gave a pattern resembling the one obtained with anti-Tac mAb. However, mAb M28 (VIM2) expressed a peculiar reactivity since the corresponding anti-

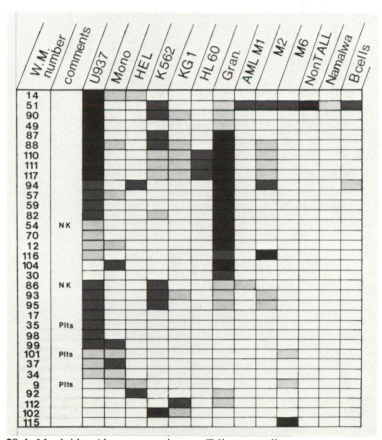

Fig. 29.4. Myeloid mAbs not reactive on T lineage cells.

gen was only detectable at day 7 of the culture. This disassociation in the expression of membrane markers during lymphocyte activation with PHA has been already reported with anti–activated T cell mAbs (8).

Figure 29.4 summarizes the general serological pattern of Workshop myeloid panel antibodies found not reactive on cells from the T lineage. Included in this group are antibodies against platelet glycoproteins (AN51, 943Dl, PL1) and NK cells and antibodies relatively specific for late granulocytic antigens. Antibodies which were found negative with the myeloid and T lineage cells selected for this study are not included in the figure (for example, NKH2, NKH1A, J15, etc.).

Discussion

The fact that molecules defined by "myeloid-specific" or myeloid-associated mAbs were expressed on cells derived from the T cell lineage is not surprising since very few differentiation antigens have been found restricted to only one lineage. For example RFB-1 has been shown to recognize an antigen common to human hematopoietic cells and T lineage cells (9). In the same manner, to our knowledge, there are no human cell surface molecules documented to be exclusively leukemia specific.

One of the classical explanations of this interlineage cross-reactivity is the recognition by a mAb of a common epitope present on a molecule that can vary in its polypeptide or carbohydrate composition (10). This has been reported for the CALLA antigen, the X-hapten, and several others (1, 7). For example, the CALLA antigen has a restricted distribution on cells from the B cell pathway, but could also be detected on kidney epithelial cells as well as on mature granulocytes, without evident relationships between the origin or the functions of such cells.

In this study, we observed that some of these cross-reactive mAbs detected antigen(s) which could be related to the general process of hematopoietic cell differentiation and maturation. For instance, the first group of cross-reactive mAbs recognized most of the hematopoietic cells. One antibody (86H1) produced in our laboratory defined a protein structure expressed on cells from all lineages except mature granulocytes, platelets, and red blood cells. The corresponding antigen was found to be expressed also on all leukemic cells isolated from AML or T-ALL patients, but not those from patients with common B-ALL. Such an antibody probably defines a common hematopoietic cell antigen which undergoes changes in its structure during the process of differentiation–maturation. This phenomenon has been widely documented with mAbs against the X-hapten mainly expressed during the myeloid pathway from the promyelocytic cells to the end stage represented by the mature granulocyte. It has been also recognized that this carbohydrate moiety could be linked to a protein or a lipidic structure and that sialic acid plays a role in the presentation or the assembly of this structure since treatment by neuraminidase greatly enhances the reactivity of such mAbs against normal or leukemic cells (data not shown).

Since this X-hapten has been also found on normal bone marrow progenitors (BFU-E, CFU-GEMM, CFU-GM, (7) and leukemic promegakaryoblasts) (J. B. Gorius *et al.,* manuscript in preparation), it is probable that this structure is widely expressed on progenitor cells from different lineages but remains expressed only in the myeloid pathway.

It is surprising to observe that so many mAbs against granulocytes or monocytes reacted also with thymocytes and not with T or activated T cells. Again these mAbs probably identify structures common to thymocytes, granulocytes, and monocytes which are probably transformed during T cell maturation but not during myeloid differentiation.

It would be now necessary to distinguish which among these common antigens could be related either to a specific function or to a precise stage of cellular differentiation. It is tempting but dangerous to assume some general function of cell regulation or cell differentiation based upon the wide expression of such antigens. This possibility has to be tested by studying inhibition of functions or interactions of these antibodies with growth factors or differentiation-inducing agents.

Considering the results obtained with this selected panel, it is not surprising that leukemia cells could express markers from apparently different lineages. This "lineage infidelity" could in fact represent an early stage of normal differentiation, when cells not completely differentiated along a specific pathway could still express membrane structures common to cells of different lineages. To assess the relevance of these cross-reactive structures in the normal differentiation process it will be necessary to study the expression of these antigens on normal bone marrow progenitor cells. In another study performed in our laboratory we found that M11 (JOAN-1) strongly inhibited the CFU-GM, BFU-E, and CFU-GEMM colonies, when normal bone marrow was preincubated with mAb and complement. Such an approach should lead to a better definition of antigen expressed on stem cells and to the possible recognition of receptors specific for the control of cell differentiation and proliferation. This approach seems to be more promising for the identification of leukemic phenotypes than the putative quest for leukemia-specific antigens.

Acknowledgments. This work was supported in part by a grant from the Alberta Heritage Foundation for Medical Research. P. Mannoni is an AHFMR scholar from the University of Alberta.

References

1. Cossman, J. 1983. Expression of CALLA antigens on normal granulocytes. *J. Exp. Med.* **157**:802.
2. Huang, L.C., C.I. Civin, J.L. Magnani, J.H. Shaper and V. Ginsburg. 1983. MY1 the human myeloid specific antigen detected by mouse monoclonal antibodies is a sugar sequence found in Lacto-*N*-Fucopentose III. *Blood* **61**:1020.

3. Tabilio, A., P.G. Pellici, G. Vinci, P. Mannoni, C.I. Civin, W. Vainchenker, U. Testa, M. Lipinski, H. Rochant and J. Breton-Gorious. 1983. Myeloid and megakaryocytic properties of K562 cell lines. *Cancer Res.* **43**:4569.
4. Greaves, M.F. 1982. Target cells, cellular phenotypes and lineage fidelity in human leukemia. *J. Cell. Physiol.* (suppl.) **1**:113.
5. McCulloch, E.A. 1983. Stem cells in normal and leukemic hemopoiesis. *Blood* **62**:1.
6. Smith, L.J., J.E. Curtis, H.A. Messner, J.S. Senn, H. Furthmayr, and E.A. McCulloch. 1983. Lineage infidelity in acute leukemia. *Blood* **61**:1138.
7. Janowska-Wieczorek, A., P. Mannoni, A.R. Turner, A. Shaw, and J.M. Turc. 1983. A monoclonal antibody specific for human stem cells and granulocytic lineage cells. *Exp. Hematol.* II (suppl. 14):204.
8. Cotner, T., J.M. Williams, L. Christenson, H.M. Shapiro, T.B. Strom, and J.L. Strominger. 1983. Simultaneous flow cytometric analysis of human T cell activation antigen expression and DNA content. *J. Exp. Med.* **157**:461.
9. Bodger, M.P., G.E. Francis, D. Delia, S.M. Granger, and G. Jannossy. 1981. A monoclonal antibody specific for immature human hemopoietic cells and T lineage cells. *J. Immunol.* **6**:2269.
10. Milstein, C. 1984. The impact of monoclonal antibodies on studies of the differentiation of lymphocytes. In: *Leucocyte typing,* A. Bernard, L. Boumsell, J. Dausset, C. Milstein, and S.F. Schlossman, eds. Springer-Verlag, Berlin, Heidelberg, pp. 3–9.

Selected Bibliography

1. Kersey, J.H., and T.W. Lebien. 1984. What happened to lineage and leukemia specific antigens? Implications for therapeutic use of monoclonal antibodies. In: *Leucocyte typing,* A. Bernard, A., L. Boumsell, J. Dausset, C. Milstein, and S.F. Schlossman, eds. Springer-Verlag, Berlin, Heidelberg, pp. 651–655.
2. Koeffler, H.P. and D.W. Golde. 1980. Human myeloid leukemia cell lines: A review. *Blood* **54**:344.
3. Marie, J.P., C.A. Izaquire, C.I. Civin, J. Mirro, and E.A. McCulloch. 1981. The presence within single K562 cells of erythropoietic and granulopoietic differentiation markers. *Blood* **58**:708.
4. Minowada, J., E. Tatsumi, K. Sagawa, M.S. Lok, T. Sugimoto, K. Minato, L. Zgoda, L. Prestine, L. Kover, and D. Gould. 1984. A scheme of human hematopoietic differentiation based on the marker profiles of cultured and fresh leukemia lymphomas. In: *Leucocyte typing,* A. Bernard, Boumsell, J. Dausset, C. Milstein, and S.F. Scholossman, eds. Springer-Verlag, Berlin, Heidelberg, pp. 519–527.
5. Tabilio, A., W. Vainchenker, D. Van Haeke, G. Vinci, J. Guichard, A. Henri, F. Reyes, J. Breton-Gorius. 1984. Immunological characterization of the leukemic megakaryocytic line at light and electron microscopic level. *Leukemia Research* **8**:769.
6. Smith, L.J., and E.A. McCulloch. 1984. Lineage infidelity following exposure of T lymphoblasts (MOLT-3 cells) to 5-Azacytidine. *Blood* **63**:1324.
7. Vinci, G., A. Tabilio, J.F. Deschamps, D. Van Haeke, A. Henri, J. Guichard, P. Tetteroo, P.M. Lansdorp, T. Hercend, W. Vainchenker, and J. Breton-

Gorious. 1983. Immunological study of *in vitro* maturation of human mega-karyocytes. *Brit. J. Haematol.* **56**:589.

8. Urdal, D.L., T.A. Brentnall, I.D. Bernstein, and S.-I. Hakomori, 1983. A granulocyte reactive monoclonal antibody, IG10, identifies the Gal beta 1-4 (Fuc-alpha 1-3) Glc NAc (X determinant) expressed HL-60 cells on both glycolipid and glycoprotein molecules. *Blood* **62**:1022.

9. Olive, D., P. Dubreuil, D. Charmot, C. Mawas, and P. Mannoni. 1984. T cell activation antigens: Kinetics, tissue distribution, molecular weights and functions; induction on non-T cell lines by lymphokines. *9th Leucocyte Culture Conference,* Cambridge, England, A. Mitchison and M. Feldman, Eds. Humana Press. Clifton, N.J.

CHAPTER 30

Different Stages of T Cell Differentiation and Microenvironments in the Thymus as Defined by Immunohistology

László Takács and Éva Monostori

Introduction

The great number of monoclonal antibodies reacting with T lymphocytes (T cell panel) or with stromal elements (myeloid panel) offers a good opportunity to study the different stages of thymic lymphocyte differentiation and the stromal elements, *in situ,* by immunohistology. The aim of this study was to define stromal cell types in specific locations, presumably acting on certain types of thymocytes, by providing appropriate microenvironments for their differentiation.

Materials and Methods

Monoclonal antibodies from the T cell panel and from the myeloid panel were received in ascites form containing sodium azide. These antibodies were used in saturating dilutions, not less than 1 : 100. Binding of the antibodies was visualized on frozen acetone-fixed tissue sections by the Vectastain ABC method.

Four different human thymuses were used. Thymuses were derived from patients undergoing cardiac surgery (1, 3, 3, and 6 years old). Sections were cut from frozen, unfixed material. 10 μm-thick sections were fixed in acetone and used in the experiments.

Rehydrated sections were covered with 15 μl of the diluted reagents for 45 min. Vectastain reagents (Vector Lab, USA) were used as described previously (1).

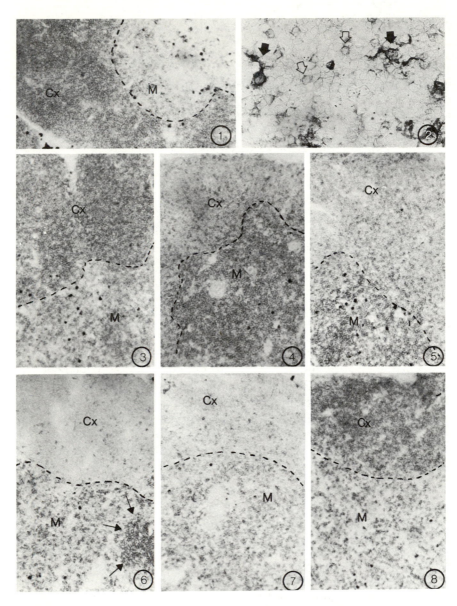

Fig. 30.1. Reaction patterns of T cell and myeloid panel antibodies described in Tables 30.1 and 30.2. Full arrows on figure panel 2 mark the nonlymphocytic cells reacting with CD1 antibodies. Open arrows mark the few positive lymphocytes present in this area of the medulla. Arrows on panel 6 label the group of positive cells where all cells react with antibody T-10; in the other regions of medullary tissue the staining pattern is heterogeneous. Cx: cortex, M: medulla, S: septum. Panels 1, 3–8, 9, 10, 12, 18: 80×; panels 11, 14, 15, 16: 200×; panels 2, 13, 17: 320×.

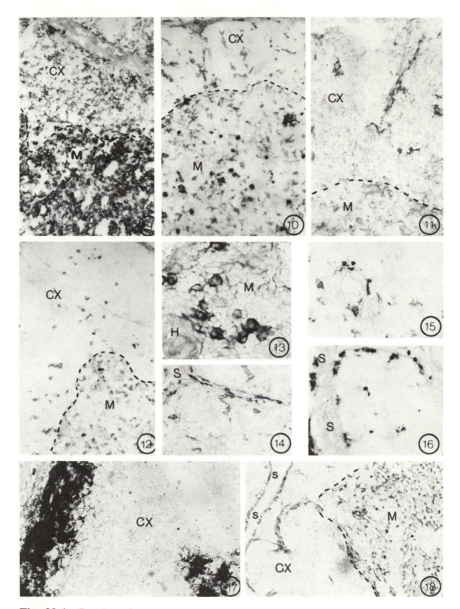

Fig. 30.1. *Continued.*

Table 30.1. Reaction patterns of T cell panel antibodies.

Group[a]	Cortex	Medulla	Exceptions	No reaction	Figure[b]
CD1	+/homogeneous	Nonlymphocytic cells	none	none	1, 2
CD2	+/homogeneous	+/homogeneous	15	23, 28	3
CD3	+/heterogeneous	+/homogeneous	119, 121, 123, 127	122, 125, 124	4
CD4	+/homogeneous	+/heterogeneous	71, 79	67, 68, 77, 78, 80	none
CD5	+/heterogeneous	+/homogeneous	10	11	5, 6
CD6	none	+			
CD7	none	+	30, 31, 33	32, 35, 36, 34	7
CD8	+/homogeneous	+/heterogeneous	96, 98	81, 85, 100	8

[a] Others: mixed reactions.
[b] Panel no. in Fig. 30.1.

Table 30.2. Reaction patterns of selected antibodies from the myeloid panel.[a]

Antibody no.	Subcapsular zone	Cortex	Cortico-medullary junction and peri-vascular zone	Medulla	Comment	Figure[b]
M40	+	+	+	+	DR-like	9
M46	+/mph.	+/mph.	+/mph.	+/mph.		10
M54	−	+/mph.	weak, +/mph.	weak, +/mph.		11
M65	−	few +	+/mph.	+/mph.		12, 13
M118	+/mph.	few +	+/mph.	+/mph.		14, 15
M111	+	few +	−	−	Large cell with processes	16, 17
M83	−	+	+	+	Similar cell as with M111	
T48	+/epithel.	−	+/epith.	+/epith.		18

[a] Macrophage-like cells : mph.; epithelial-like cells : epith.
[b] Panel no. in Fig. 30.1.

Results

Reactions on cells in general fall into two major groups. Most of the antibodies in the T cell panel reacted with thymocytes and most of the selected antibodies in the myeloid panel reacted with stromal elements. Further characterization of stromal cell types, e.g., by double-label experiments, was not performed. The majority of the reactions with T cell panel antibodies fall into the already known characteristic groups. Such a group-specific pattern was not observed with the myeloid panel antibodies. Thus in this latter case some of the antibodies presumably reacting with subtypes of stromal elements were selected and used in the study.

Group-specific reaction patterns of T cell panel antibodies are given in Table 30.1 and Fig. 30.1.

Reaction patterns of stromal element-specific reagents are given in Table 30.2 and Fig. 30.1.

Discussion

These data together with the results presented in recent publications on the heterogeneity of thymic stromal elements (2–6) indicate that there are at least four characteristic microenvironments defined by monoclonal antibodies or ultrastructural studies. These are: the subcapsular zone, cortex, cortico-medullary junction and perivascular zone, and the medulla. In all of these locations great heterogeneity of stromal elements is seen and further studies are required to analyze the overlap between the different populations characterized.

These microenvironments provide appropriate conditions for the differentiation of immature T cells to mature ones. The subcapsular blast cells entering the thymus presumably require the special subcapsular microenvironment. The murine subcapsular blast cells are characterized by their IL-2 receptors (See Ref. 7); however, monoclonal antibody reacting with the human subcapsular blasts was not observed. Cortical microenvironment induces cortex-specific antigens while in the medulla most of these are lost and new antigens appear. To elucidate the functional relationships between the heterogeneity of stromal elements and lymphocyte differentiation further *in vitro* studies are required.

References

1. Takács, L., I. Törö, H. Osawa, and T. Diamantstein. 1984. Immunohysto-chemical localization of cells reacting with monoclonal antibodies directed against interleukin-2 receptor of murine, rat and human origin. *Clin. Exp. Immunol.* **58.**
2. Wijngaert, F.P., M.D. Kendall, H. Schuurman et al. 1984. Heterogeneity of

epithelial cells in the human thymus. An ultrastructural study. *Cell Tissue Res.* **237**:227.

3. Van Vliet, A., M. Melis, and W. Van Ewijk. 1984. Monoclonal antibodies to stromal cell types of the mouse thymus. *Eur. J. Immunol.* **14**:524.

4. Savino, W., P.C. Huang, A. Corrigan, S. Berrih, and M. Dardenne. 1984. Thymic hormone containing cells. V. Immunohystological detection of metallothiocnein within the cells bearing thymulin, a zinc containing hormone in human and mouse thymus. *Histochem. Cytochem.* **32**:942.

5. Ritter, M.A., C.A. Sauvage, and S.F. Cotmore. 1981. The human thymus microenvironment: *in vivo* identification of thymic nurse cells and other antigenically-distinct subpopulations of epithelial cells. *Immunology.* **44**:439.

6. Wekerle, H.U.P., Ketelsen, and E.M. Ernst. 1980. Thymic nurse cells: lymphoepithelial cell complexes in murine thymuses: morphological and serological characterization. *J. Exp. Med.* **151**:925.

7. Takács, L., H. Osawa, and T. Diamantstein. 1984. Detection and localization by monoclonal anti-interleukin-2 receptor antibody AMT-13 of IL-2 receptor bearing cells in the developing thymus of mouse embryo and in the thymus of cortisone-treated mouse. *Eur. J. Immunol.* **14**:1152.

Index

Leucocyte Typing

Human Leucocyte Differentiation Antigens Detected by Monoclonal Antibodies: Specification, Classification, Nomenclature

Report on the First International Reference Workshop
Sponsored by INSERM, WHO, and IUIS

Edited by **A. Bernard, L. Boumsell, J. Dausset, C. Milstein, S.F. Schlossman**

The results of a large collaborative study of the best-characterized anti-leucocyte monoclonal antibodies are presented in this book. These results are vital for all scientists and clinicians in the fields of immunology, physiology, clinical biology, hematology, oncology and pathology who are interested in the technical details, pending problems and provisional conclusions about leucocyte subpopulations. The workshop was sponsored by several official organizations, in particular the WHO, forming the basis of the nomenclature recommendations of the WHO/IUIS standardization and nomenclature subcommittees at the final workshop meeting.

Contents: The impact of monoclonal antibodies on studies of the differentiation of lymphocytes. Joint report of the First International Workshop on Human Leucocyte Differentiation Antigens by the investigators of the participating laboratories. **T-Cell Antigens:** Studies involving workshop's battery of reagents. Characterization of T-cell antigens. T-cell subsets and functions. **B-Cell Antigens and Calla:** Studies involving workshop's battery of reagents. Characterization of B-cell antigens. **Monocytes-Granulocytes Antigens:** Studies involving workshop's battery of reagents. Characterization of monocytes-granulocytes antigens. **Tissue Section Studies:** Studies involving workshop's battery of reagents. **Leucocytes Common Antigens. Pathology of Leucocytes and Leucocyte Cell Lines:** Studies involving workshop's batteries of reagents. T-cell malignancies and T-cell disorders. B-cell and non-T, non-B cell malignancies. **Techniques for Detecting Leucocytes Antigens:** Editorials. Abstracts. Listings. Subject index.

1984/814 pp/158 illus/252 tables/1 color poster/$48.50/#12056-4

To order, write to: **Springer-Verlag New York, Inc.,** 175 Fifth Ave., New York, NY 10010.